RELATIONS BETWEEN STRUCTURE
AND FUNCTION
IN THE PROKARYOTIC CELL

Other Publications of the
*Society for General Microbiology**
THE JOURNAL OF GENERAL MICROBIOLOGY
THE JOURNAL OF GENERAL VIROLOGY

SYMPOSIA

1 THE NATURE OF THE BACTERIAL SURFACE
2 THE NATURE OF VIRUS MULTIPLICATION
3 ADAPTATION IN MICRO-ORGANISMS
4 AUTOTROPHIC MICRO-ORGANISMS
5 MECHANISMS OF MICROBIAL PATHOGENICITY
6 BACTERIAL ANATOMY
7 MICROBIAL ECOLOGY
8 THE STRATEGY OF CHEMOTHERAPY
9 VIRUS GROWTH AND VARIATION
10 MICROBIAL GENETICS
11 MICROBIAL REACTION TO ENVIRONMENT
12 MICROBIAL CLASSIFICATION
13 SYMBIOTIC ASSOCIATIONS
14 MICROBIAL BEHAVIOUR, 'IN VIVO' AND 'IN VITRO'
15 FUNCTION AND STRUCTURE IN MICRO-ORGANISMS
16 BIOCHEMICAL STUDIES OF ANTIMICROBIAL DRUGS
17 AIRBORNE MICROBES
18 THE MOLECULAR BIOLOGY OF VIRUSES
19 MICROBIAL GROWTH
20 ORGANIZATION AND CONTROL IN PROKARYOTIC AND
   EUKARYOTIC CELLS
21 MICROBES AND BIOLOGICAL PRODUCTIVITY
22 MICROBIAL PATHOGENICITY IN MAN AND ANIMALS
23 MICROBIAL DIFFERENTIATION
24 EVOLUTION IN THE MICROBIAL WORLD
25 CONTROL PROCESSES IN VIRUS MULTIPLICATION
26 THE SURVIVAL OF VEGETATIVE MICROBES
27 MICROBIAL ENERGETICS

* Published by the Cambridge University Press, except for the first Symposium, which
was published by Blackwell's Scientific Publications Limited.

RELATIONS BETWEEN STRUCTURE AND FUNCTION IN THE PROKARYOTIC CELL

EDITED BY

R. Y. STANIER, H. J. ROGERS AND J. B. WARD

TWENTY-EIGHTH SYMPOSIUM OF THE
SOCIETY FOR GENERAL MICROBIOLOGY
HELD AT
THE UNIVERSITY OF SOUTHAMPTON
APRIL 1978

Published for the Society for General Microbiology

CAMBRIDGE UNIVERSITY PRESS

CAMBRIDGE

LONDON · NEW YORK · MELBOURNE

Published by the Syndics of the Cambridge University Press
The Pitt Building, Trumpington Street, Cambridge CB2 1RP
Bentley House, 200 Euston Road, London NW1 2DB
32 East 57th Street, New York, NY 10022, USA
296 Beaconsfield Parade, Middle Park, Melbourne 3206, Australia

First published 1978

Printed in Great Britain
at the University Press, Cambridge

Library of Congress cataloguing in publication data

Main entry under title:
Relations between structure and function
in the prokaryotic cell.
(Symposium – Society for General Microbiology; 28)

Includes bibliographies and index.
1. Bacteria–Morphology–Congresses.   2. Bacteria–Physiology–Congresses.
I. Stanier, Roger Y.   II. Rogers, H. J.
III. Ward, J. B.   IV. Society for General Microbiology.
V. Series: Society for General Microbiology. Symposium; 28
QRI. S6233 no. 28 [QR75] 576′.08s [589.9′04]   77–21093

ISBN 0 521 21909 4

CONTRIBUTORS

BERG, HOWARD C., Department of Molecular, Cellular and Developmental Biology, University of Colorado, Boulder, Colorado 803029, USA

BRAUN, VOLKMAR, Lehrstuhl für Mikrobiologie II Universität Tübingen, West Germany

BRIMACOMBE, RICHARD, Max-Planck-Institut für Molekulare Genetik Abt. Wittmann, Berlin-Dahlem, West Germany

BROMLEY, DAVID B., Department of Microbiology, West Virginia University Medical Centre, Morgantown, West Virginia 26506, USA

BUCKINGHAM, R. H., Institut de Biologie Physico-Chimique, 13 rue Pierre et Marie Curie, 75005 Paris, France

BURDETT, I. D. J., Division of Microbiology, National Institute for Medical Research, Mill Hill, London NW7 1AA, UK

CHARON, NYLES W., Department of Microbiology, West Virginia University Medical Centre, Morgantown, West Virginia 26506, USA

COOPERMAN, B. S., Institut de Biologie Physico-Chimique, 13 rue Pierre et Marie Curie, 75005 Paris, France

ELLAR, D. J., Department of Biochemistry, University of Cambridge, Tennis Court Road, Cambridge CB2 1QW, UK

GRUNBERG-MANAGO, M., Institut de Biologie Physico-Chimique, 13 rue Pierre et Marie Curie, 75005 Paris, France

HENDERSON, RICHARD, Medical Research Council Laboratory of Molecular Biology, Hills Road, Cambridge CB2 2QH, UK

HERSHEY, J. W. B., Institut de Biologie Physico-Chimique, 13 rue Pierre et Marie Curie, 75005 Paris, France

HILMEN, M., Department of Biology, University of California, San Diego, La Jolla, California 92037, USA

KOMEDA, Y., Department of Biology, University of California, San Diego, La Jolla, California 92037, USA

LAMPEN, J. OLIVER, Waksman Institute of Microbiology, Rutgers – The State University of New Jersey, PO Box 759, Piscetaway, NJ 08854, USA

LEDERER, E., Institut de Biochemie, Université de Paris-Sud, 91405 Orsay, France

MATSUMURA, P., Department of Biology, University of California, San Diego, La Jolla, California 92037, USA

PETIT, J. F., Institut de Biochemie, Université de Paris-Sud, 91405 Orsay, France

RIDGWAY, H., Department of Biology, University of California, San Diego, La Jolla, California 92037, USA

ROGERS, HOWARD J., Division of Microbiology, National Institute for Medical Research, Mill Hill, London NW7 1AA, UK

SALTON, MILTON R. J., Department of Microbiology, New York University School of Medicine, New York, NY 10016, USA

SILVERMAN, M., Department of Biology, University of California, San Diego, La Jolla, California 92037, USA

SIMON, M. I., Department of Biology, University of California, San Diego, La Jolla, California 92037, USA

WALSBY, A. E., Marine Science Laboratories, Menai Bridge, Anglesey LL59 5EH, Wales, UK

WARD, J. BARRIE, Division of Microbiology, National Institute for Medical Research, Mill Hill, London NW7 1AA, UK

CONTENTS

Editor's preface   *page* ix

RICHARD BRIMACOMBE:
The structure of the bacterial ribosome   1

M. GRUNBERG-MANAGO, R. H. BUCKINGHAM,
B. S. COOPERMAN AND J. W. B. HERSHEY:
Structure and function of the translation machinery   27

VOLKMAR BRAUN:
Structure–function relationships of the Gram-negative
bacterial cell envelope   111

H. J. ROGERS, J. B. WARD AND I. D. J. BURDETT:
Structure and growth of the walls of Gram-positive bacteria   139

J. F. PETIT AND E. LEDERER:
Structure and immunostimulant properties of mycobacterial
cell walls   177

MILTON R. J. SALTON:
Structure and function of bacterial plasma membranes   201

RICHARD HENDERSON:
The purple membrane of halobacteria   225

J. OLIVER LAMPEN:
Phospholipoproteins in enzyme excretion by bacteria   231

J. B. WARD:
The reversion of bacterial protoplasts and L-forms   249

M. SIMON, M. SILVERMAN, P. MATSUMURA,
H. RIDGWAY, Y. KOMEDA AND M. HILMEN:
Structure and function of bacterial flagella   271

HOWARD C. BERG, DAVID B. BROMLEY AND
NYLES W. CHARON:
Leptospiral motility   285

D. J. ELLAR:
Spore specific structures and their function   295

A. E. WALSBY:
The gas vesicles of aquatic prokaryotes   327

Index   359

EDITORS' PREFACE

It is thirteen years since the Society published its Fifteenth Symposium also on Function and Structure. That volume concerned itself with both eukaryotes and prokaryotes. Yet even when the present symposium had been limited to consider only prokaryotes, many important aspects had to be omitted for reasons of time and space. We believed that after thirteen years of advances in knowledge, any attempt to deal with the whole range of microbes from viruses to protozoa would necessarily have led to an embarrassing superficiality in the treatment of all the topics. We therefore must risk the charge of 'bacterial arrogance' whether from the editors of the earlier volume or from the Society as a whole. As an aside, it might be thought that such a charge in 1977 has a somewhat hollow ring, which it would not have had in 1965, in view of the numbers of our colleagues who have deserted the microbes to study higher things. Indeed the editors of this volume hope that such of these renegades who seriously peruse the papers between these two covers, may see the error of their ways and realise that for many relevant problems bacteria may still be the necessary and most expeditious model organisms to use.

Those who edit books or organise symposia have perforce to examine more closely the meaning of titles, than the casual reader or member of an audience. We may, therefore, be excused for dilating a little on how we saw in 1977 this daunting title 'Structure and Function'. In its general biological context it is open to interpretations as widely apart as that taken by the natural historian, interested in the relation of the visual appearances of structures in living creatures to the behaviour of the creatures, all the way over to the enzymologist interested in the three-dimensional conformation of proteins interacting with small molecules of known crystal structure. We believed that the role of the present symposium should be to discuss the state of the art at levels in between these two extremes. Readers will, we think, find that the authors have opened to us fascinating vistas of macromolecular organisations functioning to accomplish the closely integrated and expanding network we please to call life. Undoubtedly, they will find many other subjects relevant to the theme that should have been included – we would agree. There is no treatment, among other subjects, of photosynthesis, nitrogen fixation, energy metabolism or molecular transport. However the time taken by symposia and printed space are both limited; moreover some of these subjects have been treated recently in reviews or symposia either organised by our own Society or by others.

In interpreting the word function, a further diversity is possible. With one very clear exception, authors have been primarily concerned with the structure, organisation and function of the component parts of bacteria, in so far as they are of importance to the micro-organisms themselves, or at most to understand how events happen within cells without prejudice as to their interaction. Quite a different complexion might have been given to a symposium involving either medical microbiologists or those interested in the economic importance of bacteria. 'Function' would primarily then have concerned function of structures interacting with the environment. Strangely such symposia would usually not have been entitled 'Structure and Function' so coloured do words become by their common usage. Nevertheless, as Luria (1960) has cogently said 'The organised features of living beings reach all the way from the macroscopic to the molecular level and one of the major tasks of biology is to understand and describe this organisation: its independent levels, its stability and its variability.' It is within this context that the present symposium has been organised.

It is inevitable that some comparison should be made between the present volume and that published in 1965. One of the striking characteristics is the enormous accretion of detailed and aesthetically pleasing knowledge in the intervening years. This is particularly true in areas such as the structure and function of the ribosomes, the general apparatus for translating the code into functional proteins and the control of flagella formation. We could say it is equally true of the organisation of the chromosomal DNA which would, of course, have been included but for the failure to receive a suitable manuscript in time to print it. Some of this increased knowledge about the parts of bacteria is due to advances in technology as in the use of biophysical techniques, such as neutron scattering to help in the study of the organisation of ribosomal proteins and the use of a variety of powerful methods to investigate membranes: some is due to advances in subjects such as genetics. However, there is also a pronounced change in the underlying philosophy. The introduction to symposium fifteen was entitled 'The architecture of the microbial cell' and was given by a man who was himself one of the architects of modern microbiological chemistry, D. D. Woods. It would probably be fair to say that such a title to an introduction to the present volume would be quite inappropriate. We are impelled to ask the question – why? Probably because the signpost written by the great pioneers who thought about form and function of living creatures, such as D'Arcy Thompson in his *Growth and Form*, led to a cross-roads. The majority of the contributors to the present symposium, in common with many biologists, have chosen to take the roads leading to explanations of the machinery that goes to make the total structure, or to study detailed parts of the building, rather than to consider the whole. Whether or not this rejection of holism is justified, only time will tell. The reasons for choosing these paths are clear. The great descriptive phase of biology is past together with our ability to stay with our aesthetic pleasure in looking at the total structure. We are only now preparing the base camp for the ascent to the Everest of understanding the derivation of the form and function of the whole integrated structure – even of bacteria – in molecular terms. Let us hope that in another thirteen years, we may see preparations for striking the base camp and beginning to move towards camp one.

It only remains for us to thank all the authors who have laboured to produce such interesting and informative manuscripts, some of our colleagues who have been prepared so kindly to supplement our limited knowledge in reviewing of the papers, and the Cambridge University Press, in particular Mrs Elizabeth Bowden, for the care and unfailing helpfulness shown to amateur Editors.

H. J. Rogers
R. Y. Stanier
J. B. Ward

THE STRUCTURE OF THE BACTERIAL RIBOSOME

RICHARD BRIMACOMBE

*Max-Planck-Institut für Molekulare Genetik, Abt. Wittmann,
Berlin–Dahlem, West Germany*

INTRODUCTION

A structure as complex as that of the bacterial ribosome cannot possibly be elucidated by any single experimental approach. In consequence, the experimenter is forced to concentrate on selected aspects of the problem, but the design of his experiments will be influenced to a considerable extent by his conception of the structure as a whole. In the past it was convenient to think of the ribosome as an arrangement of more or less globular proteins, and attempts to correlate ribosomal structure with function were based on the assumption that, on the one hand, specific functions in protein synthesis could be assigned to individual proteins, and that, on the other hand, the spatial distribution of the proteins could be deduced simply by determining which pairs of individual protein 'spheres' were able to interact with one another. The large ribosomal RNA molecules could be considered as being interspersed through this structure, but, since nucleic acids are rather unreactive, it seemed unlikely that the RNA was doing anything very interesting other than holding the particle together.

During the past few years, however, the picture has changed considerably; it has become clear for example that most functional steps in protein synthesis involve several ribosomal components, that most of the proteins are very far from being globular in shape, and also that the RNA is by no means playing a merely passive role. As a result, our conception of the ribosome has taken on a new dimension of complexity, and it is the purpose of this article to describe some of the developments which have brought about this change, and to review briefly the current status of research into bacterial ribosome structure. Ribosome function is the subject of a separate article by M. Grunberg-Manago and colleagues in this volume.

Very roughly speaking, structural research has been pursued at three levels, namely studies on isolated ribosomal components, studies on the topographical arrangement of the individual components within the ribosomal particles, and general studies on the properties of the intact ribosomes. Most of the relevant work has been concerned with the ribosome of *Escherichia coli*, a particle which has a sedimentation coefficient of 70S, and which consists of a 50S and a 30S subunit. In general, the literature cited refers to the most recent work on a particular topic, and the reader is referred to appropriate reviews for the older or original findings.

RIBOSOMAL PROTEINS

Sequences, shapes and stoichiometry

All the proteins of the *E. coli* ribosome have long since been isolated and characterized (reviewed by Wittmann, 1974). In addition to a 16S RNA molecule, the smaller subunit contains 21 proteins, named S1 to S21 according to their positions on a two-dimensional polyacrylamide gel (Kaltschmidt & Wittmann, 1970), and the larger subunit correspondingly contains 34 proteins, L1 to L34, as well as a 5S and a 23S RNA species. The amino acid sequences of about two thirds of these proteins have so far been fully determined. In addition amino acid replacements in proteins of many mutants have been analysed and some ribosomal proteins from other organisms have been studied. These data have been reviewed in detail elsewhere (Stöffler & Wittmann, 1977) and will not be discussed here. Suffice it to say that all the *E. coli* proteins have distinct non-homologous amino acid sequences, with the exception of two pairs of proteins. These are proteins L7 and L12, which differ only in the presence of an N-terminal acetyl group in L7, and proteins S20 and L26, which are identical. Very recently, investigations of the secondary structures of the ribosomal proteins have been undertaken, involving both theoretical predictions of the amounts of α-helix, extended structure, etc., and also circular dichroism measurements (Wittmann-Liebold, Robinson & Dzionara, 1977).

Previously, it was thought that the ribosomal particles were very heterogeneous, as the proteins appeared to be present in variable stoichiometry (e.g. Voynow & Kurland, 1971), but a more recent study (Hardy, 1975) has indicated that 70S ribosomes *in vivo* possess one copy of each ribosomal protein, with the exception of L7/L12 which occurs in three or four copies (see also Subramanian, 1975). The identical proteins S20/L26 appear to contribute one copy per 70S particle. While this finding is noteworthy as being the only facet of ribosome structure to have actually become simpler during the last few years, Hardy (1975) was careful to point out that most ribosomal subunits *in vitro* are indeed heterogeneous as a result of the commonly used isolation procedures, and the errors involved in this type of measurement of stoichiometry are large enough to obscure a certain degree of heterogeneity *in vivo*. It should also be added that a 1:1 stoichiometry of the ribosomal proteins by no means implies that the ribosome is a rigid structure.

The ribosome as a whole is far too complex for any attempt to be made at this time to undertake an X-ray crystallographic analysis, but progress is being made in this direction on the isolated proteins. Protein L7/L12, which is particularly interesting in view of its multiple appearance in the ribosome and its importance in a number of ribosomal functions (reviewed by Möller, 1974, and by Pongs, Nierhaus, Erdmann & Wittmann, 1974), has been recently crystallized (Liljas & Kurland, 1976), and it is not unreasonable to hope that the structural features to be revealed in an isolated protein will be at least partially preserved in the ribosomal particle.

In addition, small-angle X-ray studies have been made on several isolated proteins, and these experiments have indicated that some proteins, for example S4, L18 and L25 (Österberg, Sjöberg & Garrett, 1976; Österberg, Sjöberg, Garrett & Littlechild, 1977), have distinctly non-spherical shapes. As was mentioned in the Introduction, this question of the shape of the ribosomal proteins has had a considerable impact on our conception of ribosome structure, but the most detailed information in this context has come not from experiments with isolated proteins, but from topographical studies of the ribosomal subunits, using the technique of immune electron microscopy.

Immune electron microscopy

It has been known for some years that antibodies can be raised against individual ribosomal proteins, and that these antibodies are highly specific – no ribosomal protein will react to any significant extent with any other than its cognate antibody (Stöffler & Wittmann, 1971). Further, these antibodies are able to react with proteins within the intact subunits, and the position of attachment of the antibody can then be localized by electron microscopy, taking advantage of the fact that both ribosomal subunits have readily recognizable shapes under the electron microscope. The technique thus allows the direct visualization of the protein antigenic sites on the ribosome surface.

Data from this method have been forthcoming from two research groups (Tischendorf, Zeichhardt & Stöffler, 1975; Lake, 1976, and earlier papers), and the topographical distribution of some of the protein antibody binding sites obtained by both groups on the two

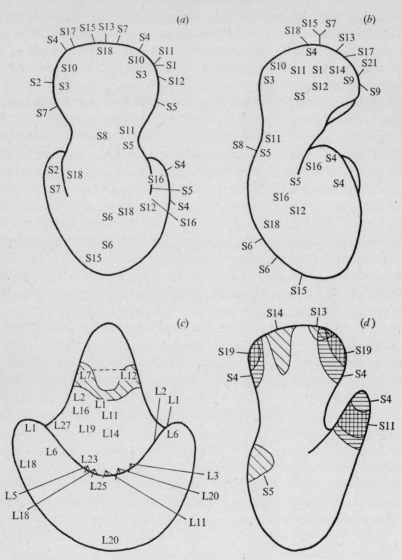

Fig. 1. Comparative sketches of the positions of antigenic sites for the ribosomal proteins on the 30S and 50S subunits: (a) & (b) two views of the 30S subunit, according to Stöffler & Wittmann (1977), (c) a view of the 50S subunit, according to Stöffler & Wittmann (1977), (d) a view of the 30S subunit, according to Lake (1976).

ribosomal subunits is illustrated in Fig. 1. There is no space here to describe the results in detail (see Stöffler & Wittmann, 1977, for review), but for the purpose of this discussion the most important feature is that several proteins (e.g. S4, S7, L6 (Fig. 1)) have multiple antibody binding sites which are situated at widely separated points on the subunit surface. Since each protein occurs in only one copy per ribosome, this result can only be interpreted as an indication that the proteins concerned have highly extended conformations. In particular, two proteins (S15 and S18) appear to stretch the whole length of the 30S subunit, and it has been possible to demonstrate in the case of S18 that one of the sites corresponds to the N-terminal region of the protein and the other to the C-terminal region, by using antibodies raised against specific protein fragments (G. Tischendorf, K. H. Rak, G. Stöffler & M. Yaguchi, personal communication). It is important to remember here too that the technique gives a minimum estimate of the number of elongated proteins; further antibody binding sites may well be found later for those proteins which at present only exhibit a single site.

There is a disagreement between the two research groups as to the degree of asymmetry exhibited by the subunits under the electron microscope (Stöffler & Wittmann, 1977, Lake 1976, and cf. Fig. 1), and this difference, which affects some of the protein placements, has yet to be resolved. On the other hand, both groups are in substantial agreement as to the relative placing of several of the protein sites as well as to the tendency for elongated protein conformations.

Preparation of the samples for electron microscopy involves a fairly drastic drying-down procedure, and this undoubtedly accounts at least partly for the difference in overall dimensions of the ribosomal subunits, when measured by electron microscopy as opposed to physical methods such as X-ray scattering (Hill & Fessenden, 1974). However, although this drying procedure must inevitably cause some distortion of the particles, it is reasonable to expect that the relative topographical distribution of the proteins will remain substantially unaltered. This conclusion is supported to some extent by data obtained from another new technique, namely neutron scattering, which makes measurements on ribosomes in solution. This method also indicates that some proteins are elongated within the subunits.

Neutron scattering

The advantage of a scattering technique is that not only does it measure properties of the whole particle (and not just the particle surface), but it enables the activity of the ribosome preparation to be tested at the end of the experiment. The danger of artefact is thus reduced, but on the other hand the neutron technique is in its early stages of development and does not yet have the resolving power of the electron microscope measurements. Neutron scattering measurements have primarily been designed to obtain the distance between centres of mass of pairs of ribosomal proteins, and for this purpose a reconstituted ribosome must

be prepared in which the two proteins under consideration are deuterated. Distances between several pairs of proteins have so far been
estimated by this method, in the first instance S2–S5, S5–S8 and S3–S7
(Engelman, Moore & Schoenborn, 1975). Of these distances, S3–S7 was
found to be short (3.5 nm), and the others were both long (of the order
of 10 nm). A further analysis of the data suggested that S2 was elongated
(cf. Fig. 1), but that S8 and S5 were compact, the latter result being at
variance with the electron microscope data, where S5 is clearly elongated.
More recently, these studies have been extended (Moore, Langer,
Schoenborn & Engelman, 1977) to include six other protein pairs
(S3–S4, S3–S5, S3–S8, S4–S7, S4–S8 and S5–S6), and it was concluded
that most if not all of these proteins are elongated, with protein S7
having a particularly unusual shape. In another neutron study, the
distance between the centres of mass of L7/L12 and L10 was estimated
at 10 nm (Hoppe *et al.*, 1975), which is difficult to reconcile with the
observation that L7/L12 and L10 form a stable complex in solution
(Pettersson, Hardy & Liljas, 1976), unless of course these proteins are
also highly elongated.

Neutron scattering has also been used to determine the relative
positions of the centres of mass of RNA and protein within the subunits. Here it appears that in the 50 S particle (but not in the 30 S) these
centres of mass are not coincident (Moore, Engelman & Schoenborn,
1974), indicating an RNA-rich and a protein-rich region in the 50 S
subunit, but there is some disagreement as to the magnitude of the
displacement (Stuhrmann *et al.*, 1976).

Science is supposed to thrive on controversy, and there are clearly a
number of points in all the studies described above which need to be
resolved. Nevertheless, it should be emphasized that the disagreements
reflect the sophistication of the questions being asked, rather than the
inadequacy of the methods, and the central thesis is clear, viz. that
many ribosomal proteins have highly extended conformations. We must
now turn to the question of how this concept affects the more classical
methods of investigating the spatial distribution of the ribosomal
proteins.

Protein–protein cross-linking

The most widely used method for investigating protein topography
within the ribosomal subunits has been chemical cross-linking. In this
approach, pairs of proteins are covalently linked by a suitable bifunctional reagent, the total protein is extracted and fractionated, and
the components of the cross-linked pairs are identified. Several research

Table 1. *Protein–protein cross-linking data from the 30 S subunit*

S2–S3	S4–S5–S8	S5–S13	S11–S18–S21
S2–S5	S4–S6	S6–S14–S18	S12–S13
S2–S8	S4–S8	S6–S18	S12–S20
S3–S4	S4–S9	S7–S8	S12–S21
S3–S5	S4–S12	S7–S9	S13–S17
S3–S9	S4–S13	S7–S13	S13–S19
S3–S10	S4–S17	S8–S11	S14–S19
S3–S12	S5–S8	S8–S13	S18–S21
S4–S5	S5–S9	S11–S13	

IF2–S1, S2, S11, S12, S13, S14, S19.
IF3–S1, S7, S11, S12, S13, S19, S21.

Cross-links between ribosomal proteins are according to the summary by Sommer & Traut (1976), and the data for the initiation factors IF2 and IF3 are from van Duin, Kurland, Dondon & Grunberg-Manago (1975) and the summary by Heimark *et al.* (1976).

groups have been involved in this work (e.g. Sommer & Traut, 1976; Lutter, Kurland & Stöffler, 1975; Clegg & Hayes, 1974; Peretz, Towbin & Elson, 1976) and a variety of different reagents have been used, the most successful of which have been imido ester derivatives. The proteins in the isolated cross-linked complex can be identified by Ouchterlony double diffusion tests using protein-specific antibodies (e.g. Lutter *et al.*, 1975), but in the more recent work techniques involving the use of reversible cross-linking reagents have been developed, coupled with identification of the proteins by gel electrophoresis. One such reagent is mercaptobutyrimidate, which is allowed to react with protein amino groups; cross-linking is then effected by a mild oxidation between the mercapto groups thus introduced, and the reaction can subsequently be reversed by mild reduction (Sommer & Traut, 1976, and earlier papers). Other workers have used *bis*-azide derivatives of tartaric acid as reversible cross-linkers, which can be cleaved by periodate oxidation (Lutter, Ortandel & Fasold, 1974).

Many protein pairs have been cross-linked by this approach in both subunits, and the results from the 30 S subunit are summarized in Table 1. If these are compared with the electron microscopy data of Fig. 1, it is clear that several (but not all) of the elongated proteins (e.g. S2, S3, S4, S5) are involved in relatively large numbers of cross-links, and there is also a measure of agreement between the cross-link data and the corresponding antibody binding sites. On the other hand the fact that many of the proteins have extended conformations reduces the significance of the cross-linking data to a considerable degree; the effect of such conformations, as opposed to globular shapes, is that the proteins are potentially able to share a large number of possibly widely dispersed contacts with one another, and it also follows that any particular point on the ribosome may be in the direct neighbourhood of a number of different 'long thin' proteins. This is shown, for instance, in the rather large number of proteins which have been cross-linked to the initiation factors of protein synthesis (see Table 1).

The consequence is that, in order to be able to interpret this type of cross-linking data fully, it will be necessary to pursue the analysis to a more intimate level. In other words, identification of the proteins contained in a cross-linked pair is no longer sufficient; the amino acid residues, or at least the oligopeptides, involved in such a cross-link will need to be identified in order to define the particular neighbourhood between two polypeptide chains in a meaningful way. This will not be easy; but, fortunately, the lion's share of the task, which is the analysis of the amino acid sequences of the proteins (see above), is already well under way.

The same argument applies to a number of studies on ribosomal function, in particular affinity labelling studies (see Brimacombe *et al.*, 1976, for review), which are designed to identify those ribosomal components concerned with a particular function in protein synthesis (such as peptidyl transferase activity, or tRNA binding). The number of proteins implicated in individual functions by such studies is continually increasing, and in some cases specific RNA regions have also been implicated (e.g. Wagner *et al.*, 1976). This has led to the concept of 'functional domains' in the ribosome (Kurland, 1977), in which parts of several different ribosomal components contribute to a particular functional site. The situation is analogous to the active site of an enzyme, to which amino acids are contributed by widely separated parts of the amino acid chain. It is worth noting here that there is already one example of a precise identification of the point of reaction in an affinity labelling study; an affinity analogue of the initiator codon AUG was found to react with a cysteine residue at position 10 in protein S18 (O. Pongs, E. Lanka & M. Yaguchi, personal communication).

Chemical modification

A number of chemical modification studies have been made on the ribosome (for summary see Benkov & Delihas, 1974). The object of this approach in earlier experiments was to treat the ribosome with different protein reagents, and analyse for the degree of reactivity of the individual proteins, with a view to discovering which proteins were 'outside' and which were 'inside' the subunits. The extended conformations of the proteins, coupled with the uncertainty as to how far a small reagent could penetrate into the ribosomal particle, makes these early studies virtually impossible to interpret, and more recent work has concentrated on establishing differences in protein reactivity between ribosomes in different functional states.

To give just two recent examples, reaction with N-ethyl-maleimide (Ginzburg & Zamir, 1975), and lactoperoxidase catalysed iodination (Litman, Beekman & Cantor, 1976) have both been used to probe conformational changes accompanying inactivation of subunits, and also to try to identify proteins located at the 30S–50S interface. In the latter case the experimental strategy is to detect those proteins which are accessible in isolated subunits but not in the 70S ribosome. It was clear from both these studies (together with earlier work from the same groups) that a considerable conformational change takes place in the 30S particle upon 70S formation, since several 30S proteins become actually more accessible to modification in the 70S particle. Unfortunately, it follows that those proteins which become correspondingly less accessible upon 70S formation cannot be unambiguously assigned as being located at the subunit interface.

It should be added at this juncture that direct evidence concerning the topography of the subunit interface has proved extremely difficult to come by. Most of the experiments are rather indirect and suffer from the same ambiguity as those just mentioned. Protein S16 can however be cross-linked directly to 50S proteins by reaction with aldehydes (Sun, Traut & Kahan, 1974), and protein S20/L26, which occurs on both subunits but in only one copy per 70S ribosome (see above) is also presumably at the interface. Further, proteins S11 and S12 appear to be able to interact specifically with the 23S RNA from the large subunit (Morrison, Garrett, Zeichhardt & Stöffler, 1973). Here again however the extended conformations of the proteins make it likely that a large number of proteins will contribute to the subunit interface, and, again, a more detailed analysis will be required. A new technique which has promising potential in this direction is the measurement of energy transfer between fluorescent markers.

Singlet–singlet energy transfer

This method has been developed by Cantor and his co-workers and involves the preparation of reconstituted ribosomes in which the proteins are labelled with two different fluorescent dyes. The singlet–singlet energy transfer from the donor to the acceptor dye is dependent on a number of parameters, one of which is the distance between the two labelled proteins, and this distance can be estimated by appropriate fluorescence measurements. Up to now the technique has been applied using individual proteins which are labelled as randomly as possible, and so has effectively given measurements of the relative distances between centres of mass of the protein pairs concerned (Huang, Fairclough & Cantor, 1975). Many such interprotein distances in the 30S subunit were measured, and the results compare favourably with those of the cross-linking and electron microscopy data, bearing in mind the reservations already discussed. In another similar study the distance between the binding site of the antibiotic erythromycin on the 50S subunit and proteins L7/L12 was measured (Langlois *et al.*, 1976), but, unfortunately, data are not yet available for the distances between those protein pairs discussed above, which have been determined independently by neutron scattering, and so a direct comparison of the two methods cannot at the moment be made. As with the neutron scattering technique, the fluorescence method is in its very early stages of development but, as the authors point out, it has the potential that the fluorescent dyes could in principle be placed at specific sites on a protein, which should enable measurements to be made of the distance between precise points on two polypeptide chains within the ribosome.

The neutron scattering and fluorescence techniques both rely very heavily on a fairly old principle which one tends nowadays to take rather for granted, viz. that active 30S ribosomal subunits can be reassembled from their component proteins and RNA (Traub & Nomura, 1969), and will fortunately even allow a certain degree of modification of the proteins. There have, however, been some interesting recent developments in this field of subunit reassembly, which are worthy of comment.

RECONSTITUTION

30S subunit

Early studies on the in-vitro reconstitution of 30S subunits indicated that a few proteins (S4, S7, S8, S15, S17, S20 and possibly S13 (reviewed by Zimmermann, 1974)) were able to interact specifically with 16S RNA, and that the remaining proteins could be added to complexes containing RNA and one or more of these 'primary binding proteins' in a very specific manner. The interactions thus found between the proteins were incorporated into an 'assembly map' (Held, Ballou, Mizushima & Nomura, 1974), which it seemed reasonable to assume was a direct reflection of the topographical neighbourhoods of the ribosomal proteins. Indeed several groups have combined the assembly map with protein–protein cross-linking and other data to produce 'ping-pong ball' models of the protein arrangement in the 30S subunit (reviewed by Brimacombe *et al.*, 1976). While the futility of this type of model-building is now self-evident from the preceding discussion, it has to be added that the data do show a considerable degree of consistency when incorporated into these models.

However, it has been suggested by Kurland (1974) that interactions during assembly are probably not quite so simple. Two thirds of the mass of the ribosome is after all RNA, and the suggestion was that the steps in the assembly process might be reflecting steps in the organization of the RNA necessary for the incorporation of further proteins, rather than merely reflecting protein–protein interactions. In other words, the assembly process would involve interactions between 'mixed neighbourhoods' of protein and RNA, and while this process would be likely to be related to the topographical arrangement of the proteins, there could also be long-range effects mediated through alterations in the RNA structure. This concept had the additional implication that most if not all of the ribosomal proteins would have substantial direct contact with the RNA in the finished particles.

Strong support for this way of thinking has been provided by some new data of Craven and his co-workers (Hochkeppel, Spicer & Craven, 1976), which rest on the very simple observation that the protein-binding properties of the ribosomal RNA are drastically dependent upon the way in which the RNA is prepared. When 16S RNA is prepared by a phenol extraction procedure, only the six or seven proteins named above are able to bind individually and specifically to it. In contrast, RNA prepared by an acetic acid–urea technique is able to bind an additional six or seven proteins in an apparently specific manner (namely S3, S5, S9, S12, S13, S18 and possibly S11). While several important criteria have not yet been tested in these experiments (such as the stoichiometry of binding of all of the proteins), the central point is well established, viz. that protein binding sites can become either obscured or activated according to the conformation of the isolated RNA, and in consequence at least a part of the effect of the primary binding proteins may be the reorganization of the RNA to make other protein sites available. Another very recent study has demonstrated that the protein conformation is also important for RNA binding, as, by using proteins prepared under very mild conditions, a further new set of 'binding proteins' was found, namely S2, S5, S13 and S19 (Littlechild, Dijk & Garrett, 1977). The same authors also found two new 23S RNA binding proteins, in addition to the eleven already described.

50 S subunit

Studies on the detailed assembly of the 50S subunit lag far behind those on the 30S, for the simple reason that a successful reconstitution procedure for *E. coli* 50S subunits has only recently been established (Nierhaus & Dohme, 1974). This is incidentally in contrast to the case of the 50S subunit from the thermophilic organism. *Bacillus stearothermophilus* which was first reconstituted several years ago (Nomura & Erdmann, 1970). The procedure developed for reassembly of the *E. coli* 50S subunit involves a two-step incubation procedure, and has been shown to proceed through formation of a number of intermediate particles, which parallel fairly closely the corresponding in-vivo intermediates. Nierhaus and his co-workers have identified those proteins which are essential for one of the early steps in this reassembly chain, which is a conformational change from a 33S to a 41S particle (S. Spillmann, F. Dohme & K. H. Nierhaus, personal communication). Interestingly, all of the proteins concerned (L4, L13, L20, L22 and L24) have RNA binding sites in the 5'-proximal region of the 23S RNA (Chen-Schmeisser & Garrett, 1976, and see later), which is a clear indication that the assembly proceeds in a general 5' to 3' direction along the ribosomal RNA.

The binding proteins and the significance of the term 'RNA binding site' will be dealt with in more detail later, in the section on RNA–protein interaction; but first the ribosomal RNA will be discussed.

RIBOSOMAL RNA

Functional importance

It is perhaps unfortunate that the term 'translation' in protein synthesis has come to be associated with the function of the ribosome. In fact the 'translation' of a nucleotide sequence into an amino acid takes place not on the ribosome, but on an aminoacyl tRNA synthetase molecule, and the function of the ribosome is one of 'reading' rather than of translation; the nucleotide codon sequences of the messenger RNA are recognized by the corresponding anticodon sequences of the tRNA, and the polypeptide chain is then generated between the adjacent aminoacyl and peptidyl tRNA molecules by a relatively simple enzymatic reaction, namely the peptidyl transferase activity.

In other words, the primary job of the ribosome is to supervise specific interactions between nucleic acids, and it would therefore be very strange if the nucleic acid components did not themselves play a decisive role in this function. However, as suggested in the Introduction, this has not in the past been a popular idea, and while some authors (e.g. Crick, 1968; Kurland, 1974) have stressed the probable functional importance of the ribosomal RNA, it was not until fairly recently that strong evidence began to appear in support of this contention.

The first case concerns the binding of tRNA to the 50S subunit. Here, several lines of evidence suggest that this binding involves at least in part a base-pairing between the TψCG loop of the tRNA and a complementary sequence on the 5S RNA molecule (reviewed by Erdmann, 1976). There is some conflicting evidence (Noller & Herr, 1974), as the relevant G-residues in the 5S RNA within the ribosome are not exposed to attack by kethoxal, a reagent which binds specifically to non-paired guanine bases (Litt, 1969), but this difficulty can be overcome by postulating appropriate conformational changes (Schwarz, Lührmann & Gassen, 1974).

The second case concerns the proposal made by Shine and Dalgarno (1974) that the 3'-terminal sequence of 16S RNA is involved in messenger RNA recognition. In this instance, Steitz and Jakes (1975) have been able to isolate a dissociable oligonucleotide complex consisting of the 3'-terminus of the 16S RNA and an RNA fragment from a protein initiator region of R17 phage RNA; the two fragments contained a seven-base complementary sequence, the 3'-terminus of the 16S RNA being released from the 30S subunit as a 49-nucleotide fragment under very mild conditions with the help of the antibiotic colicin E3 (Senior & Holland, 1971).

A number of other attractive hypotheses (van Knippenberg, 1975; Herr & Noller, 1975; Branlant, Sriwidada, Krol & Ebel, 1976c) involving complementary sequences of between 6 and 12 base-pairs between various nucleic acid components of the protein-synthesizing system have been put forward, but, while there is circumstantial evidence in favour of these ideas, hard facts are very difficult to obtain, and it must be remembered that, in molecules the size of ribosomal RNA, any given nucleotide sequence of up to six or seven bases would be expected to appear on a statistical basis. It is noteworthy however that 5S RNA and the 'colicin fragment' are at the moment the only two readily identifiable RNA species which can be removed from the ribosome under mild conditions, and since both of them are strongly implicated in separate functions, it seems highly probable that many other regions of RNA within the bulk of the 16S and 23S molecules will ultimately reveal a similar functional importance.

Primary, secondary and tertiary structure

As with the amino acid sequences of the ribosomal proteins, a large amount of information is already available concerning the nucleotide sequences of the *E. coli* ribosomal RNA species. The total sequence of 5S RNA (120 nucleotides) has been known for many years (Brownlee, Sanger & Barrell, 1968), and the primary structure of the 16S RNA (1600 nucleotides) is now almost fully established (Ehresmann *et al.*, 1975), although there is still a measure of disagreement as to the detailed sequences of some of the oligonucleotide degradation products (Magrum, Zablen, Stahl & Woese, 1975). Progress has also been made on the sequence of the 23S molecule (3200 nucleotides, e.g. Branlant, Sriwidada, Krol & Ebel, 1977). A comparison with the corresponding RNA molecules from other organisms indicates that the sequence of 5S RNA has been strongly conserved throughout evolution (see Erdmann, 1976, for review), and the oligonucleotide catalogues of 16S RNA from different bacteria show a considerable degree of homogeneity (Woese *et al.*, 1975), again suggesting that regions of the sequence have been conserved.

As with the ribosomal proteins, however, a knowledge of the primary sequences of the RNA molecules is but one step along the road to understanding how they are fitted into the ribosome. A meaningful discussion of the various complementary interactions which have been proposed during protein synthesis (see above) demands a knowledge of the secondary structure of the nucleic acid components involved, and here the evidence is unfortunately very thin. More than a dozen different secondary structures have so far been proposed for the 5S RNA (see Erdmann, 1976, for review), and while a secondary structure has been put forward as a working hypothesis for the 16S RNA (Ehresmann *et al.*, 1975), there is at the moment, very little evidence to support it. In particular, chemical modification studies with kethoxal (Chapman & Noller, 1977), which as already mentioned attacks non-paired guanine residues, do not agree with this proposed secondary structure.

The kethoxal results are also noteworthy because they were pursued to the level of identifying the actual nucleotide residues involved in the reaction. The experiments showed that the kethoxal-accessible sites on 16S RNA within the 30S subunit are distributed along the RNA in distinct groups, and are mostly in regions of the RNA suspected to contain conserved nucleotide sequences (Woese *et al.*, 1975, see above), including the 3'-terminal region already mentioned (Steitz & Jakes, 1975). These kethoxal-sensitive sites are therefore likely candidates for functionally important regions. Further, when 70S ribosomes were the target for attack, the reactivity towards kethoxal was altered, as compared to that found in the 30S subunit; some groups of kethoxal sites became less sensitive, while others became more sensitive (Chapman & Noller, 1977). This should be compared with the results mentioned earlier, obtained by lactoperoxidase catalysed iodination or N-ethylmaleimide modification of the 30S proteins (Litman *et al.*, 1976; Ginzburg & Zamir, 1975), and gives support to the idea that a pronounced conformational change takes place in the 30S particle upon 70S formation.

Conformational changes in the RNA have been observed in a variety of circumstances. It was previously thought that the secondary structure of isolated ribosomal RNA was the same as that of the RNA within the subunits (e.g. Miall & Walker, 1969), but a more recent study suggests that the helical content of the RNA is lower inside the subunits than in the isolated RNA molecules (Araco, Belli & Onori, 1975). The base-pairing of the RNA within the subunits also changes upon going from a compact to an 'unfolded' conformation induced by EDTA treatment, as indicated by optical rotatory dispersion measurements (reviewed by Spitnik-Elson & Elson, 1976), although the overall degree of base-stacking is not affected. Circular dichroism studies showed that the protein conformations are not altered to any significant extent by this 'unfolding' process (Spitnik-Elson & Elson, 1976), but it is known that the unfolding allows random exchange of the ribosomal proteins to occur (Newton, Rinke & Brimacombe, 1975). These results, taken together with the importance of the RNA conformation during 30S reassembly already discussed earlier, indicate that the proteins are able to influence the secondary and/or tertiary structure of the RNA, and that maintenance of this secondary and tertiary structure is important for the integrity of the subunit structure as a whole.

Such studies of the overall properties of the RNA inevitably only lead to general conclusions, but some specific tertiary structural features of the 16S RNA have recently been described. That is to say, stable interactions have been observed between regions of the RNA which are widely separated in the primary sequence (see later). This suggests that the topological arrangement of the RNA within the subunits is very complex, involving tertiary RNA–RNA interactions as well as RNA–protein interactions, and in fact specific RNA–RNA crosslinks can be generated between remote regions of the 16S RNA in the 30S subunit by irradiation with ultraviolet light (C. Zwieb & R.

Brimacombe, unpublished results). The question of the RNA–protein
interactions is the subject of the next and last section.

RNA–PROTEIN INTERACTIONS

The three-dimensional arrangement of the ribosomal proteins (as
exemplified in Fig. 1) is complicated enough, without the RNA. If one
now adds the RNA to this picture (remembering that RNA constitutes
two-thirds of the mass of the particle), and concedes that the RNA is
also likely to contribute significantly to the 'functional domains'
already discussed, then it follows that the interaction between protein
and RNA becomes a vital part of, and perhaps the key to an understanding of ribosome structure and function.

Unfortunately, the study of protein–nucleic acid interactions in
general has been one of the most disappointingly slow-moving fields in
molecular biology. This has not been due to any lack of effort, but
reflects rather the experimental difficulties and also the difficulty of
formulating general rules for such interactions. Indeed there is no
reason why there should be any 'rules'; an interaction between say
DNA and an 'unwinding protein' is unlikely to have much in common
with the interaction between tRNA and an aminoacyl synthetase, or
with the interaction between a ribosomal protein and ribosomal RNA.
It is already clear that even interactions between individual ribosomal
proteins and RNA differ considerably from one another; the considerable variation in the ability of the proteins to bind to RNA has already
been mentioned, and one protein (S1) has been shown to be capable of
'unwinding' double-helical RNA (Szer, Hermoso & Boublik, 1976).
Questions which have been asked in the past to try to establish general
rules, such as 'do ribosomal proteins bind to single- or double-stranded
regions of the RNA?' (e.g. Cotter, McPhie & Gratzer, 1967), are
therefore very probably irrelevant.

Most recent studies on RNA–protein interactions in the ribosome
have concentrated on the topographical aspects of the problem, that is
the question of which regions of the RNA are involved in interaction
with individual ribosomal proteins. Here, two different approaches have
been used, one involving the study of complexes between single proteins
and RNA, and the other concerned with controlled nuclease digestion
of the ribosomal subunits. The experiments can be further subdivided
into those where an RNA–protein complex is isolated which is itself
stable, and those where the complex is stabilized by a covalent cross-
link between protein and RNA.

Stable protein–RNA complexes

In the 'single protein' approach, a pre-formed complex between protein and RNA is subjected to mild digestion with nuclease, following which the RNA region 'protected' by the protein can be identified. Alternatively, the binding of the protein to isolated RNA fragments of known nucleotide sequence (also produced by mild nuclease digestion) can be studied. A considerable amount of data has been accumulated using this approach, mostly by Garrett and Zimmermann and their co-workers, and the results have been reviewed in detail elsewhere (Zimmermann, 1974; Brimacombe *et al.*, 1976).

To summarize briefly, 'binding sites' on 16S RNA have been found for proteins S4, S7, S8, S13, S15, S17 and S20, which constitute the original set of RNA binding proteins (see the reviews just cited). So far no data are available on the 'new' binding proteins (Hochkeppel *et al.*, 1976; Littlechild *et al.*, 1977) which were mentioned above in the section on reconstitution of the 30S subunit. The size and complexity of the RNA regions identified in these experiments varies considerably. Proteins S8 and S15, for example, are found associated with quite short RNA sequences, of the order of 50 nucleotides long, whereas, in contrast, S4 has a very large 'binding site' comprising almost one third of the 16S molecule. It is important to realize here that the nature of the RNA fragments which are isolated is limited both by the availability of nuclease sensitive sites in the neighbourhood of the protein and by the structure of the RNA itself, and it by no means follows that the protein is in intimate contact with the whole of its RNA 'binding site'. This point is underscored by the recent finding that two regions of RNA within the S4 binding site, separated by about 100 nucleotides, are able to form a stable complex in the absence of protein (Mackie & Zimmermann, 1975; Ungewickell, Ehresmann, Stiegler & Garrett, 1975). This type of tertiary interaction has been mentioned earlier, and a similar instance has also been observed in the RNA region corresponding to protein S7 (Rinke, Yuki & Brimacombe, 1976).

In the case of 23S RNA, most of the available data have been obtained by making use of the fact that 23S RNA can be readily split into a 13S and 18S fragment (Allet & Spahr, 1971), to which individual binding proteins (Spierer, Zimmermann & Mackie, 1975) or groups of proteins (Chen-Schmeisser & Garrett, 1976) can be re-bound. A more precise localization has been made of the 'binding sites' of some of the 23S RNA binding proteins (reviewed by Brimacombe *et al.*, 1976), as well as of the three proteins, L5, L18 and L25, which bind to 5S RNA

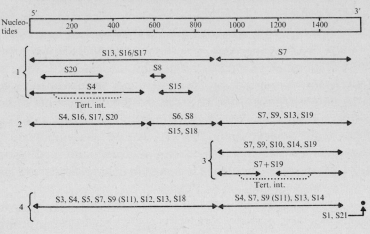

(a) 16 S RNA and 30 S proteins

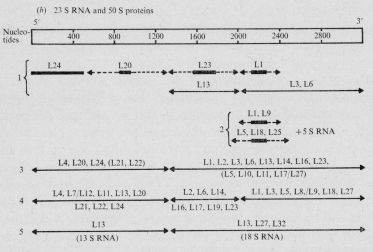

(b) 23 S RNA and 50 S proteins

Fig. 2. Summary of data on interactions of proteins with ribosomal RNA (see text for details). The scales indicate number of nucleotides, from the 5′-end. (a) 16 S RNA. 1. Protein 'binding sites' determined by nuclease digestion studies on single protein–RNA complexes (reviewed by Zimmermann, 1974; Brimacombe, Nierhaus, Garrett & Wittman, 1976). The dotted line indicates RNA fragments which form a tertiary interaction in the absence of protein (see text). 2. Distribution of proteins in fragments of a reconstitution intermediate particle (Zimmermann, 1974). 3. Distribution of proteins in fragments of the 30S subunit (Rinke *et al.*, 1976). 4. Distribution of proteins found by RNA–protein cross-linking within the 30S subunit (Czernilofsky, Kurland & Stöffler, 1975; Möller *et al.*, 1977). (b) 23S RNA. 1. Protein 'binding sites' determined by nuclease digestion studies (see text for individual references). 2. Distribution of proteins in fragments of the 50S subunit (Branlant, Krol, Sriwidada & Brimacombe, 1976a). 3. Distribution of proteins which rebind to the 13S and 18S RNA fragments (Spierer *et al.*, 1975). Proteins in brackets require the presence of other proteins in order to bind (Spierer & Zimmermann, 1976). 4. Distribution of proteins by reconstitution experiments with RNA fragments (Chen-Schmeisser & Garrett, 1976). 5. Distribution of proteins found by RNA–protein cross-linking within the 50S subunit (Möller *et al.*, 1977).

(reviewed by Erdmann, 1976), and, very recently, detailed analyses have been published of the binding sites on 23S RNA of L1 (Branlant *et al.*, 1976*b*) and L24 (Branlant *et al.*, 1977). All these results, for both 16S and 23S RNA, are summarized in Fig. 2.

For those proteins which are unable to bind individually to RNA, the alternative approach involving controlled hydrolysis of ribosomal subunits has yielded some information. Here, the object is to isolate fragments of ribonucleoprotein, and to establish topographical groupings of protein and RNA by analysing the protein and RNA moieties of the fragments. This method has been usefully applied both to the intact 30S subunit and also to a reconstituted intermediate particle, but has had only a limited success in the case of the 50S subunit (reviewed by Brimacombe *et al.*, 1976). Again, the results are included in Fig. 2.

The disadvantage of all these methods is that they are limited both by the ease with which the RNA–protein complexes dissociate, and by the ease with which the specificity of the interaction is lost. In consequence, a number of laboratories have turned their attention to the covalent cross-linking of RNA to protein.

RNA–protein cross-linking

Cross-linking studies yield a completely different type of information to those outlined above. In the latter, a strong physical association is required to hold the RNA–protein complex together, whereas cross-linking (as is the case in protein–protein cross-linking work) is only a measure of topographical neighbourhood. For this very reason, cross-linking studies on intact subunits are particularly interesting, as they have the potential of revealing RNA–protein contacts or neighbourhoods in regions where no strong physical association exists, and the complex nature of the protein arrangement (see Fig. 1) would predict that many such contacts will occur for any given protein, possibly at widely separated points along the RNA chain.

One specialized example of RNA–protein cross-linking within the 30S subunit has been the linking of the 3′-terminus of the 16S RNA to proteins S1 and S21 (Czernilofsky *et al.*, 1975) via periodate oxidation of the 3′-terminal ribose followed by borohydride reduction. For cross-linking within the bulk of the 16S and 23S RNA molecules, more general methods are required, and while some bifunctional chemical RNA–protein cross-linking reagents have been described (Fink & Brimacombe, 1975; Oste, Parfait, Bollen & Crichton, 1977), they have as yet yielded no detailed information on RNA–protein contacts within the subunits. At present the most widely used method for inducing

RNA–protein cross-links is direct irradiation with ultraviolet light. This has the advantage that it is a 'zero-length' cross-linking technique, in contrast to the chemical methods just mentioned (where the reagents involved are rather long). Gorelic (1976, and earlier publications) has shown that almost all the ribosomal proteins can be cross-linked to their respective RNA in the subunits by a simple ultraviolet irradiation procedure. On the other hand, it has also been shown that high levels of irradiation cause 'unfolding' of the ribosomal subunits (Möller & Brimacombe, 1975), in a manner likely to lead to protein exchange (Newton *et al.*, 1975), and under milder irradiation conditions the reaction yields a very specific cross-linking of only S7 within the 30S subunit, and of L2 and L4 in the 50S (Möller & Brimacombe, 1975; Baca & Bodley, 1976).

As with protein–protein cross-linking studies, the ultimate objective in this type of experiment should be the identification of the nucleotide and amino acid residues involved in the cross-link. A partial location of the 16S RNA region concerned has been made in the case of the ultraviolet induced cross-linking of protein S7 to 16S RNA (Rinke *et al.*, 1976), and studies on the peptide residues involved in all three cases just mentioned (S7, L2 and L4) suggest that a single cross-link has occurred, as evidenced by the appearance in each case of a single oligonucleotide–peptide complex after trypsin and nuclease digestion (K. Möller & R. Brimacombe, unpublished results). This is in contrast to the findings of other workers, who have cross-linked S7 and S4 individually to 16S RNA, by irradiation of the protein–RNA complex (e.g. Ehresmann *et al.*, 1976). Here, several peptides were found to be missing from each cross-linked protein after trypsin digestion, and this was interpreted as evidence of involvement of these peptides in the RNA–protein cross-links.

In another recent study, formaldehyde has been used to cross-link proteins to RNA within the 30S and 50S subunits (Möller *et al.*, 1977). While formaldehyde is not an ideal reagent for this purpose, as the cross-linking reaction is readily reversible, it was possible partially to locate a number of cross-linking points on the RNA. The results are included in Fig. 2, and it is noteworthy that several proteins showed formaldehyde cross-linked 'contacts' in both halves of their respective ribosomal RNA molecules. In each case these proteins (S4, S7, S13 and L13) are either elongated (Fig. 1) or else have shown ambiguous binding properties in other experiments (Fig. 2). This experiment therefore supports the contention made above that, within the subunits, some (and perhaps many) proteins do not possess a single RNA 'binding site', but rather have a number of contacts with dispersed regions of the RNA chain. A precise analysis of such contacts will obviously play an important part in working out the topological organization of the RNA.

CONCLUSION

The purpose of this article has been to review some of the recent developments in research into bacterial ribosome structure, and to show how these developments have brought with them a new generation of fascinating problems needing solution. While we obviously still have a very long way to go before ribosome structure and function can be correlated at the molecular level, preoccupation with these problems should not be allowed to overshadow what has in fact already been achieved. Ten years ago the ribosome was just a 'black box'. Since then we have seen the isolation and characterization of the individual ribosomal components, and the determination of the primary structures of a large proportion of the various protein and RNA species. We have sufficient data, on the shapes and the distribution in space of the individual proteins, to allow preliminary models to be constructed of the three-dimensional protein arrangement. There is detailed information on the in-vitro assembly of active ribosomal particles from their constituent protein and RNA molecules, and we know in many cases which regions of RNA interact with individual proteins. A comparable level of sophistication has not been reached in research into any other cellular organelle of similar complexity.

The author is very grateful to Dr H. G. Wittmann for his critical reading of the manuscript.

REFERENCES

ALLET, B. & SPAHR, P. F. (1971). Binding sites of ribosomal proteins on two specific fragments derived from *E. coli* 50S ribosomes. *European Journal of Biochemistry*, **19**, 250–5.

ARACO, A., BELLI, M. & ONORI, G. (1975). The secondary structure of *E. coli* ribosomes and ribosomal RNAs; a spectrophotometric approach. *Nucleic Acid Research*, **2**, 373–81.

BACA, O. G. & BODLEY, J. W. (1976). Ultra-violet induced covalent cross-linking of *E. coli* ribosomal RNA to specific proteins. *Biochemical and Biophysical Research Communications*, **70**, 1091–6.

BENKOV, K. & DELIHAS, N. (1974). Analysis of kethoxal bound to ribosomal proteins from *E. coli* 70S reacted ribosomes. *Biochemical and Biophysical Research Communications*, **60**, 901–8.

BRANLANT, C., KROL, A., SRIWIDADA, J. & BRIMACOMBE, R. (1976a). RNA sequences associated with proteins L1, L9 and L5, L18, L25 in ribonucleoprotein fragments isolated from the 50S subunit of *E. coli* ribosomes. *European Journal of Biochemistry*, **70**, 483–92.

BRANLANT, C., KROL, A., SRIWIDADA, J., EBEL, J. P., SLOOF, P. & GARRETT, R. A. (1967b). The binding site of protein L1 on 23S ribosomal RNA of *E. coli. European Journal of Biochemistry*, **70**, 457–69.

BRANLANT, C., SRIWIDADA, J., KROL, A. & EBEL, J. P. (1976c). Extensions of the known sequences at the 3′ and 5′ ends of 23S ribosomal RNA from *E. coli. Nucleic Acid Research*, **3**, 1671–87.

BRANLANT, C., SRIWIDADA, J., KROL, A. & EBEL, J. P. (1977). RNA sequences in ribonucleoprotein fragments of the complex formed from ribosomal 23S RNA and ribosomal protein L24 of *E. coli. European Journal of Biochemistry*, **74**, 155–70.

BRIMACOMBE, R., NIERHAUS, K. H., GARRETT, R. A. & WITTMANN, H. G. (1976). The ribosome of *E. coli. Progress in Nucleic Acid Research*, **18**, 1–44.

BROWNLEE, G. G., SANGER, F. & BARRELL, B. G. (1968). The sequence of 5S ribosomal RNA. *Journal of Molecular Biology*, **34**, 379–412.

CHAPMAN, N. M. & NOLLER, H. F. (1977). Protection of specific sites in 16S RNA from chemical modification by association of 30S and 50S ribosomes. *Journal of Molecular Biology*, **109**, 131–49.

CHEN-SCHMEISSER, U. & GARRETT, R. A. (1976). Distribution of protein assembly sites along the 23S ribosomal RNA of *E. coli. European Journal of Biochemistry*, **69**, 401–10.

CLEGG, C. & HAYES, D. (1974). Identification of neighbouring proteins in the ribosome of *E. coli. European Journal of Biochemistry*, **42**, 21–8.

COTTER, R. I., MCPHIE, P. & GRATZER, W. B. (1967). Internal organization of the ribosome. *Nature, London*, **216**, 864–8.

CRICK, F. H. C. (1968). The origin of the genetic code. *Journal of Molecular Biology*, **38**, 367–79.

CZERNILOFSKY, A. P., KURLAND, C. G. & STÖFFLER, G. (1975). 30S ribosomal proteins associated with the 3′-terminus of 16S RNA. *FEBS Letters*, **58**, 281–4.

EHRESMANN, B., REINBOLT, J., BACKENDORF, C., TRITSCH, D. & EBEL, J. P. (1976). Studies of the binding sites of *E. coli* ribosomal protein S7 with 16S RNA by ultraviolet irradiation. *FEBS Letters*, **67**, 316–19.

EHRESMANN, C., STIEGLER, P., MACKIE, G. A., ZIMMERMANN, R. A., EBEL, J. P. & FELLNER, P. (1975). Primary sequence of the 16S ribosomal RNA of *E. coli. Nucleic Acid Research*, **2**, 265–78.

ENGELMAN, D. M., MOORE, P. B. & SCHOENBORN, B. P. (1975). Neutron scattering measurements of separation and shape of proteins in 30S ribosomal subunit of *E. coli. Proceedings of the National Academy of Sciences, USA*, **72**, 3888–92.

ERDMANN, V. A. (1976). Structure and function of 5S and 5.8S RNA. *Progress in Nucleic Acid Research*, **18**, 45–90.

FINK, G. & BRIMACOMBE, R. (1975). Synthesis and applications of a two-step reagent forming cross-links between RNA and protein. *Biochemical Society Transactions*, **3**, 1014–15.

GINZBURG, I. & ZAMIR, A. (1975). Characterization of different conformational forms of 30S ribosomal subunits in isolated and associated states. *Journal of Molecular Biology*, **93**, 465–76.

GORELIC, L. (1976). Photo-induced cross-linkage *in situ* of *E. coli* 30S ribosomal proteins to 16S RNA. *Biochemistry*, **15**, 3579–90.

HARDY, S. J. S. (1975). The stoichiometry of the ribosomal proteins of *E. coli, Molecular and General Genetics*, **140**, 253–74.

HEIMARK, R. L., KAHAN, L., JOHNSTON, K., HERSHEY, J. W. B. & TRAUT, R. R. (1976). Cross-linking of initiation factor IF-3 to proteins of the *E. coli* 30S ribosomal subunit. *Journal of Molecular Biology*, **105**, 219–30.

HELD, W. A., BALLOU, B., MIZUSHIMA, S. & NOMURA, M. (1974). Assembly mapping of 30S ribosomal proteins from *E. coli*. *Journal of Biological Chemistry*, **249**, 3103–11.

HERR, W. & NOLLER, H. F. (1975). A fragment of 23S RNA containing a nucleotide sequence complementary to a region of 5S RNA. *FEBS Letters*, **53**, 248–52.

HILL, W. E. & FESSENDEN, R. J. (1974). Structural studies on the 30S ribosomal subunit from *E. coli*. *Journal of Molecular Biology*, **90**, 719–26.

HOCHKEPPEL, H. K., SPICER, E. & CRAVEN, G. R. (1976). A method of preparing *E. coli* 16S RNA possessing previously unobserved 30S ribosomal protein binding sites. *Journal of Molecular Biology*, **101**, 155–70.

HOPPE, W., MAY, R., STÖCKEL, P., LORENZ, S., ERDMANN, V. A., WITTMANN, H. G., CRESPI, H. L., KATZ, J. J. & IBEL, K. (1975). Neutron scattering measurements with the label triangulation method on the 50S subunit of *E. coli* ribosomes. In *Neutron scattering for the analysis of biological structures; Brookhaven Symposium* no. 27; pp. iv, 38–48.

HUANG, K. H., FAIRCLOUGH, R. & CANTOR, C. R. (1975). Singlet energy transfer studies of the arrangement of proteins in the 30S *E. coli* ribosome. *Journal of Molecular Biology*, **97**, 443–70.

KALTSCHMIDT, E. & WITTMANN, H. G. (1970). Number of proteins in small and large ribosomal subunits of *E. coli* as determined by two-dimensional gel electrophoresis. *Proceedings of the National Academy of Sciences, USA*, **67**, 1276–82.

KURLAND, C. G. (1974). Functional organization of the 30S ribosomal subunit. In *Ribosomes*, ed. M. Nomura, A. Tissières & P. Lengyel, pp. 309–31. Cold Spring Harbor Laboratory, USA.

KURLAND, C. G. (1977). Structure and function of the bacterial ribosome. *Annual Review of Biochemistry*, **46**, 173–200.

LAKE, J. A. (1976). Ribosome structure determined by electron microscopy of *E. coli* small subunits, large subunits, and monomeric ribosomes. *Journal of Molecular Biology*, **105**, 131–59.

LANGLOIS, R., LEE, C. C., CANTOR, C. R., VINCE, R. & PESTKA, S. (1976). The distance between two functionally significant regions of the 50S *E. coli* ribosome; the erythromycin binding site and proteins L7/L12. *Journal of Molecular Biology*, **106**, 297–313.

LILJAS, A. & KURLAND, C. G. (1976). Crystallization of ribosomal protein L7/L12 from *E. coli*. *FEBS Letters*, **71**, 130–2.

LITMAN, D. J., BEEKMAN, A. & CANTOR, C. R. (1976). Further studies on the identity of proteins at the subunit interface of the 70S *E. coli* ribosome. *Archives of Biochemistry and Biophysics*, **174**, 523–31.

LITT, M. (1969). Labelling of exposed guanine sites in yeast phenylalanine tRNA with kethoxal. *Biochemistry*, **8**, 3249–53.

LITTLECHILD, J., DIJK, J. & GARRETT, R. A. (1977). The identification of new RNA-binding proteins in the *E. coli* ribosome. *FEBS Letters*, **74**, 292–4.

LUTTER, L. C., KURLAND, C. G. & STÖFFLER, G. (1975). Protein neighbourhoods in the 30S ribosomal subunit of *E. coli*. *FEBS Letters*, **54**, 144–50.

LUTTER, L. C., ORTANDEL, F. & FASOLD, H. (1974). The use of a new series of cleavable protein cross-linkers on the *E. coli* ribosome. *FEBS Letters*, **48**, 288–92.

MACKIE, G. A. & ZIMMERMANN, R. A. (1975). Characterization of fragments of 16S RNA protected against pancreatic ribonuclease digestion by ribosomal protein S4. *Journal of Biological Chemistry*, **250**, 4100–12.

MAGRUM, L., ZABLEN, L., STAHL, D. & WOESE, C. (1975). Corrections in the catalogue of oligonucleotides produced by digestion of *E. coli* 16S RNA with T_1 ribonuclease. *Nature, London*, **257**, 423–6.

MIALL, S. H. & WALKER, I. O. (1969). Structural studies on ribosomes. *Biochimica Biophysica Acta*, **174**, 551–60.

MÖLLER, K. & BRIMACOMBE, R. (1975). Specific cross-linking of proteins S7 and L4 to ribosomal RNA, by ultra-violet irradiation of *E. coli* ribosomal subunits. *Molecular and General Genetics*, **141**, 343–55.

MÖLLER, K., RINKE, J., ROSS, A., BUDDLE, G. & BRIMACOMBE, R. (1977). The use of formaldehyde in RNA-protein cross-linking studies with ribosomal subunits from *E. coli*. *European Journal of Biochemistry*, **76**, 175–87.

MÖLLER, W. (1974). The ribosomal components involved in EF-G and EF-Tu dependent GTP hydrolysis. In *Ribosomes*, ed. M. Nomura, A. Tissières & P. Lengyel, pp. 711–31. Cold Spring Harbor Laboratory, USA.

MOORE, P. B., ENGELMAN, D. M. & SCHOENBORN, B. P. (1974). Assymmetry in the 50S ribosomal subunit of *E. coli*. *Proceedings of the National Academy of Sciences, USA*, **71**, 172–6.

MOORE, P. B., LANGER, J. A., SCHOENBORN, B. P. & ENGELMAN, D. M. (1977). Triangulation of proteins in the 30S ribosomal subunit of *E. coli*. *Journal of Molecular Biology*, **112**, 199–234.

MORRISON, C. A., GARRETT, R. A., ZEICHHARDT, H. & STÖFFLER, G. (1973). Proteins occurring at, or near, the subunit interface of *E. coli* ribosomes. *Molecular and General Genetics*, **127**, 359–68.

NEWTON, I., RINKE, J. & BRIMACOMBE, R. (1975). Random exchange of ribosomal proteins in EDTA sub-particles. *FEBS Letters*, **51**, 215–18.

NIERHAUS, K. H. & DOHME, F. (1974). Total reconstitution of functionally active 50S ribosomal subunits from *E. coli*. *Proceedings of the National Academy of Sciences, USA*, **71**, 4713–17.

NOLLER, H. F. & HERR, W. (1974). Accessibility of 5S RNA in 50S ribosomal subunits. *Journal of Molecular Biology*, **90**, 181–4.

NOMURA, M. & ERDMANN, V. A. (1970). Reconstitution of 50S ribosomal subunits from dissociated molecular components. *Nature, London*, **228**, 744–8.

OSTE, C., PARFAIT, R., BOLLEN, A. & CRICHTON, R. R. (1977). A new nucleic acid–protein cross-linking reagent. *Molecular and General Genetics*, **152**, 253–7.

ÖSTERBERG, R., SJÖBERG, B. & GARRETT, R. A. (1976). Small-angle X-ray scattering study of the 5S RNA binding proteins L18 and L25 from *E. coli* ribosomes. *FEBS Letters*, **65**, 73–6.

ÖSTERBERG, R., SJÖBERG, B., GARRETT, R. A. & LITTLECHILD, J. (1977). Small-angle-X-ray scattering study of the 16S RNA binding protein S4 from *E. coli* ribosomes. *FEBS Letters*, **73**, 25–8.

PERETZ, H., TOWBIN, H. & ELSON, D. (1976). The use of a cleavable cross-linking reagent to identify neighbouring proteins in the 30S ribosomal subunit of *E. coli*. *European Journal of Biochemistry*, **63**, 83–92.

PETTERSON, I., HARDY, S. J. S. & LILJAS, A. (1976). The ribosomal protein L8 is a complex of L7/L12 and L10. *FEBS Letters*, **64**, 135–8.

PONGS, O., NIERHAUS, K. H., ERDMANN, V. A. & WITTMANN, H. G. (1974). Active sites in *E. coli* ribosomes. *FEBS Letters*, **40**, S28–37.

RINKE, J., YUKI, A. & BRIMACOMBE, R. (1976). Studies on the environment of protein S7 within the 30S subunit of *E. coli* ribosomes. *European Journal of Biochemistry*, **64**, 77–89.

SCHWARTZ, U., LÜHRMANN, R. & GASSEN, H. G. (1974). On the mRNA induced conformational change of aa-tRNA, exposing the T-ψ-C-G sequence for binding to the 50S ribosomal subunit. *Biochemical and Biophysical Research Communications*, **56**, 807–14.

SENIOR, B. W. & HOLLAND, I. B. (1971). Effect of colicin E3 upon the 30S ribosomal subunit of *E. coli. Proceedings of the National Academy of Sciences, USA*, **68**, 959–63.

SHINE, J. & DALGARNO, L. (1974). The 3′-terminal sequence of *E. coli* 16S RNA; complementarity to nonsense triplets and ribosome binding sites. *Proceedings of the National Academy of Sciences, USA*, **71**, 1342–6.

SOMMER, A. & TRAUT, R. R. (1976). Identification of neighbouring protein pairs in the *E. coli* 30S ribosomal subunit by cross-linking with methyl-4-mercaptobutyrimidate. *Journal of Molecular Biology*, **106**, 995–1015.

SPIERER, P. & ZIMMERMANN, R. A. (1976). RNA–protein interactions in the ribosome: co-operative interactions in the 50S subunit of *E. coli. Journal of Molecular Biology*, **103**, 647–53.

SPIERER, P., ZIMMERMANN, R. A. & MACKIE, G. (1975). RNA-protein interactions in the ribosome: binding of 50S subunit proteins to the 5′ and 3′ terminal segments of the 23S RNA. *European Journal of Biochemistry*, **52**, 459–68.

SPITNIK-ELSON, P. & ELSON, D. (1976). Studies on the ribosome and its components. *Progress in Nucleic Acid Research*, **17**, 77–98.

STEITZ, J. A. & JAKES, K. (1975). How ribosomes select initiator regions in mRNA. *Proceedings of the National Academy of Sciences, USA*, **72**, 4734–8.

STÖFFLER, G. & WITTMANN, H. G. (1971). Sequence differences of *E. coli* 30S ribosomal proteins as determined by immunochemical methods. *Proceedings of the National Academy of Sciences, USA*, **68**, 2283–7.

STÖFFLER, G. & WITTMANN, H. G. (1977). Primary structure and three-dimensional arrangement of proteins within the *E. coli* ribosome. In *Protein Synthesis*, ed. H. Weissbach & S. Pestka, pp. 117–202. New York: Academic Press.

STUHRMANN, H. B., HAAS, J., IBEL, K., DE WOLF, B., KOCH, M. H. J., PARFAIT, R. & CRICHTON, R. R. (1976). New low resolution model for 50S subunit of *E. coli* ribosomes. *Proceedings of the National Academy of Sciences, USA*, **73**, 2379–83.

SUBRAMANIAN, A. R. (1975). Copies of L7 and L12 and heterogeneity of the large subunit of the *E. coli* ribosome. *Journal of Molecular Biology*, **95**, 1–8.

SUN, T. T., TRAUT, R. R. & KAHAN, L. (1974). Protein–protein proximity in the association of ribosomal subunits of *E. coli. Journal of Molecular Biology*, **87**, 509–22.

SZER, W., HERMOSO, J. M. & BOUBLIK, M. (1976). Destabilisation of the secondary structure of RNA by ribosomal protein S1 from *E. coli. Biochemical and Biophysical Research Communications*, **70**, 957–64.

TISCHENDORF, G. W., ZEICHHARDT, H. & STÖFFLER, G. (1975). Architecture of the *E. coli* ribosome as determined by immune electron microscopy. *Proceedings of the National Academy of Sciences, USA*, **72**, 4820–4.

TRAUB, P. & NOMURA, M. (1969). Structure and function of *E. coli* ribosomes; mechanism of assembly of 30S ribosomes studied *in vitro. Journal of Molecular Biology*, **40**, 391–413.

UNGEWICKELL, E., EHRESMANN, C., STIEGLER, P. & GARRETT, R. A. (1975). Evidence for tertiary structural RNA–RNA interactions within the protein S4 binding site at the 5′-end of 16S ribosomal RNA of *E. coli. Nucleic Acid Research*, **2**, 1867–93.

VAN DUIN, J., KURLAND, C. G., DONDON, J. & GRUNBERG-MANAGO, M. (1975). Near neighbours of IF-3 bound to 30S ribosomal subunits. *FEBS Letters*, **59**, 287–90.

VAN KNIPPENBERG, P. H. (1975). A possible role of the 5′-terminal sequence of 16S ribosomal RNA in the recognition of initiation sequences for protein synthesis. *Nucleic Acid Research*, **2**, 79–85.

VOYNOW, P. & KURLAND, C. G. (1971). Stoichiometry of the 30S ribosomal proteins of *E. coli. Biochemistry*, **10**, 517–24.

WAGNER, R., GASSEN, H. G., EHRESMANN, C., STIEGLER, P. & EBEL, J. P. (1976). Identification of a 16S RNA sequence located in the decoding site of 30S ribosomes. *FEBS Letters*, **67**, 312–15.

WITTMANN, H. G. (1974). Purification and identification of *E. coli* ribosomal proteins. In *Ribosomes*, ed. M. Nomura, A. Tissières & P. Lengyel, pp. 93–114. Cold Spring Harbor Laboratory, USA.

WITTMANN-LIEBOLD, B., ROBINSON, S. M. L. & DZIONARA, M. (1977). Prediction of secondary structures in proteins from the *E. coli* 30S ribosomal subunit. *FEBS Letters*, **77**, 301–7.

WOESE, C. R., FOX, G. E., ZABLEN, L., UCHIDA, T., BONEN, L., PECHMAN, K., LEWIS, B. J. & STAHL, D. (1975). Conservation of primary structure in 16S ribosomal RNA. *Nature, London*, **254**, 83–6.

ZIMMERMANN, R. A. (1974). RNA–protein interactions in the ribosome. In *Ribosomes*, ed. M. Nomura, A. Tissières & P. Lengyel, pp. 225–69. Cold Spring Harbor Laboratory, USA.

STRUCTURE AND FUNCTION OF THE TRANSLATION MACHINERY

M. GRUNBERG-MANAGO, R. H. BUCKINGHAM
B. S. COOPERMAN AND J. W. B. HERSHEY

Institut de Biologie Physico-Chimique,
13 rue Pierre et Marie Curie, 75005 Paris, France

INTRODUCTION

Brimacombe's review opening this volume shows strikingly the complexity of organization of the organelles on which protein synthesis occurs. The complexity is even greater when one considers the translation machinery as a whole which consists, in addition to ribosomes, of aminoacyl–tRNA (aatRNA), mRMA, numerous ligands (monovalent and divalent ions, nucleotides) and proteins transiently associated with the ribosomes at the different steps of protein synthesis.

STRUCTURE AND FUNCTION OF PROTEIN SYNTHESIS INITIATION COMPLEX

The steps preceding the formation of the first peptide bond are particularly important in synthesizing correct polypeptides. This process, called initiation, consists of a series of events ensuring the recognition of particular regions of the RNA chain. These regions, called the initiation signals, specify the place where decoding of the template RNA begins. During this step, the tRNA providing the N-terminal aminoacyl residue of the nascent chain is positioned opposite the initiation triplet present within the initiation signal. The formation of such an initiation complex is critical in phasing the 'readout' of the RNA and is probably the rate-limiting step in protein synthesis. Initiation also represents an obvious and important control point in translation. For instance, alteration of any one of the components involved in the formation of the initiation complex could theoretically eliminate recognition of a particular initiation region, cause a latent site to become active, or alter the relative efficiency of ribosome binding to various initiation triplets in a mRNA.

In the case of prokaryotes and mainly in *Escherichia coli*, the components participating in the initiation step are now well characterized. For a certain number, we know their site of attachment on the ribosome and we have an overall hypothesis to explain the selection of initiation sites as well as the role of protein factors involved in this step.

STRUCTURE OF THE COMPONENTS
OF THE INITIATION COMPLEX

A brief summary of the initiation sequence is shown in Fig. 1 and will be detailed below. Step (1) is required to generate the small ribosomal subunit on which the initiation complex containing mRNA and tRNA is first formed. Under physiological conditions, as will be discussed further (p. 50), the equilibrium is towards the 70S, and two proteins,

$$70S \rightleftharpoons 50S + 30S \tag{1}$$

$$30S + fMet\text{-}tRNA^{fMet} + mRNA + GTP \rightleftharpoons 30S \cdot fMet\text{-}tRNA^{fMet} \cdot mRNA \cdot GTP \tag{2}$$

$$50S + 30S \cdot fMet\text{-}tRNA^{fMet} \cdot mRNA \cdot GTP \rightleftharpoons \underbrace{70S \cdot fMet\text{-}tRNA^{fMet}(P) \cdot mRNA}_{\text{initiation complex}} + GDP + P_i \tag{3}$$

Fig. 1. Initiation sequence.

called initiation factors IF1 and IF3, play a role in altering the equilibrium position. IF1 serves to increase the rate constant for the dissociation of the 70S ribosome while IF3 shifts the equilibrium strongly towards dissociation by binding to the smaller 30S particle.

The second step requires a third initiation factor, IF2, to direct fMet–tRNA^{fMet} binding. The third step is accompanied by the release of all three factors; whereas IF1 and IF3 are released before or during the junction of the 50S subunit, IF2 release occurs after this junction and requires GTP hydrolysis. The three steps result in the formation of the initiation complex which contains a 70S ribosome, mRNA and fMet–tRNA^{fMet}.

Initiator tRNA

The first residue incorporated at the NH$_2$ terminus into a growing protein chain is always a methionine residue, whatever the nature of the protein (Grunberg-Manago & Gros, 1977). This residue is bound to the messenger-ribosome complex as a derivative of fMet–tRNA which is different from the tRNA positioning methionine internally in the protein chain. Thus, two species of methionine-accepting tRNAs are found in prokaryotes: an initiator tRNA (tRNA^{fMet}) and an elongation tRNA (tRNA^{Met}). They are both acylated by the same synthetase and recognize the triplet AUG, although the initiator tRNA will also recognize GUG *in vitro* and *in vivo* as well as, under some

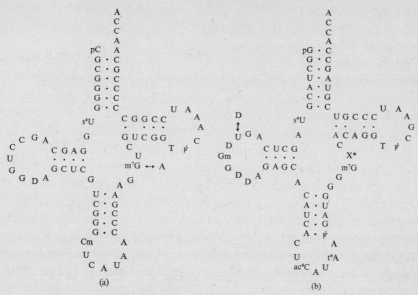

Fig. 2. Sequences of *E. coli* Methionyl-tRNAs, (*a*) *E. coli* CA 265 tRNA^fMet, (*b*) *E. coli* CA 265 tRNA^Met. From Barrell & Clark (1974).

conditions, CUG and UUG. Initiation of protein synthesis by methionyl–tRNA^fMet requires the aminoacyl residue to be *N*-substituted by a formyl group. Formylation is catalysed by a special transformylase (Marcker, 1965) which transfers a formyl residue from an *N*-formyl-tetrahydrofolate donor to Met–tRNA^fMet. It cannot, however, formylate the methionine of Met–tRNA^Met. *In vivo*, the formyl group, initially present on the growing chain, is rapidly removed by a 'deformylase' (Adams, 1968). Once the methionine from the growing polypeptide is deformylated, other enzymes may split off the methionine residue itself.

Two particular properties of the initiator tRNA have emerged from a comparison of the primary sequences of tRNAs (Fig. 2; Barrell & Clark, 1974) that may be significant for the discrimination, shown by enzymes, between the initiator and other tRNAs. No Watson–Crick base-pair is found at the 5'-terminal position of the acceptor stem (base-pair 1–73) and there is an adenine in position 11 in the dihydro-uridine stem, whereas all other tRNA species of known sequence have a cytidine or uridine in that position. Bisulphite modification of the tRNA has demonstrated that the missing base-pair is important in the resistance of fMet–tRNA^fMet to hydrolysis by peptidyl-hydrolase (Schulman & Pelka, 1975) and in the lack of interaction between Met–tRNA^fMet and EF–Tu·GTP (Schulman, & Her, 1973). On the other hand the modified tRNA^{fMet} is still recognized by the transformylase and shows initiation-factor dependent binding to ribosomes (L. H. Schulman, unpublished data; see Ofengand, 1977).

While formylation of Met–tRNA^{fMet} is generally required for its incorporation into the 70S initiation complex, an interesting exception has been established in *Streptococcus faecalis* that sheds some light on the role of *N*-formylation (Delk & Rabinowitz, 1974). *S. faecalis* grown in the absence of folic acid cannot formylate its Met–tRNA^{fMet} but unformylated tRNA initiates at a normal rate (Delk & Rabinowitz, 1974). A second alteration of the initiator tRNA structure, the failure to convert uracil to pseudo-uracil in the GTψC sequence, may somehow compensate for the lack of formylation (Delk & Rabinowitz, 1975). A similar situation was observed in two other Gram-positive bacteria, *Bacillus subtilis* and *Micrococcus luteus* (Arnold, Schmidt & Keroten, 1975); here, trimethoprim (an inhibitor of dihydrofolate reductase) caused the same modifications as those described above; no formylated Met–tRNA could be detected. Nevertheless, both organisms synthesize proteins at normal rates under these conditions provided that the low MW products of folate metabolism are supplied in the medium. Finally in a halophilic bacterium lacking transformylase, non-formylated Met–tRNA^{fMet} serves as the initiator (White & Bayley, 1972). Thus initiation in Gram-positive bacteria differs in some respect from that in Gram-negative species (see p. 41) and it would be interesting to know whether the ability to initiate in the absence of formylation could be extended to all Gram-positive bacteria.

Ribosomes

The morphology of the ribosomal subunit has been reviewed by Brimacombe in this volume. In ribosome preparations, some structural herterogeneity exists (Voynow & Kurland, 1971; see p. 62); proteins such as S1 and S21 are easily removed during the isolation process (Hardy, 1975). In addition, ribosomes can undergo conformational changes. The 30S and the 50S subunits exist in active and inactive conformations (McLaughlin *et al.*, 1968; Kaufman & Zamir, 1972) and sometimes the two conformations are reversible. However, an apparently irreversible heterogeneity of 30S and 50S also exists in most preparations of ribosomes (Noll & Noll, 1976; Van Diggelen & Bosch, 1973; Debey *et al.*, 1975). Two classes of ribosomes can thus be discriminated by their difference in sensitivity to hydrostatic pressure during high speed centrifugation in sucrose gradients. If the ribosomes are prepared by centrifugation in 5 mM Mg^{2+}, the tight couples (or type-A ribosomes) can be selected which are the most active and most resistant to hydro-static pressure. This preparative technique leads to the selection of ribosomes forming an initiation complex with an efficiency of over 50% (Noll, Noll, Hapke & van Dieijen, 1973). When special precautions are taken in growing the bacteria (harvesting at a point one-third along the exponential phase of growth) and when the ribosomes are prepared immediately after harvesting the bacteria and without freezing them, a homogeneous preparation of type-A ribosomes is obtained (Debey *et al.*, 1975).

The subunit association curves, as a function of Mg^{2+}, differ widely according to the type of ribosomes used (Debey *et al.*, 1975; Noll & Noll, 1976). At a given concentration of monovalent cations, the interval between the Mg^{2+} concentrations required for 10% and 90% association is narrower, and the half-saturation lower, for type-A ribosomes. Furthermore, when exposed to 1 mM Mg^{2+} they showed no hysteresis effect in their reassociation curve, in contrast to type-B.

The structural differences between the two types of ribosomes are not clear. The conformational difference usually appears to be in the 50S subunit. Type-B ribosomes are more flexible and undergo conforma-tional changes that result in their being unfit for reassociation. These conformational changes can sometimes be reversed by thermal treat-ment.

The kinetics of association and dissociation of *E. coli* 30S and 50S ribosomal subunits of type-A appear to fit the simple scheme:

$$30S + 50S \underset{k_1}{\overset{k_2}{\rightleftharpoons}} 70S$$

over a wide range of Mg^{2+} and ribosome concentrations (Wishnia *et al.*, 1975). Therefore, the association of free 50S and 30S is not an association reaction limited by complex conformational changes in the subunits, but a second-order reaction governed by the rate of collision of these particles. Both rate constants strongly depend on Mg^{2+}, monovalent ions, and on the preparation of the particles. When particles are washed with high concentrations of salt, k_2 (at 25 °C and 50 mM NH_4Cl) ranges from 0.04×10^6 to 21×10^6 M^{-1} s^{-1} as Mg^{2+} is increased from 1.5 mM to 8 mM; k_1 from 150 to 2.5 $\times$ 10^{-3} s^{-1} for concentrations of 1.0 to 3.0 mM Mg^{2+}. The kinetics are very similar in the presence of other divalent ions such as Ca^{2+}, Sr^{2+}, Ba^{2+}, whereas Mn^{2+} causes an increase in the rate of association k_2, while k_1 is lowered (A. Wishnia, unpublished experiments). The highest rate may not be far from the diffusion-controlled limit.

The equilibrium of ribosomal particles is temperature-dependent (Hui Bon Hoa, Graffe & Grunberg-Manago, 1977). The association is exothermic with an enthalpy change of:

$$\Delta H^\circ = -19.9 \text{ kcal mol}^{-1}$$

The entropy change is also negative. The enthalpy term is predominant for the reaction. However, type-A ribosomes exhibit less negative enthalpy and entropy changes than type-B ribosomes where

$$\Delta H^\circ = -38 \text{ kcal mol}^{-1}.$$

This probably indicates that type-A couples undergo less conformational change upon association than type-B couples or have a smaller difference in the amount of bound water between the free and associated subunits.

These thermodynamic data for the overall equilibrium association are quite different from those obtained for reactions involving hydrophobic interactions, which are generally endothermic processes. This seems to eliminate any preponderant contribution of hydrophobic-binding between the subunits. On the other hand, the enthalpy change of the association reaction between ribosomal subparticles is close to that for complementary oligonucleotide interactions, but the difference between these two types of interactions is shown in the entropy change. In accordance with the discussion (see pp. 58–9), the thermodynamic data suggest that interaction between ribosomal subunits is mainly due to interactions between RNA chains. The effect of Mg^{2+} concentration, although strong enough in terms of rate to have important physiological consequences, is relatively weak in terms of free energy (3.6 to 2.1 kcal). The system is well balanced and other factors may easily tip it one way or the other.

The primary effect of Mg^{2+} (as calculated from the rather large changes in binding to the ribosomal particles as a function of concentration) could be explained by a decrease in the contribution of electrostatic repulsion to the free energy of activation. However, whether in addition specific interactions of Mg^{2+} with some RNA components or proteins contribute to the Mg-dependence is an unsettled question, but there is some indication that this is indeed so (Kliber, Hui Bon Hoa, Douzou & Grunberg-Manago, 1976).

mRNA

Synthetic polynucleotides having an AUG or GUG codon at their 5′-termini direct the synthesis of polypeptides initiated with *N*-formylmethionine. For natural mRNA, most of our knowledge of structure comes from the RNA bacteriophages. The ease with which the single-stranded RNA genomes of these viruses can be labelled to high specific activities and purified from the phage particles has made them ideal subjects for sequence analysis. The complete primary sequence of MS2 RNA, 3569 nucleotides long, has recently been achieved by Fiers *et al.* (1975, 1976) and we shall undoubtedly gain a lot from this knowledge.

The RNA of R17, MS2 or f2 phage genomes encodes only three proteins (Lodish & Robertson, 1969; Weissmann *et al.*, 1973). These are the coat protein (MW 14000) which is synthesized in largest amounts both *in vivo* and *in vitro*; the phage specific subunit of the replicase (MW 60000–67000, depending on the phage); and the 'A' or maturation protein (MW about 40000) which is normally translated in smallest amounts and is a minor constituent of the virus particle. The particle contains 1 A protein molecule and approximately 180 coat protein molecules (Steitz, 1968). The MS2 genome starts from the 5'-end with an untranslated leader sequence, 129 nucleotides long, followed by the A protein gene, an intercistronic region of 26 nucleotides, the coat gene, an intercistronic region of 36 nucleotides, the replicase gene, and finally, an untranslated segment at the 3'-terminus 174 nucleotides long (De Wachter *et al.*, 1971; Fiers *et al.*, 1976). In Qβ RNA, a fourth protein (MW 36000) appears, which is the product of natural read-through at the UGA termination signal of the Qβ coat cistron, and is therefore not considered as a discrete protein (Weiner & Weber, 1971).

Tentative models (Fiers *et al.*, 1975, 1976) for the secondary structural folding of the viral MS2 RNA propose a secondary structure for each gene, resembling a flower with similar folding in other parts of the molecule, the secondary structure of the whole viral RNA constituting a bouquet (Fig. 3). The existence of the proposed RNA secondary structure is supported both by the results of partial ribonuclease digestion experiments, and by the observed specificity of chemical mutagenesis performed on whole phage particles. The early physico-chemical measurements on phage RNA molecules indicated not only a high helix content (60–80%) but an unusually compact overall structure (Boedtker & Gesteland, 1975).

Before the brilliant achievement by Fiers's group in elucidating the total primary sequence of MS2 RNA, several groups (Steitz, 1969, 1977) isolated and sequenced the ribosomal binding sites of the RNA bacteriophages. The sequencing of these binding sites was based on results (Takanami, Yan & Jukes, 1965) which demonstrated that regions of

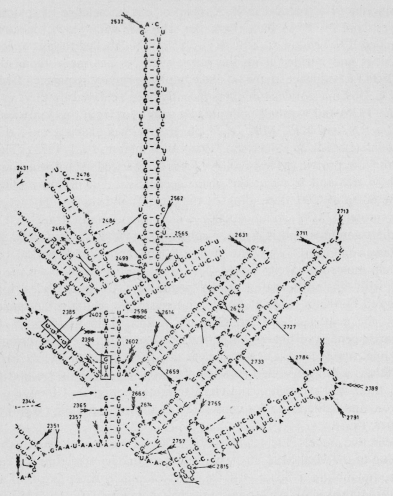

Fig. 3. A portion of the nucleotide sequence with secondary structure of the replicase gene. Primarily the sequence is read from the 5'-end to the 3'-end. The numbering starts at the first 5'-nucleotide of MS2 RNA and ends at the 3'-terminal residue 3569. Arrows point to sites easily split by nucleases during partial digestion of MS2 RNA: solid arrows, T1 ribonuclease: dashed arrows, carboxymethylated ribonuclease. The feathers on the arrows indicate sensitivity of the bond under the conditions used; (0) split very seldom, (2) rather often, (3) very often, (4) always. The AUG and the Shine and Dalgarno sequence, GGAG, enclosed in rectangles denote initiation type sequences not recognized as such by ribosomes. From Fiers *et al.* (1976).

bacteriophage RNA are protected from ribonuclease digestion by association with ribosomes. When ^{32}P-labelled mRNA is mixed with washed *E. coli* ribosomes plus initiation factors and initiator tRNA, the ribosomes bind selectively at the initiation sequence and are unable to move from this site due to the lack of aatRNAs and elongation factors. Upon digestion of the complex, either by pancreatic ribonuclease or by T1 ribonuclease, the whole mRNA is hydrolysed, except for the portions associated with the ribosomes. The protected fragments can be liberated upon treatment with a denaturing agent such as sodium dodecylsulphate, and the sequence determined by Sanger's procedure (Barrell, 1971). Steitz (1969, 1977) was thus able to establish the sequence of initiation signals preceding the three cistrons of different phage RNAs. In each example the amount of ^{32}P-labelled RNA remaining bound to ribosomes after fractionation from the bulk of the degraded RNA depended on the presence of fMet–tRNA^{fMet} in the original initiation reaction. Sites protected by ribosomes are exclusively at the beginning of genes and in general, the protected fragments are about 30–40 nucleotides long. Ribosomes can bind, even in the absence of translation, to internal initiator regions on a polycistronic mRNA. Under normal conditions, *E. coli* ribosomes bind much more strongly to the initiator site of the coat protein cistron, other sites, especially that of the replicase, being hidden by secondary structure of the mRNA.

Analysis of initiation signals showed for the first time, that these sequences all include an initiator triplet preceding the first translated codons. The initiator triplet appears approximately in the middle of the protected sequence. Indeed, knowing the N-terminal peptides from the maturation and coat proteins and from the replicase, it was easy to verify that the triplets to the right of the initiator triplet in the protected fragments were those predicted by the code. The initiation codon is AUG except for the MS2 maturation protein and the *E. coli* lactose repressor protein (*lacI*), where it is GUG. Of particular interest is the fact that the protected sequences sometimes contain more than one AUG triplet. This is especially clear, for example, for the initiation sequence of $Q\beta$ replicase where two AUGs are present in the reading phase, and one out of phase. This is not unique, a similar situation being encountered in the initiation signal of the replicase gene of R17 or MS2.

A number of signals contain the sequence AGGAGGU, or at least GAGG preceding the intiator triplet at a certain distance, and this will be discussed at length (see pp. 42–5). Normally, even the first cistron of a polycistronic mRNA is located some distance from the 5′-end of the mRNA. The true initiation triplets are often preceded by termination triplets that stop the reading (they are generally found among the 15 bases preceding the initiator AUG). This is an expected consequence of the translation mechanism, since ribosomes have to stop reading the preceding cistron before protein synthesis is reinitiated. However, in phage mRNA, none of the initiator AUG triplets is directly preceded by a UAG, UAA or UGA codon, the three known terminators in *E. coli*, thus non-translated intercistronic regions exist in polycistronic mRNA, as is clear from the total sequence determinations. Besides being implicated in ribosome recognition, such untranslated nucleotides presumably eliminate possible interference between ribosomes involved in termination and initiation.

It is particularly noteworthy that the region preceding AUG often contains more than one termination triplet. This is shown by the sequence preceding the initiation signal of $Q\beta$ replicase, which contains one stop signal (UAA) in phase, and two (UAA and UGA) out of phase. Thus, even if the reading frame happens to be scrambled, that is to say, even if the ribosome is reading out-of-phase, growth of the polypeptide chain will stop and this will cause the system to reinitiate at the next AUG. The probable reason for this safety device is to prevent ribosomes from sliding and translating two adjacent cistrons into a single protein, a situation that would be lethal. However, this is not always so; in the $Q\beta$ coat cistron only one UGA triplet functions as the termination triplet and suppression by UGA-suppressor leads to read-through into the subsequent nucleotide sequence. In cellular mRNA the situation might sometimes be different from that of the phage mRNA. In tryptophan mRNA the intercistronic region between *trpB* and *trpA* (the last two genes in the operon) consists of only two untranslated nucleotides, the UGA terminator codon of *trpB* overlaps the *trpA* AUG initiation codon (Platt & Yanofsky, 1975).

Table 1 presents a complete list of currently known sequences for ribosome binding sites from coliphage and *E. coli* mRNAs. Each of these has been identified as a starting site for protein synthesis, either by ribosome protection after initiation of complex formation in vitro, or by comparison with the known N-terminal amino acid sequence of the protein specified. All sites isolated from native mRNA using *E. coli* ribosomes have proved to be initiators for translation.

Initiation factors

E. coli initiation factors are now reasonably well characterized; three general types, IF1, IF2 and IF3, are known. They have been purified to homogeneity and IF1 has even been crystallized. Recently Hershey, Yanov, Johnson & Fakunding (1977) set up a very reproducible procedure to prepare the three factors simultaneously in good yield.

IF1 is a basic protein (MW 9500), which is not retained on DEAE-cellulose. Earlier it was reported that the amino acid composition of

Table 1. *Initiation sequences recognized by* E. coli *ribosomes* (*from Steitz, 1977*)

mRNA	Ribosome binding sites [a]
R17 A	GAU UCC UAG GAG GUU UGA CCU AUG CGA GCU UUU AGU G
MS2 A	GAU UCC UAG GAG GUU UGA CCU GUG CGA GCU UUU AGU G
Qβ A	UCA CUG AGU AUA AGA GGA CAU AUG CCU AAA UUA CCG CGU
R17 coat	CC UCA ACC GGG GUU UGA AGC AUG GCU UCU AAC UUU
Qβ coat	AAA CUU UGG GUC AAU UUG AUC AUG GCA AAA UUA GAG ACU
f2, MS2 coat	CC UCA ACC GAG GUU UGA AGC AUG GCU UCC AAC UUU ACU
R17, MS2 replicase	AA ACA UGA GGA UUA CCC AUG UCG AAG ACA ACA AAG
Qβ replicase	AG UAA CUA AGG AUG AAA UGC AUG UCU AAG ACA G
f1 coat	UUU AAU GGA AAC UUC CUC AUG AAA AAG UCU UU
f1 gene 5	A AGG UAA UUC ACA AUG AUU AAA GUU GAA AU
f1 gene?	A AAA AAG GUA AUU CAA AUG AAA UU
T3 in vitro[b]	AAC AUG AGG UAA CAC CAA AUG AUU UUC ACU AAA GAG
T7 gene 0.3 site a	AAC UGC ACG AGG UAA CAC AAG AUG GCU AUG UCU AAC AUG
T7 gene 0.3 site b	GUA CGA GGA GGA UGA AGA GUA AUG U
λPR	ppp AUG UAC UAA GGA GGU UGU AUG GAA CAA CGC
λCI	TTG CGG TGA TAG ATT TAA CGT ATG AGC ACA AAA AAG
φX174 G	TTT CTG CTT AGG AGT TTA ATC ATG TTT CAG ACT TTT ATT
φX174 F	CCT ACT TGA GGA UAA AUU AUG UCU AAU AUU CAA ACU
φX174 D	ACC ACT AAT AGG TAA GAA ATC ATG AGT CAA GTT ACT
φX174 H	ACT TAA GTC AGG TGA TTT ATG TTT GGT GCT ATT
φX174 B	UAA AGG UCC AGG AGC UAA AGA AUG GAA CAA CUC ACU
φX174 J	ACG TGC GGA AGG AGT GAT GTA ATG TCT AAA GGT AAA
araB	UUU UUU GGA UGG AGU GAA ACG AUG GCG AUU GCA AUU
trp leader	CAC GUA AAA AGG GUA UCG ACA AUG AAA GCA AUU UUC GUG
trpE	GAA CAA AAU UAG AGA AUA ACA AUG CAA ACA CAA AAA CCG
trpA	GAA AGC ACG AGG GGA AAU CUG AUG GAA CGC UAC GAA UCU
lacz	AAU UUC ACA CAG GAA ACA GCU AUG ACC AUG AUU ACG GAU
lacI	AGU CAA UUC AGG GUG GUG AAU GUG AAA CCA GUA ACG
galE	AUA AGC CUA AUG GAG CGA AUU AUG AGA GUU CUG GUU ACC
galT	TAT CCC GAT TAA GGA ACG ACC ATG ACG CAA TTT AAT CCC

16S RNA 3'-end HOAUUCCUCCACUAGs.

[a] Underlining indicates contiguous bases complementary to the 3'-oligonucleotide of E. coli 16S rRNA. Dots indicate G·U base pairs. Initiator triplets are indicated by bold type.
[b] What was originally thought to be bacteriophage T7 has turned out to be T3.

IF1 is similar to that of most ribosomal proteins (Lee-Huang, Sillero & Ochoa, 1971). However, the IF1 preparations of Hershey *et al.* (1977) and C. Gualerzi (personal communication) show an amino acid composition quite different from that reported by Lee-Huang *et al.* (1971).

IF2 exists in two forms, $IF2_a$ and $IF2_b$. $IF2_b$ is the smaller with MW of 85000, while $IF2_a$ has a MW between 95000 and 117000, depending on the isolation technique used (Grunberg-Manago & Gros, 1977). $IF2_b$ is less thermostable than $IF2_a$ but otherwise no functional difference has been found between the two factors. The sedimentation constant of

$IF2_a$ is about 3.5 S, suggesting that the relatively high MW compound has an elongated shape. *E. coli* contains twice as much $IF2_a$ as $IF2_b$ and $IF2_b$ may result from a proteolytic cleavage of $IF2_a$ (Hershey *et al.*, 1977; Grunberg-Manago & Gros, 1977).

IF3 (MW 22000) is a thermostable protein. The sedimentation constant, 2.2 S, (Dubnoff & Maitra, 1971) agrees with the MW determination by acrylamide–dodecylsulphate gel electrophoresis. However, the value by gel filtration on Sephadex was approximatively 32000 (Gualerzi, Pon & Kaji, 1971) which may indicate that this factor does not behave as a typical globular protein. It has been reported that *E. coli* contains at least two forms of IF3 with almost identical MWs (Lee-Huang & Ochoa, 1971, 1973). $IF3_\alpha$ (MW 22500) is supposed to catalyse the translation of MS2 RNA preferentially to late T4 messenger; $IF3_\beta$ (MW 21000) shows the inverse specificity. More recently, Suryanarayana & Subramanian (1977) have been studying the protein specifically associated with native 30S ribosomal proteins by two-dimensional gel electrophoresis and they obtained purified factor preparations from several laboratories for identification in this system. They found that different laboratories purified two different molecular forms of IF3, and in addition native 30S from different strains of *E. coli* yielded at least two forms of IF3. The primary structures of the two forms have been determined (Brauer & Wittmann-Liebold, 1977) and they differ at their N-terminal regions, one being 6 amino acids shorter than the other. It is, therefore, likely that they are both products of the same gene, a finding consistent with the results of Springer, Graffe & Hennecke (1977*b*) (see p. 60). The long form consists of 181 amino acids and contains one cysteine at position 66; the short form has only 175 amino acids. The MWs determined by chemical analysis are 20668 and 19997. In addition to these two main forms, at least one minor one has been detected which differs from the other IF3s by being methylated on the N-terminal amino group of methionine. All three forms were active in phage RNA protein synthesis and in ribosome dissociation. Many workers now believe that highly purified IF3 has no mRNA discriminating ability *per se* (Schiff, Miller & Wahba, 1974) (see pp. 55–7). In *B. stearothermophilus* and *Caulobacter crescentus* only two initiation factors have been found and purified: they correspond to IF2 and IF3 (Leffler & Szer, 1974*a, b*; Kay, Graffe & Grunberg-Manago, 1976). This situation might reveal the somewhat dispensable nature of IF1 in the other prokaryotic systems.

FUNCTION OF THE COMPONENTS
OF THE INITIATION COMPLEX

Initiator tRNA – the role of formylation

It is believed that the initiator tRNA has the same general conformation as a polypeptidyl–tRNA. It is located at the operationally defined P site, a specific 70S territory made by certain regions of both particles. When located in this site, an aatRNA derivative reacts with puromycin, an antibiotic analogue of aatRNA, which occupies a portion of the acceptor, or 'A' site. Entry of the initiator tRNA is thought to be directly into the P site, in contrast to aatRNAs for internal amino acids, which are first bound in the A site. The definition of the A site is also operational; aatRNA bound in the A site does not react with puromycin although its binding is inhibited by low concentrations of the antibiotic tetracycline.

Polypeptide chain initiation proceeds via a mechanism involving fMet–tRNA^{fMet}. One could ask whether the structure of the initiator tRNA or the formylation of the methionine is responsible for its entry into the P site. That formylation *per se* is not necessary has been discussed on p. 30, and *E. coli* can grow to a certain extent in a medium where the level of formylation is strongly depressed (Danchin, 1973; Harvey, 1973). Moreover, when Met–tRNA^{fMet} is used *in vitro*, formation of the 30S–tRNA initiation complex is strongly stimulated by all initiation factors and is messenger-dependent, as with fMet–tRNA^{fMet} (Petersen, Danchin & Grunberg-Manago, 1976a). Therefore, formylation does not affect formation of the preinitiation complex on the 30S subunit. The requirements for binding Met–tRNA^{fMet} are somewhat different from fMet–tRNA^{fMet}, having a higher Mg^{2+} optimum and not involving GTP. When the unformylated tRNA is first bound to the 30S particles and the 50S subunits are subsequently added, most of the bound Met–tRNA^{fMet} reacts with puromycin, indicating that it is correctly positioned in the P site. The situation is quite different with the 70S ribosomes (Petersen, Danchin & Grunberg-Manago, 1976b) where only the attachment of the formylated species is stimulated by the initiation factors. Presence of the factors strongly inhibits the binding of Met–tRNA^{fMet}. In their absence both species are bound to the 70S ribosomes, but only the formylated tRNA reacts completely with puromycin. Experiments with tetracycline support the idea that, although bound in a manner different from that of fMet–tRNA^{fMet}, the Met–tRNA^{fMet} is not located preferentially at the A site under these conditions. Met–tRNA^{Met}, whether formylated or not, is not able to bind to the P site when added to 70S ribosomes or to 30S subunits followed by addition of 50S subunits. Neither are Val–tRNA^{Val} or fVal–tRNA^{Val} capable of direct entry into the P site. However, fVal–tRNA^{fMet} (initiator tRNA mischarged with Val and chemically formylated) binds to the puromycin-reactive P site (Giégé, Ebel, Springer & Grunberg-Manago, 1973). Phe–tRNA^{Phe} and Trp–tRNA^{Trp}, when formylated or acetylated, also bind to the P site in the presence of initiation factors and poly(U) or poly(UG) respectively, but the binding is less stable than that of fMet–tRNA^{fMet} (Grunberg-Manago et al., 1969; R. H. Buckingham, J. Dondon and M. Grunberg-Manago, unpublished results). Nevertheless, this suggests that their structures are similar to that of the true initiator tRNA, because other species that have been examined, including tRNA^{Met}, tRNA^{Lys} and tRNA^{Val} (Rudland, Whybrow, Marcker & Clark, 1969), are not bound under these conditions.

In conclusion, it appears that 70S ribosomes are able to distinguish not only between Met–tRNA^{fMet} and Met–tRNA^{Met}, but also between fMet–tRNA^{fMet} and Met–tRNA^{fMet}; and between the latter pair even in the absence of initiation factors. Presumably since no energy is required in this recognition step, although the presence of IF2 and GTP enhances the ability to discriminate between the two initiator tRNA species, either two sites or two 70S ribosome conformations exist for the binding of initiator tRNA, depending on whether it is formylated or not. Binding experiments, and experiments on the formation of aminoacylpuromycin, as a function of Mg^{2+}, K^+, or initiation factors, have led Danchin and co-workers (Petersen et al., 1976b) to propose a two-state equilibrium for 70S particles, involving a minor, active conformation, and a major one that is not readily active. The formyl group would act as a specific trigger to select the active conformation. For this model, equilibrium parameters and kinetic constants of the peptidyl-transferase activity have been tentatively derived.

Finally, this leads to the proposition of a general function for the regulatory action of the formyl group. Initiation of polypeptide synthesis generally proceeds via a complex involving fMet–tRNA^{fMet}, messenger RNA, and the 30S ribosomal subunit. Since this subunit is unable to discriminate between the formylated and unformylated initiator tRNA species, one may wonder whether, in special cases, the 70S particle may be used as such in initiation complexes. In particular, for polycistronic messengers, the first cistron may be translated via dissociated ribosomes, but when passing to the next cistron an appreciable proportion of the ribosomes would remain as 70S particles, requiring formylation for the initiation step. Formylation might, therefore, be used as a regulatory device in the translation of polycistronic messengers, since lowering the formylation level would increase the polarity effect. This model has been tested *in vivo*. Studies on the polarity of the expression of the lactose operon genes showed that the polarity strongly increased when the level of formylation decreased (H. U. Petersen, A. Ullmann & A. Danchin, manuscript in preparation). The observation that a termination factor releases the polypeptide chain without releasing the ribosome from mRNA also suggests that after reaching the termination codon, the ribosome can slide across the intercistronic region until it reaches the AUG triplet at the beginning of the next cistron (Noll & Noll, 1972).

Ribosomes

The brilliant discovery of mRNA overshadowed the part played by ribosomes and particularly by ribosomal RNA. The whole specificity of protein synthesis was thought to proceed from codon–anticodon interactions and ribosomes were considered to provide only a surface for the positioning of the major components of protein synthesis. However, AUG directs not only the binding of Met–tRNA^{fMet}, but also the binding of internal Met–tRNA^{Met} on the ribosome. Thus some other feature should distinguish the correct initiator codon from the many other AUG triplets bound either in or out of phase in the mRNA.

It was first believed that protein synthesis started simply with the AUG triplet closest to the 5'-end of an mRNA molecule (see Steitz, 1977). However, as was discussed in the section on mRNA, translation of naturally occurring mRNA molecules usually does not begin near the 5'-end. Moreover ribosomes can select different initiation sites in polycistronic mRNA and finally ribosomes can initiate protein synthesis very efficiently on a circular messenger DNA molecule (Bretscher, 1969). Thus *E. coli* ribosomes are capable of selecting and binding to true initiation signals found in internal regions of an mRNA.

It was then generally believed that ribosome recognition of initiator regions was, as for tRNA-synthetase or methylase recognition, a specific protein–nucleic acid interaction. One of the several ribosomal proteins, or initiation factor IF3 or some other protein factor was presumed to bind selectively to some unidentified, important sequence in mRNA.

Species specificity

Lodish (1970) was the first to demonstrate clearly that the specificity is not due to initiation factors but lies in the small ribosomal subunit. *E.*

coli ribosomes recognize the beginning of the three cistrons of the phage f2 RNA; protein A, coat, and replicase, with coat being the most efficient initiation site. In contrast *B. stearothermophilus* translates only cistron A, either at 65 °C, its physiological temperature, or at 47 °C where *E. coli* translates the three cistrons. In general it appears that Gram-positive and Gram-negative bacteria have different specificities.

Crossing the various elements from the translation machinery of *E. coli* and *B. stearothermophilus* at 47 °C, where both organisms are active, Lodish (1970) showed that the specificity of action of their ribosomes is not altered when heterologous initiation factors are used and that cistron specificity lies in the 30S particle. Subsequently, the components of the 30S ribosomal particles, responsible for the specific recognition of initiation signals on mRNA, have been investigated (Held, Gette & Nomura, 1974; Goldberg & Steitz, 1974). Using the techniques of Nomura for the reconstitution of 30S particles they found that the ability to discriminate between the initiation signals present on the phage mRNA lies in protein S12 and the 16S RNA. Finally, Isono & Isono (1975) have found that S1 also appears to be involved.

The interaction of R17 RNA with the *B. stearothermophilus* system is similar to that of f2 RNA whereas with Qβ RNA ribosomes at temperatures above 65 °C ignore all three normal sites and bind instead to two other regions of the RNA molecule (Steitz, 1973). These binding sites lack initiator triplets and accordingly do not direct the binding of fMet–tRNA although they are both very rich in polypurine tracks (Table 2).

Role of the 16S ribosomal RNA (*Shine and Dalgarno's hypothesis*)

In 1974 Shine & Dalgarno (1974, 1975) presented a revolutionary hypothesis to explain the specificity of selecting the initiation regions. This was at once generally accepted by scientists working on initiation, and surprisingly it focused the attention on and renewed interest in ribosomal RNA, which was assumed until then to have only a structural role. Shine and Dalgarno suggested that the 3'-OH terminal dodecanucleotide of *E. coli* 16S ribosomal RNA directly participates in the selection of initiation sites by forming several Watson–Crick base-pairs with the polypurine-rich sequence located in the vicinity of the initiator triplet in the mRNA. It is significant that all coliphage RNA ribosome binding sites contain part or the whole purine-rich sequence, AGGAGGU, at a relatively similar position with respect to the initiator triplet AUG, on its 5'-side (see the section on mRNA and Table 1). Seven base-pairs could possibly be involved in the interaction

Table 2. *RNA phage sites bound by* B. stearothermophilus *ribosomes*
(*from Steitz, 1977*)

Binding temperatures (°C)	Site	Sequence[b]	Theoretical T_m of mRNA.rRNA complex[c] (°C)
65	R17 A	CCUAGGAGGUUUGACCUAUG	38
	Qβ non-initiator *a*	CUGAAAGGGGAGAUUACUCG	29
	Qβ non-initiator *b*	GGAAGGAGC	29
49	R17 A	CCUAGGAGGUUUGACCUAUG	38
	R17 replicase	AUGAGGAUUACCCAUG	−7
	Qβ replicase	AACUAAGGAUGAAAUG	−7
	Qβ coat	AAACUUUGGGUCAAUUUGAUCAUG	−25
Not bound	R17 coat	CAACCGGGGUUUGAAGCAUG	−46

B. stearothermophilus 16S 3′-end   HO*AUCU*UUCCUCCACUAG₃,

[a] Note that only the R17 A site is bound strongly by *B. stearothermophilus* ribosomes at 49 °C; the other regions are recognized much more weakly by thermophilic than by *E. coli* ribosomes at 49 °C.

[b] Underlining indicates phage RNA regions which are complementary to the 3′-end sequence of *B. stearothermophilus* 16S rRNA aligned below. Dots represent G·U base pairs. Initiator AUGs are in bold type. Italics indicate that portion of the 16S terminal sequence which differs in *E. coli* and *B. stearothermophilus*.

[c] T_ms calculated according to Gralla & Crothers (1973*b*). Note the correlation between the T_m of the mRNA·rRNA complex and the temperature at which each site binds stably to *B. stearothermophilus* ribosomes.

between 16S RNA and the initiation site for the A protein cistron, whereas four or five pairs could form at the initiation regions of the other cistrons. The lengths of the complementary regions vary between three and nine nucleotides, the average being 4–5. The number of nucleotides separating the complementary region from the initiator triplet varies, with an average of 10 nucleotides separating the middles of each. Finally, wherever two or more initiator triplets appear in a single initiator region, that preceded by the strongest and most appropriately positioned complementary region functions in polypeptide chain initiation. Thus *lacI* repressor mRNA has three GUG in phase but only one acts as the initiator. The *trpA* ribosome binding site is not preceded by intercistronic sequences but contains, however, the Shine and Dalgarno sequence in the *trpB* segment.

In addition to the appearance of a sequence complementarity to 16S RNA in each known ribosome binding site, other data made the suggestion of Shine and Dalgarno quite attractive. Evidence has accumulated that an intact 3′-terminus of 16S RNA is necessary for initiation, as shown by the inhibitory action of colicine E3 (Bowman *et al.*, 1971; Boon, 1971; Dahlberg *et al.*, 1973), which results in the removal of about 50 nucleotides from the 3′-terminus of the 16S RNA by a single endonucleolytic cleavage. Other inhibitors of initiation (streptomycin (Crépin, Lelong & Gros, 1973) and kasugamycin (Melser, Davies & Dahlberg, 1971)) have their site of action in the vicinity of the 3'-OH terminus of 16S RNA. Cross-linking experiments (see pp. 61–8) also suggest that the 3'-OH terminus of 16S RNA is near or part of the binding site for initiation factors and ribosomal proteins involved in initiation. Random co-polymers rich in A and G are the best competitive inhibitors for initiation on phage mRNA (Revel & Greenshpan, 1970). Moreover, the long polypurine stretch of the fragment of $Q\beta$ RNA which is bound by *B. stearothermophilus* ribosomes at 65 °C and does not possess an initiator triplet has a sequence complementary to the 3'-OH end of 16S RNA (Steitz, 1973). Finally, the 3'-terminus of 16S RNA is located on the interface between the two ribosomal subunits where the mRNA is presumed to bind (Chapman & Noller, 1977; Santer & Shane, 1977).

More direct evidence for the validity of Shine and Dalgarno's hypothesis comes from experiments by Steitz & Jakes (1975), who treated initiation complexes formed by *E. coli* ribosomes with colicin E3. They used as messenger the fragment of the initiation cistron for the maturation protein A (this fragment was ^{32}P-labelled). After removing proteins by exposure to sodium dodecylsulphate, they fractionated the components on a polyacrylamide gel. An mRNA–rRNA hybrid, containing approximately equimolar amounts of the 30-nucleotide mRNA fragment and the 49- to 50-nucleotide colicin fragment, was detected (Fig. 4). This hybrid, which exhibits an electrophoretic mobility different from each of the above two fragments, appears only in the presence of all the components necessary for complex formation. Furthermore, it does not appear if the colicin treatment is omitted, since in that case the mRNA co-sediments in a sucrose gradient with intact 16S RNA. Moreover, aurintricarboxylic acid, an inhibitor of mRNA binding to ribosomes, lowers the amount of complex formation. The hybrid can be dissociated by heat ($T_m = 32$–37 °C). With ^{32}P-labelled ribosomes in addition to the ^{32}P-labelled initiator region, chromatographic analysis reveals that the portion of gel containing the complex consists of oligonucleotides assigned to the colicin E3 fragment and to the R17 protein initiator fragment. These observations strongly support the hypothesis of Shine and Dalgarno that base-pairing between the mRNA and the 3'-terminus of 16S RNA occurs during the formation of a functional initiation complex.

Fig. 4 suggests that a specific secondary structure is assumed by the 3'-terminal region of the 16S RNA and that, upon mRNA binding, some of the intramolecular base-pair interactions may be exchanged

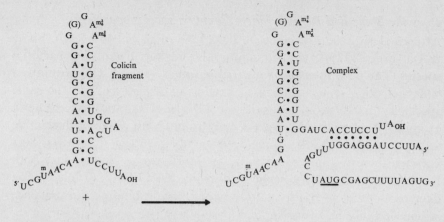

5' AUUCCUAGGAGGUUUGACCUAUGCGAGCUUUUAGUG 3'

R 17 A protein initiator region

Fig. 4. Postulated hydrogen-bonding between the colicin fragment of 16S RNA and the R17 A protein initiator region. The secondary structure drawn, involving internal base-pairing, according to Shine and Dalgarno, is predicted to be stable under physiological conditions. From Steitz & Jakes (1975).

for intermolecular hydrogen bonds. Studies by temperature jump relaxation methods (Yuan, Steitz & Crothers, 1976) and NMR techniques (Baan *et al.*, 1977) give results which are consistent with the structural conformation of the isolated colicin E3 fragment. Thus a T_m of 80 °C indicates that there is a quite stable helical region containing 9 base-pairs and temperature jump studies also reveal a second transition at 21 °C which could correspond to the melting of additional base-pairs below the loop. This is in agreement with the results from high resolution proton magnetic resonance spectra suggesting that the rRNA fragment exists at 25 °C as a hair-pin consisting of 8 intramolecular base-pairs, the 3'-terminal dodecanucleotide being unpaired (Baan *et al.*, 1977).

Since protein synthesis can begin both in vivo and in vitro at temperatures far exceeding the T_m of the most stable isolated hybrids, mRNA–rRNA interactions must be supplemented by other mRNA–ribosome contacts, particularly for initiation sites which exhibit many fewer mRNA–rRNA base-pairs. We know that this is so for the codon–anticodon interaction on the ribosome where T_m far exceeds that of the most stable codon–anticodon interaction in solution (McLaughlin *et al.*, 1968).

Does the Shine and Dalgarno theory explain species specificity?

After analysing the 12 residues at the 3'-end of *E. coli* 16S RNA, Shine & Dalgarno (1975) examined comparable regions from other bacterial species (Table 3). *Pseudomonas aeruginosa* has at the 3'-OH terminus a hexanucleotide sequence identical to that of *E. coli* and has the same specificity for cistron recognition. All the Gram-positive bacteria examined have pyrimidine-rich 3'-termini but with different nucleotide sequences. Nevertheless, some four complementary base-pairs exist between these sequences and those of the ribosome binding sites both for the A protein and the replicase cistrons of R17 RNA (Table 4). The degree of complementarity with the coat-protein binding site sequence is considerably less (one or two base-pairs). As discussed earlier *B. stearothermophilus* binds significantly only to the A protein site on native f2 or R17 RNA. While appreciable translation of the replicase, in addition to the A protein cistron, occurs when the RNA is unfolded by formaldehyde treatment, no coat protein is synthesized under these conditions (Lodish, 1971). The 3'-OH sequence of the 16S RNA from *C. crescentus* is similar to that from *B. stearothermophilus* and does not bind to the coat-protein initiation site. It was therefore concluded that the extent of possible interaction between the 3'-OH terminus of 16S RNA from several different bacteria and the ribosome binding sites of coliphage RNA is consistent with the available data on translational specificity in these bacteria.

However, in order directly to correlate the mRNA–rRNA complementarity with translation specificity, Sprague, Steitz, Grenley & Stocking (1977) determined the nucleotide sequences of a longer portion of the 3'-terminus of 16S rRNA from *B. stearothermophilus* (Table 3) and of the two polypurine stretches in the RNA from Qβ bacteriophage that is bound at high temperature by ribosomes from this species. Their results confirmed that specific recognition of the two Qβ regions can be

Table 3. *3'-Terminal sequences of 16S rRNA*
(*from Shine & Dalgarno, 1975*)

E. coli B	GAUCACCUCCUUA$_{OH}$
Ps. aeruginosa	G(X)$_2$YCUCUCCUU(A)$_{OH}$[a,b]
B. stearothermophilus	G(X)$\sim_5$YUCCUUUCU(A)$_{OH}$[a,b]
B. stearothermophilus	GAUCACCUCCUUUC(U)A$_{OH}$[b,c]
B. subtilis	G(X)$\sim_7$YCUUUCU$_{OH}$[a]
C. crescentus (ATCC 15252)	G(X)$_3$YUCCUUUCU$_{OH}$[a]

[a] X represents any nucleoside other than guanosine; Y = pyrimidine nucleoside.
[b] The variable presence of the 3'-terminal adenosine in 16S RNA is found in a variety of bacteria and depends on the culture conditions.
[c] Sequence as determined by Sprague *et al.* (1977).

Table 4. *R17 ribosome-binding site sequences and proposed pairing with the 3'-terminus of 16S RNA (from Shine & Dalgarno, 1975)*

The sequences shown represent all or part of the conserved sequence [$_5$'AGGAGGU$_3$'] from the A-protein, replicase, and coat-protein initiator region of R17 RNA. The initiator AUG is located eight to nine bases from this sequence. Apart from the A transition (see below), these sequences are identical to those available for the corresponding regions of f2 and MS2 RNA. Two sequences have been reported for this region of the coat-protein binding site in the two different R17 stocks used, the G→A transition represents a spontaneous mutation that occurred after the two stocks were separated. Such a transition may have some selective advantage, since it would increase the stability of interaction between the coat-protein initiation site and the 3'-end of 16S RNA. The corresponding sequence from MS2 and f2 RNA contains the A substitution. The in-vitro ribosome-binding data were obtained with R17 RNA containing the GGGG sequence and with f2 or MS2 RNA presumably containing the GGAG sequence. The apposition of a G and a U residue in the proposed helical region formed between the coat-protein binding site and 16S RNA would cause little, if any, destabilization of the base-paired structure.

Bacteria	R17 cistron	Possible pairing of 3'- terminus with appropriate region of ribosome-binding site sequence		Number of base-pairs possible	Ribosome binding to unfolded R17, MS2 or f2 RNA
E. coli B		$_{OH}$A U U C C U C C A C Py	16 S RNA		
	A protein	$_5$'C U [A G G A G G U] U U$_3$'		7	
	Replicase	A C A U [G A G G] A U U		4	
	Coat protein	A C C [G G] G [G] U U U G		3(4)	+
		↓ [A]			
Ps. aeruginosa		$_{OH}$A U U C C U C U C Py	16 S RNA		
	A protein	$_5$'C U [A G G A G] G U U$_3$'		5	+
	Replicase	U G [A G G A] U U A C		4	+
	Coat protein	A C C [G G] G [G] U U U		3(4)	+
		↓ [A]			
B. stearothermophilus		$_{OH}$A U C U U U C C U Py	16 S RNA		
	A protein	$_5$'A U C C U [A G G A G]$_3$'		>4	+
	Replicase	A C A U G [A G G A] U		4	+
	Coat protein	A A C C G G [G G] U U		2(1)	−
		↓ A			
C. crescentus		$_{OH}$U C U U U C C U Py	16 S RNA		
	A protein	$_5$'U C C U [A G G A G]$_3$'		>4	
	Replicase	C A U G [A G G A] U		4	
	Coat protein	A C C G G [G G] U U		2(1)	
		↓ A			

explained by an extensive mRNA–rRNA pairing (Table 2). Masking of the two non-initiator $Q\beta$ regions by RNA secondary structure probably explains the absence of these sites among the fragments protected by *B. stearothermophilus* ribosomes at lower temperatures. However, the comparative translational specificities of *B. stearothermophilus* and *E. coli* were more difficult to explain. The 3'-OH termini of both rRNAs are identical, except that the 3'-terminal adenosine of *E. coli* rRNA is replaced by $UCUA_{OH}$ in the bacillus. Thus, *B. stearothermophilus* ribosomes have the same potential for binding to R17 or f2 coat-protein initiator regions as do *E. coli* ribosomes, unless one postulates that the four extra nucleotides at the 3'-OH end prevent that sequence from being available for mRNA recognition.

We have already discussed the importance of proteins S12 and S1 as selectivity determinants. They are involved in facilitating the mRNA–rRNA interaction by stabilizing the bonds of the RNA or by opening the intramolecular base-pairs of the colicin E3 fragment from the 16 S RNA ends. Nevertheless, the possibility exists that proteins can create other interaction sites between mRNA and ribosomes, and we must await further evidence before drawing conclusions as to the particular interactions determining prokaryotic translational specificity. In conclusion both proteins and rRNA must be regarded as potential contributors to the specificity of initiation of protein synthesis with different cistrons.

Is the calculated strength of the mRNA–rRNA interaction related to the efficiency of initiation?

This question may be tested by considering the 'reinitiation' of protein synthesis which occurs *in vivo* after polypeptide chain termination at the site of a nonsense mutation. In the lactose repressor gene (*lacI* gene) Weber, Miller and their colleagues (Platt, Weber, Ganem & Miller, 1972; Ganem *et al.*, 1973; Files, Weber & Miller, 1974; Files, Weber, Coulondre & Miller, 1975) have identified three discrete restarting sites which give rise to C-terminal repressor fragments. The codons of valine 23 (GUG), methionine 42 (AUG) and leucine 62 (CUG) of the repressor mRNA were used to reinitiate translation beginning with *N*-formyl-methionine. A single nonsense mutation can activate more than one reinitiation site. Steege (1977) sequenced the 5'-terminus of the *lacI* mRNA (Fig. 5). The wild type repressor protein is initiated at a GUG codon preceded by a sequence exhibiting substantial complementarity to the 16S RNA. Of the three in-phase GUG triplets clustered in this region, the real initiator is that codon which is

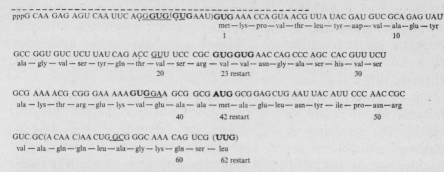

Fig. 5. Sequences of *lacI* mRNA (Steege, 1977) and the repressor protein (Beyreuther, Adler, Geisler & Klemm, 1973), arranged so that the initiator codons for the three restart proteins are aligned under that for the wild-type repressor. All in-phase initiator triplets are shown in bold type. Sequences which are complementary to the 3′-end of 16S rRNA and precede the utilized initiators are indicated by underlying (with G·U pairs denoted by dots). The ribosome-protected region (Steege, 1977) is indicated by a dashed line. From Steitz (1977).

most appropriately situated relative to the polypurine stretch (Steitz, 1977).

Relative to the true *lacI* initiator region, mRNA complementarity to rRNA is weak at the restart sites, and is non-existent at valine 23. Apparently several additional in-phase GUG codons in the sequence also do not serve as reinitiation signals in the nonsense mutants. It is clear that, following termination, ribosomes do not restart simply by moving to the next available codon. Reinitiation at valine 23 (GUG) occurs concomitantly with that at methionine 42 (AUG), binding at the accessible methionine 42 destabilizing the site at valine 23 and making it also available.

mRNA: its secondary structure as control

In the previous sections only regions of mRNA adjacent to authentic initiator codons have been considered. However, recognition by ribosomes is influenced by intramolecular interactions within the mRNA molecule. For instance, if the initiator triplet or the polypurine tract are made unavailable, a potential initiation signal could be rendered inactive. When one examines the sequence of MS2 RNA, many regions in the viral RNA fulfill the criteria of Shine & Dalgarno (1974, 1975) base-pairing and yet do not bind to ribosomes. For example, the sequence of the replicase gene segment 2381–2396 (Fig. 3) is UAAGGA-GCCUGAUAUG and according to the criteria should interact with 16S rRNA from both *E. coli* and *B. stearothermophilus* (Fiers *et al.*, 1976). That this is not observed may mean that its accessibility is masked by secondary and tertiary structure. We can see (Fig. 3) that the sequences surrounding this additional initiator triplet are sequestered in RNA secondary structures. Accordingly, when phage mRNA is treated with formaldehyde, additional initiation sites are revealed (Lodish, 1971).

Obviously the potential for base-pairing plus the presence of an initiator triplet is not sufficient to describe a true mRNA initiator region. The secondary structure plays an active negative control in limiting initiation to the proper regions. It can sometimes play a positive role as well. The RNA fragment derived from the R17 coat-protein initiator site binds to the ribosomes less well than does the same site when it is part of the intact RNA (Steitz, 1977). Possibly the fragment containing the coat-protein site falls into a stable secondary structure not assumed in the intact molecule, or the overall tertiary structure of the viral RNA may be designed so that this region is exposed in such a way that it forms a more stable interaction with ribosomes.

Initiation factors

The general requirement for initiation factors can be shown operationally, by the use of an in-vitro system which contains the RNA from an RNA phage as messenger, NH_4Cl-washed 70S ribosomes, GTP, an energy-generating system, the appropriate ions (Mg^{2+}, K^+, etc.), and a crude supernatant fraction containing the elongation factors EF–Tu, EF–Ts, and EF–G, the termination factors, and aatRNAs. In such a reconstituted system, no protein synthesis occurs unless the three initiation factors are present. By contrast, when a synthetic messenger, such as poly(U) is used, and provided that appropriate ionic conditions are met (high Mg^{2+} concentrations), translation occurs at a maximal rate in the absence of initiation factors. These types of experiment indicate that the initiation factors from the ribosomal wash are not involved in the elongation process *per se* but rather that natural messengers carry particular signals, the recognition of which requires the presence of both a special tRNA and initiation factors.

Reversible dissociation of 70S ribosomes

At physiological Mg^{2+} concentrations (4–5 mM), type-A ribosomes are nearly all associated. Since each initiation event is preceded by dissociation of the ribosomes, this explains the need for protein factors. The part played by each factor has been shown by studying its influence on ribosomal equilibria as well as on the dissociation kinetics, as analysed by scattered light (Godefroy-Colburn *et al.*, 1975). The IF3 dissociating action is purely passive. The factor binds to the 30S subunit, which then becomes unfit for reassociation. The affinity of IF3 for the 30S subunit, determined under various conditions (25 °C to 37 °C; 2–5 mM Mg^{2+}), is 2.5 to 4×10^7 M^{-1}. With type-A ribosomes IF3 does not bind significantly to the 70S couple but with type-B ribosomes the affinity for the 70S is significant and cannot be neglected. The presence of IF3 causes a decrease in the rate constant k_2 for associa-

Table 5. *Effect of IF1 and IF3 on the kinetic constants of ribosome dissociation (from Godefroy-Colburn et al., 1975)*

The constants were determined at 2.5 mM Mg^{2+}, 100 mM NH_4^+, by second order analysis of recordings.

$$70S \underset{k_2}{\overset{k_1}{\rightleftharpoons}} 30S + 50S$$

$$K_{d,app.} \frac{(50S)(30S)}{70S}$$

Concentrations (nM)			$k_{1,app.}$ $(s^{-1} \times 10^3)$	$k_{2,app.}$ $(\mu M^{-1} s^{-1})$	$K_{d,app.}$ (nM)
Ribosomes	IF1	IF3			
39.6	—	—	10	0.24	42
50.0	—	—	10.9	0.21	52
50.1	—	—	8.0	0.19	42
51.9	30	—	17	0.21	81
51.3	31	—	15	0.21	72
51.0	59	—	20	0.21	96
51.3	65	—	21	0.27	79
51.7	88	—	24	0.28	88
41.8	116	—	21	0.32	65
51.0	144	—	33	0.36	90
41.8	—	117	10.7	0.077	139
38.7	—	117	7.5	0.053	143
63.5	—	146	7.0	0.040	173
51.2	87	58	36	0.19	188
41.0	114	114	19.5	0.060	325

tion of the subunits, but not in the dissociation rate constant k_1 (Table 5). In contrast, IF1 alone has little effect on the proportion of associated ribosomes whereas it drastically increases the two rate constants k_1 and k_2 (the ratio $k_1/k_2 = k_d$ varies by a factor of only 2 in the presence of excess IF1). Moreover, IF1 does not affect the affinity of the 30S particles for IF3.

The results can therefore be interpreted by the following scheme, under conditions where the factors bind independently (i.e. no excess of IF1 or IF3).

$$
\begin{array}{ccc}
70\,\text{S} & \xrightleftharpoons[\quad]{K_1'} & 70\,\text{S·IF1} \\[1mm]
k_2 \Updownarrow k_1 & & k_2' \Updownarrow k_1' \\[1mm]
50\,\text{S}+30\,\text{S} & \xrightleftharpoons[\quad]{K_1} & 30\,\text{S·IF1}+50\,\text{S} \\[1mm]
\Updownarrow K_3 & & \Updownarrow K_3 \\[1mm]
30\,\text{S·IF3} & \xrightleftharpoons[\quad]{K_1} & 30\,\text{S}\begin{array}{l}\text{·IF3}\\\text{·IF1}\end{array}
\end{array}
$$

The need for both factors is explained: one increases the rate of dissociation of the 70S particles, and the other prevents their reassociation. Similar results have been obtained by studying the rate of subunit exchange when the factors are added to tight 70S couples in the presence of a radioactive subunit (Noll & Noll, 1972; Noll *et al.*, 1973).

By the light-scattering technique it has been shown that IF2 has an association effect. In the presence of GTP (or of $GMPP(CH_2)P$) this effect is specific and the factor has a stronger affinity for the 70S ribosome couple than for the 30S subunit. The binding constant, K_2, for the 30S subunit is around 5×10^5 M^{-1} and K_2', for the 70S particles is 1.0 to 1.5×10^7 M^{-1}. In the absence of the nucleotide, the association effect is much less pronounced and, furthermore, aggregates are formed. These results might indicate that GTP is necessary for a correct positioning of IF2.

The sequence of initiation complex formation

After the equilibrium of ribosomes is shifted towards the 30S+50S state with the help of IF1 and IF3, two routes of initiation-complex formation have been suggested, depending on the mRNA structure (Gualerzi, Risulea & Pon, 1977). First, under the influence of IF3, the 30S subunit interacts with the messenger initiation signals *before* attachment of the initiator tRNA. Alternatively the fMet–tRNA first binds to the ribosome under the influence of IF2, this attachment being a prerequisite for the correct binding of RNA.

A number of workers regard IF2 as the 'fMet–tRNA recognition or carrier protein'. It is tempting to establish a parallel with EF–Tu, which is responsible for the recognition and positioning of internal aatRNAs in the presence of GTP (see pp. 68–72). However, whereas EF–Tu forms a stable ternary complex with GTP and aatRNA in the absence of ribosomes, interactions between highly purified IF2, GTP, and/or fMet–tRNA^{fMet}, are not stable enough, in the absence of ribosomes, to allow separation of the corresponding complexes. Recently, with purified IF2 in the absence of Mg^{2+}, a binary complex, IF2·fMet–tRNA^{fmet} has been detected on nitrocellulose filters. However, only a small fraction of input IF2 becomes bound to fMet–tRNA^{fMet}. The complex formation is GTP-independent, (see Grunberg-Manago & Gros, 1977; Van der Hofstad *et al.*, 1977).

The existence of weak sites for GTP binding to IF2 can, nevertheless be demonstrated indirectly. GTP protects IF2 against inactivation by sulphydryl reagents (Mazumder, Chae & Ochoa, 1969). Furthermore, GTP, and to a lesser extent GDP, protect IF2 against thermal inactivation (around 50 °C) while ATP, GMP, or GMPP(CH$_2$)P are without effect (Lelong *et al.*, 1970). Calculations show that GTP and IF2 associate with an apparent K_a of 10^4 M^{-1}, which indicates that the interaction is very weak. There is some parallel here with EF–G which also has a rather weak interaction with GTP and shows an active un-coupled GTPase activity (see pp. 79–80). The melting denaturation profile spectrum of the initiator tRNA as a function of temperature is modified in the presence of IF2 plus GTP, the other initiation factors having no effect (Rudland, Whybrow & Clark, 1971).

These results all indicate that IF2, GTP and fMet–tRNA interact with each other and suggest that ternary complex formation may occur prior to binding to the 30S ribosome. An alternative interaction of fMet–tRNA and IF2 is possible, however, where IF2 binds to 30S sub-units in the absence of fMet–tRNA, and is greatly stabilized there when IF1 and IF3 are also present (Fakunding & Hershey, 1973). Such a tri-initiation factor complex may be a common intermediate in both pathways postulated above, accepting either mRNA or fMet–tRNA as the next macromolecular component in the nascent initiation com-plex. To determine which of the various alternate pathways operate more sophisticated kinetic analyses are required.

Recycling of initiation factors

A number of experimental approaches indicate that initiation factors are only transiently associated with ribosomes, and leave them following completion of the formation of the initiation complex. First, initiation factors are found predominantly associated with native 30S ribosomes when cell extracts are fractionated by sucrose gradient centrifugation (Eisenstadt & Brawermann, 1966; Dubnoff, Lockwood & Maitra, 1972). The polysome fraction, which contains the bulk of the ribosomes active in protein synthesis, does not contain the factors. Secondly, the molar ratio of initiation factors to ribosomes in crude cell lysates has been estimated to be 0.03–0.10 (Van Knippenberg, Van Duin & Lentz, 1973; Krauss & Leder, 1975), which implies that the initiation factors must recycle from one ribosome to another. Thirdly, in-vitro assays of initiation of protein synthesis have been performed where more initiation events occurred than the number of molecules of factor added.

The ejection of the initiation factors from the ribosome has been shown directly by analysis of initiation complexes formed with radio-active factors (Fakunding & Hershey, 1973; Sabol & Ochoa, 1971; Hershey, Dewey & Thach, 1969; Benne & Voorma, 1972; Vermeer, de Kievit, Van Alphen & Bosch, 1973a). IF3 appears to leave the 30S subunit when fMet–tRNA binds. IF1 is displaced from the 30S initiation complex during junction with the 50S subunit, and does not require the hydrolysis of GTP. IF1 has, however, some affinity for the 70S. The recycling of IF2, in contrast, requires both the junction of the 50S particle to the 30S initiation complex and GTP hydrolysis. Indeed, after the 70S couple is formed, a GTPase is activated by the combination of 50S particle and IF2. The hydrolysis of GTP into $GDP+P_i$ probably causes a change in the conformation of the ribosome or of IF2, allowing the release of IF2 (Lelong et al., 1970; Kolakofsky, Dewey, Hershey & Thach, 1968; Dubnoff & Maitra, 1972; Dubnoff et al., 1972). However, in the absence of GTP, release can also occur, albeit less efficiently; but in the presence of $GMPP(CH_2)P$, IF2 is not released and remains bound to the 70S complex (Fakunding & Hershey, 1973). The role of GTP hydrolysis appears to be that of providing for the rapid release of IF2 so that the elongation cycle of protein synthesis can proceed.

IF1 and IF3 markedly stimulate the recycling of IF2 (Dondon, Godefroy-Colburn, Graffe & Grunberg-Manago, 1974). Nevertheless, attachment of fMet–tRNA to 70S ribosomes can occur in the absence of IF1 and IF3, but is then stoichiometric to fMet bound. The reaction is catalytically stimulated by IF2 only in the presence of either one of the other two factors. The precise mechanism whereby IF1 or IF3 promote the recycling of IF2 is not yet known.

mRNA selectivity and the role of protein factors

The rate of mRNA binding or the stability of the RNA–ribosomal complex may be influenced by the presence or absence of proteins such as IF3 or S1. In order to understand better, the roles of such proteins in initiation we shall review certain experiments which shed some light on these problems.

Requirements for IF3 with different mRNAs There was a general contention that IF3 is required only for initiation in the presence of naturally occurring mRNAs and not when synthetic messengers were used. It has been suggested that IF3 is involved in the recognition of some specific messenger starting-signals present on phage mRNA, which contain an initiator sequence (possibly also with a specific structure) which is more complex than that of an initiator codon alone (Iwasaki, Sabol, Wahba & Ochoa, 1968; Revel, Aviv, Groner & Pollack, 1970). Such a sequence may be that postulated by Shine and Dalgarno (1974, 1975), since the binding of phage MS_2 RNA to ribosomes requires the presence of IF3, whereas the binding of synthetic messenger RNAs does not (Vermeer, Van Alphen, Knippenberg & Bosch, 1973*b*). The reported isolation of two mRNA-discriminating species of IF3 lends support to the role proposed for IF3 in the process of messenger recognition (Lee-Huang & Ochoa, 1973).

Conflicting with this view, other results suggested that IF3 is essential for initiation-complex formation, not only when phage RNAs are involved, but also with synthetic mRNAs (Miller & Wahba, 1973; Wahba *et al.*, 1969; Bernal, Blumberg & Nakamoto, 1974; Suttle, Haralson & Ravel, 1973). IF3 stimulates poly(U)-directed polyphenylalanine synthesis with 18 mM Mg^{2+} present, i.e. under conditions where ribosomes are 100% associated (Schiff *et al.*, 1974). This effect constitutes the basis for a specific test of IF3 activity, for it does not depend on the presence of other initiation factors and IF1 and IF2 have no substantial effect by themselves. However, we have seen (pp. 50–2) that IF3 exhibits no affinity for type-A 70S couples, although it has some affinity for type-B ribosomes, an observation which might indicate that in the presence of all components for translation, ribosomes are in the type-B conformation. Since there is no stimulation by IF3 of the binding of phenylalanyl–tRNA to 70S ribosomes, it is unclear whether this stimulation by IF3 is at the initiation step or at another step of protein synthesis (see Grunberg-Manago & Gros, 1977).

Other experiments have shown that IF3 also stimulates the initiation step promoted by synthetic polynucleotides containing AUG or GUG initiation codons (Miller & Wahba, 1973; Wahba *et al.*, 1969; Bernal *et al.*, 1974). However, most of these experiments have been performed with 70S ribosomes, and the stimulatory activity of IF3 could have resulted from its effect on the 70S-subunit equilibrium rather than from a specific influence on initiation complex formation. This interpretation appears at first glance to be strengthened by the finding that IF3 stimulates IF1- and IF2-dependent fMet–tRNA binding in the

Table 6. *Stimulation of fMet–tRNA binding to 30S subunits by initiation factors*

Mixtures contained in 100 μl: 50 mM Tris-HCl, pH 7.4; 100 mM NH_4Cl; 5 mM Mg^{2+} acetate; 7 mM 2-mercaptoethanol; 1 mM GTP; 0.1 A_{260} units poly(A,G,U) or 0.5 A_{260} units AUG; 1.0 A_{260} units [^{3}H]fMet-tRNA, sp.act. 3000 cpm pmole^{-1}; 0.35 A_{260} units 30S subunits; and initiation factors as indicated: 15 pmole IF1; 10 pmole IF2; and 10 pmole IF3. Mixtures were incubated 10 min at 37 °C, diluted with cold buffer and filtered through Millipore HA nitrocellulose filters; the filters were dried and immersed in scintillation fluid for counting. The results are reported as counts per min, following the subtraction of blanks (minus all factors) of 855 cpm for AUG and 2677 cpm for poly(A,U,G).

Additions	poly(A, G, U)	AUG
IF2	511	553
IF2+IF3	3167	956
IF2+IF1	2597	2923
IF1+IF2+IF3	5517	2102

Hershey, J. W. B., Dondon, J., Grunberg-Manago, M., unpublished experiments.

presence of 70S ribosomes, but not in the presence of 30S+50S particles (under conditions where these subunits, probably of the type-B, do not spontaneously associate) nor in the presence of 30S alone (Meier, Lee-Huang & Ochoa, 1973).

To elucidate the requirements for the initiation factors in initiation complex formation, the levels of binding of fMet–tRNA to either 30S or 70S ribosomes were compared with different mRNAs. IF2-dependent binding of fMet–tRNA to 70S ribosomes was greatly stimulated by either IF1 or IF3, while maximum levels of binding were obtained when both factors were added to the incubation mixtures (Dondon *et al.*, 1974). This was true when either poly(A,G,U) or triplet AUG were used and was most pronounced with small amounts of IF2. Thus both IF1 and IF3 promote the recycling of IF2, regardless of the type of mRNA. Such results are also obtained under conditions where 70S dissociation is not a limiting factor. However, when binding to the 30S subunit was measured, a different pattern was observed. With poly-(AGU) the result was similar to that for 70S ribosomes (stimulation occurred with either IF1 or IF3 and maximum binding is seen with both factors) whereas, with triplet AUG, IF1 greatly stimulated the IF2-dependent fMet–tRNA binding, while IF3 did so only slightly (Table 6). In the presence of both IF1 and IF2, IF3 inhibits complex formation. In conclusion, IF1-stimulated binding with either poly(A,G,U,) or AUG may occur by stimulating IF2 binding to 30S, as was discussed (p. 53). IF1 stabilizes IF2 on the 30S ribosome (Stringer, Sarkar & Maitra, 1977; Fakunding & Hershey, 1973). IF3 appears to cause two different and opposite effects. It stimulates the formation of initiation complex on 30S ribosomes with those mRNAs containing polypurine regions

Table 7. *Effect of S1 and factors on recognition of R17 initiator regions (from Steitz et al., 1977)*

Expt. No.	Ribosomes	Factors	Ratio A:coat:replicase
I	30S	IF2, IF3	1:2.0:0.5
	30S	—	1:0.2:0.05
	30S(−S1)	IF2, IF3	1:0.3:0.04
	30S(−S1)	—	1:0.4:0.03
II	30S	Crude	1:7.1:2.7
	30S	IF2, IF3	1:3.6:0.8
	30S	—	1:0.2:0.2
	30S(−S1)	IF2, IF3	1:0.8:0.2
	30S(−S1)	—	1:0.1:0.06

near the initiator codon and also weakens the binding of fMet–tRNA to 30S subunits but not to 70S ribosomes since the stability of the 70S initiation complex is not affected by IF3. The requirement for IF3 is more drastic with natural mRNA (e.g. phage RNA) than with synthetic polymers. With natural mRNA, the different cistrons vary in their dependence on the presence of IF3. Thus of the three R17 initiator regions, ribosome recognition of the coat or replicase sites is several-fold more dependent on the presence of initiation factors (Steitz, Wahba, Laughrea & Moore, 1977) and of fMet–tRNA than is binding to the A protein initiator region (Table 7). Similarly, fMet–tRNA binding to 70S ribosomes is more dependent on IF3 with R17 RNA than with AUG or poly (A,G,U) (Hershey *et al.*, 1977).

Role of ribosomal protein S1 There is evidence that protein S1 may interact specifically with the pyrimidine-rich terminal dodecanucleotide of the RNA molecule (Steitz, 1977). Gassen (1977) found that S1 interacts with all single-stranded polynucleotides which have a flexible structure, whatever the nature of the bases but it has less affinity for stacked bases as in poly(A). S1 will thus react with the polypyrimidine-rich 3'-OH end of 16S RNA and may correctly position that sequence for subsequent base-pairing with the complementarity region on the mRNA. Other possibilities for S1 action should, however, not be disregarded (Szer, Hermoso & Boublik, 1976; Steitz, 1977).

Conclusions The binding of mRNA to 30S subunits is probably stabilized primarily by two independent RNA–RNA interactions: (1) the anti-codon of fMet–tRNA with the initiator codon, usually AUG; and (2) the 3'-terminus of 16S RNA with a portion of the mRNA as postulated by Shine and Dalgarno and already discussed. The first interaction depends on IF2, while in the second, IF3 and S1 play critical roles. Thus the efficiency of translation of a given mRNA species could depend not only on the primary sequence of the initiator region of the mRNA, but also on the levels of initiation factors and of S1 in the organism.

As will be discussed on pages 63–4, Van Duin et al. (1976) found that IF3 can be cross-linked to the 3'-terminus of 16S RNA, and also to the 3'-terminus of 23S RNA. This suggests that the two RNA ends are both near to IF3 and to each other, and leads to the postulate that, in the 70S couple, 16S RNA forms complementary base-pairs with the 23S RNA, possibly at their 3'-termini. The proposal is supported both by the reasonably complementary nature of the two sequences (Fig. 6) and by observed differences in the nuclease susceptibility of the 16S RNA in the 30S, as opposed to the 70S, ribosomal particle (Chapman & Noller, 1977; Santer & Shane, 1977). The thermodynamic data on the ribosomal subunit association are also consistent with this hypothesis (see p. 32). Upon dissociation, the base-pairs between 16S and 23S RNA are replaced by self-complementary base-pairs, existing in both the 16S RNA and 23S RNA. The resulting 30S conformation is strengthened in the presence of IF3. Thus IF3 acts as an anti-association factor by preventing the interaction of 16S RNA with 23S RNA. Ribosomal protein S1 is thought to stabilize a single-stranded conformation of a small section of 16S RNA near the 3'-terminus. This section contains the Shine & Dalgarno (1974, 1975) sequence, which becomes available for mRNA binding. Following hydrogen bond formation between the mRNA and the 16S RNA, IF3 may in turn stabilize this interaction as well. The two reactions, 30S–mRNA and 30S–50S association, are thus mutually exclusive (Fig. 6).

In this model, the different in-vitro effects of IF3 on the dissociation of 70S couples and on the binding of mRNA can be seen as two different aspects of a single function. In effect, IF3 could, by binding to the sequence involved in base-pairing between 16S and 23S RNA, be either indirectly or directly responsible for a series of transitions of complementary nucleotide interactions, from those between 16S RNA and 23S RNA to those between the self-complementary interactions of 16S RNA, and finally to those between 16S RNA and mRNA. Initiation of protein synthesis may occur more efficiently on cistrons having more complementary pairing between the mRNA initiator region and the 16S rRNA. mRNAs which are poorly complementary should bind weakly and be more dependent on stabilizing factors such as IF3. The weaker dependence on IF3 observed with poly(A,G,U) compared to natural mRNAs may be due to the presence of long oligopurine

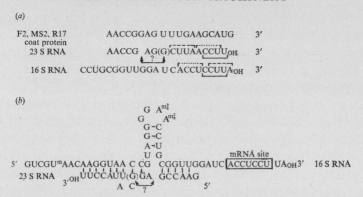

Fig. 6. Possible base-pairing between 3′-OH termini of 16S RNA and 23S or mRNA. (a) The nucleotide sequences of the 3′-end of the 16S RNA and 23S RNA are compared with those of the bacteriophage RNA ribosomal binding site associated with the coat-protein cistron. The hexanucleotide of the 23S RNA distal to the 3′-end has been sequenced by Branlant & Ebel (1977). (b) Hydrogen-bonding schemes are depicted for the 16S and 23S RNA interaction. From Van Duin et al. (1976).

sequences in the synthetic polymer compared to the weak Shine and Dalgarno complementary binding of the exposed coat cistron in the phage RNA. Possibly phage infection could result in a decrease of IF3, leading to the preferential utilization of initiator regions with relatively low initiation factor or S1 requirements. Benne & Pouwels (1975) have obtained evidence that mRNA transcribed from T7 DNA is optimally translated in an in-vitro system in the complete absence of IF3.

Specific mutations affecting initiation

Information bearing upon the theory of Shine & Dalgarno (1974, 1975), as well as about the roles of S1 and initiation factors *in vivo* and their possible regulatory roles in translation, could best be obtained with the help of specific mutants. Genetic loci for many ribosomal proteins and for all elongation factors are known but mutants affecting initiation of translation have been lacking until recently.

One class of ribosome binding site mutations comprises those that affect the triplet interaction between the initiator codon and the fMet–tRNA anti-codon sequence. Weissmann *et al.* (1977) obtained mutants of the Qβ coat-protein cistron in which the third position of the initiator AUG and the next residue (i.e. the first position of the succeeding alanine codon) are altered. They compared the potential base-pairing between the initiator tRNA anti-codon region and the relevant Qβ RNA sequences with the relative efficiencies of in-vitro ribosome attachment to the same four initiator regions. The base adjacent at the 3′-end of the AUG triplet has an important influence on the interaction with the messenger. Moreover the mutant RNAs which lack the AUG triplet retain considerable ribosome binding activity. It can be concluded that other ribosome–mRNA interactions suffice for the formation of a correctly located initiation complex.

Another very interesting mutant has recently been obtained and studied (F. W. Studier & J. J. Dunn, personal communication; Steitz, 1977). The mutation affects the T7 gene 0.3 initiator region by reducing the complementarity in the Shine and Dalgarno sequence from a 5 base-pair interaction (GAGGU) to a less stable interaction (GAAGU). A decrease of about ten-fold is observed in the efficiency of synthesis of the gene 0.3 protein in T7 infected cells. This important mutant will provide direct evidence, when controls such as mutations in regions other than in the Shine and Dalgarno sequence are found, that mRNA–rRNA base-pairing makes a significant contribution to the binding energy required for ribosome recognition of true initiation signals.

Another interesting thermosensitive mutant with a thermolabile initiation factor IF3 has been isolated by Springer, Graffe & Grunberg-Manago (1977a). Genetic data show that the mutation is located at about 38 min on the new E. coli map and is 68% co-transducible with the aroD marker. A λ hybrid transducing phage (Springer et al., 1977b) carrying the 38 min region of the E. coli genome was prepared in vitro and used to infect the thermosensitive strain. Thermoresistant trans-ductants were isolated and shown to have a normal IF3 activity. Final proof that this λ phage carries the structural gene for IF3 was obtained by sodium dodecylsulphate–polyacrylamide gel electrophoretic analysis of labelled proteins synthesized in UV irradiated cells after infection with the transducing phage. The following criteria were used to demon-strate that the phage carries the information for IF3: synthesis of a protein of MW 22000; specific cross reaction with anti-IF3-antibodies; co-migration with pure IF3 on sodium dodecylsulphate–polyacrylamide gel; and co-migration on a two-dimensional gel system separating proteins by charge in the first dimension and by molecular weight in the second. The availability of both a bacterial mutant modified in IF3 and a phage carrying the IF3 gene opens the field of the initiation of trans-lation to genetic and physiological studies.

LOCATION OF THE INITIATION COMPONENTS ON THE 30S SUBUNIT

Introduction

We now turn to the evidence for the location on the 30S subunit of the components involved in initiation – first to the interaction of each of the major components with specific 30S proteins and to a more limited extent with 16S RNA, and then to a comparison of these results with those from structural studies discussed by Brimacombe in this volume. We shall emphasize location at the level of individual ribosomal proteins, despite the fact that RNA constitutes two-thirds by weight of the ribosomal subunit. This reflects the situation that protein location can be found relatively easily by using two-dimensional gel electrophoresis and immunological techniques, and thus has proceeded rapidly, whereas RNA location is currently more lengthy.

The methods used to locate initiation components may be conveniently divided into two categories. These are direct methods, in which covalent bonds are formed between the component (usually used in radioactive form) and portions of its site on the 30S subunit, with subsequent isolation and identification of the ribosomal sites of incorporation; and indirect methods, in which the functional consequences of defined structural variations in the 30S subunit are examined. The former include two techniques; affinity labelling and cross-linking. With affinity labelling, a chemically (or photochemically) reactive derivative of the component in question is used to form covalent bonds with the receptor site, whereas in cross-linking double-headed reagents or photochemical techniques are used to form such bonds. The great advantage of affinity labelling is that covalent bond formation is restricted to reactions of the modified component. The cross-linking approach on the other hand allows utilization of the native component, thus avoiding the problems of chemical synthesis of derivatives and the demonstration that such derivatives retain biological activity.

Of the indirect methods, several are based on the ribosome reconstitution techniques of Traub & Nomura (1968). Functional studies have been carried out on reconstituted ribosomes either lacking a single protein (single omission experiments) (Nomura et al., 1969), or in which a single ribosomal protein isolated from a different bacterial strain has been substituted for its homologous protein from the strain under consideration (heterologous reconstitution experiments) (Held et al., 1974). Protein addition experiments have also been performed in which the effect of adding back proteins, present in only fractional amounts in the isolated ribosome, is studied (Randall-Hazelbauer & Kurland, 1972). Two other important indirect methods are functional studies of ribosomal mutants (Stöffler & Wittmann, 1977) and the examination of the inhibitory effects of antibody or Fab fragments for specific 30S proteins on various ribosomal functions (Lelong et al., 1974).

In presenting the evidence on site location, we have weighted the results obtained using direct methods most heavily, and for the most part have used results from indirect methods only when they bear on results obtained by direct methods. We have done this because of the overwhelming evidence that ribosomes are conformationally labile, as a result of which even small changes in structure can have consequences in regions of the ribosome far removed from where the change has been made (Ginzburg & Zamir, 1975; Kurland, 1977). Thus indirect methods of location have, by themselves, only very limited usefulness. Because of this heavy reliance on the direct methods, it is appropriate to mention potential pitfalls associated with their application. More complete discussions of these topics may be found elsewhere (Cooperman, 1977; Kurland, 1977) and we should like to note here only those problems which have arisen in the application of these methods to studies of the ribosome. Two of the problems reflect inherent difficulties in ribosomal studies. First, there is no standard method for preparing homogeneous, active ribosomes. Thus, ribosomes prepared by harvesting bacteria at different stages of the cultural growth phase, after different methods of cell-breaking, or by including or excluding high salt washing, show some variation in both protein composition and functional properties (Deusser, 1972; Weber, 1972; Noll et al., 1973; Debey et al., 1975; Van Diggelen & Bosch, 1973; Van Diggelen, Oostrom & Bosch, 1973). Secondly, ribosomal conformation appears to be highly dependent on such factors as monovalent cation and Mg^{2+} concentrations, pH, and temperature (Zamir, Miskin & Elson, 1971; Zamir, Miskin, Vogel & Elson, 1973). Hence comparison of results obtained in different laboratories, where preparative methods and reaction media vary, is difficult, particularly as little systematic study of the dependence of results from direct methods on these parameters has been made. In affinity-labelling studies insufficient attention has been paid to the problem of site multiplicity for a given ligand. Thus, the dependence of affinity-labelling results on component concentration is seldom studied, leaving open the possibility that the results obtained reflect secondary rather than primary binding sites. A major problem with the cross-linking studies is the determination of whether a given cross-linked product reflects the native or denatured 30S structure.

Finally, many of the affinity-labelling experiments and all of the cross-linking experiments utilize electrophilic reagents to effect covalent bond formation. Thus labelling results reflect not only the proximity of ribosomal proteins to the component binding site but also their nucleophilic properties. With these reservations we shall consider what is known about the various sites.

Component sites

mRNA

Affinity-labelling studies with electrophilic derivatives of four oligonucleotides (AUG, UGA, AUGU and GUUU), with photolabile poly(U) derivatives, and by using the intrinsic photolability of poly(U) itself, implicate S1, S4, S12, S18 and S21 as being at or near the mRNA binding site (Fiser, Scheit, Stöffler & Kuechler, 1974; Fiser, Margaritella & Kuechler, 1975a; Fiser, Scheit, Stöffler & Kuechler, 1975b; Schenkman, Ward & Moore, 1974; Pongs, Lanka & Bald, 1975a; Pongs, Stöffler & Lanka, 1975b; Pongs, Stöffler & Bald, 1976; Pongs & Rossner, 1975; Pongs & Lanka, 1975a, b; Pongs & Rossner, 1976; Luhrmann, Gassen & Stöffler, 1976). These results have a striking degree of overlap and correlate well with those obtained from indirect studies. Thus, mRNA binding has been shown to be dependent on added S1 (Van Duin & Kurland, 1970). Heterologous reconstitution and single omission experiments have shown S12 and S21 to be important in AUG-dependent fMet–tRNA^{fMet} binding (Nomura et al., 1969; Held et al., 1974), and Fab inhibition studies have implicated S1, S12, S18 and S21 in AUG-dependent fMet–tRNA^{fMet} binding.

Initiation factors

Essentially all the information about the binding site for initiation factors comes from cross-linking studies. These results are summarized in Table 8. For IF3, the results reported reflect experiments on binary 30S–IF3 complexes. Results for IF1 and IF2 were obtained in solutions containing 30S subunits and all three factors, either IF1 or IF2 being radioactive. Use of a cleavable cross-linking reagent permitted the demonstration that the cross-linking between S7 and IF3 was direct. All the other cross-links were obtained with non-cleavable reagents, thus introducing the ambiguity that, in some cases, the cross-links reported may be indirect and mediated by one or more other proteins. Nevertheless, the results do suffice to define protein neighbourhoods at or near the initiation-factor binding sites. The high degree of overlap among the proteins which are cross-linked either to all three factors

Table 8. *Cross-linking of initiation factors*

	Factor		
S protein	IF1[a]	IF2[b]	IF3[c]
1	+	+	+
2		+	?
3			?
4			?
7			++[d]
9			++[f]
11		++	++
12	++	++	++[c,e]
13	+	++	?
14	+	+	
18			?
19	++	++	++
20			?
21			+

++ Major cross-linked protein, + minor cross-linked protein, ? not always found.
[a] Langberg *et al.* (1977), [b] Bollen *et al.* (1975), [c] Heimark *et al.* (1976b) except as noted, [d] Van Duin *et al.* (1975), [e] Hawley, Slobin & Wahba (1974), [f] R. A. Hawley, J. E. Sobura & A. J. Wahba, unpublished results.

(S1, S12, S13, S19) or to two (IF1 and IF2, S14; and IF2 and IF3, S11) is strong evidence for the proximity of these factors to one another on the 30S particle. In fact, Langberg, Kahan, Traut & Hershey (1977) have recently found evidence for formation of an IF1–IF2 cross-link in the trifactor complex with the 30S particle. Additional evidence for the proximity of IF1, IF2 and IF3 on the 30S subunit is that all three factors cross-link to the 3′-terminus of 16S RNA on periodate oxidation and subsequent sodium borohydride reduction of factor–30S complexes (S. Langberg & J. W. B. Hershey, unpublished results). Treatment of *E. coli* with either colicin E3 or cloacin DF13 inhibits protein synthesis owing to a single cleavage of 16S RNA, approximately 50 nucleotides from the 3′-terminus (Bowman *et al.*, 1971; Senior &. Holland, 1971; Boon, 1971; De Graaf, Planta & Stouthamer, 1971), This inhibition appears to result from a loss of IF1 function (Baan *et al.*, 1976).

fMet–tRNA^{fMet}

N-acetyl-Phe–tRNA^{Phe} has been shown to substitute for fMet–tRNA^{fMet} in both the formation of an initiation complex and initiation of protein synthesis (Lucas-Lenard & Lipmann, 1967). Thus it is not unreasonable to use the results of affinity-labelling studies with electrophilic *N*-acetyl–Phe–tRNA^{Phe} derivatives to define a binding site for the 3′-terminus of fMet–tRNA^{fMet}. Two such derivatives have been found by Girshovich *et al.* (1974*a*, *b*) to react with isolated 30S subunits in poly(U)-dependent

Table 9. *Summary of site location data in the initiation reaction*

Site S protein	mRNA	Initiation factors	3'-end of tRNA	Present in region I	II
1	++	++		v	v
2		+		–	–
3			++	–	–
4	++			v	v
7		++	+	v	v
9		+		v	v
11		++		v	v
12	++	++		v	v
13		++	++	v	v
14		+	++	v	–
18	++			v	v
19		++		?ᵃ	v
21	++	+		v	v
relevant cross-links	4–12 12–21 18–21	7–9 7–13 11–13 11–18–21	12–13 12–21 13–19 14–19	7–13		

++ Strongly implicated, + Marginally implicated. *a* See text.

reactions when S3, S7, S13 and S14 were labelled. Single omission experiments implicate S13 in fMet–tRNA^{fMet} binding (Nomura *et al.*, 1969) and Fab inhibition studies implicate S3, S13 and S14 in AUG-dependent fMet–tRNA^{fMet} binding (Lelong *et al.*, 1974).

Summary of location results

The proteins implicated at the various ribosomal binding sites are listed in Table 9. There is a great deal of overlap between the proteins cross-linked to initiation factors and those found at the other two types of sites. For the mRNA site the overlap includes proteins S1, S12, and S21. Interestingly, there is no overlap between the mRNA and the fMet–tRNA^{fMet} sites. There is also evidence (see pp. 42–6; 64) that the 3'-terminal region of 16S RNA is important for mRNA recognition and the binding of initiation factors.

Correlation of site location and structural studies

We next wish to correlate the location results with knowledge obtained from other structural studies of 30S particles, emphasizing in particular the results of protein–protein cross-linking and of immune electron microscopy. The protein cross-linking results offer important support for the data summarized in Table 9. Thus, for the proteins at the mRNA site the following pairs of proteins have been found to be cross-linked: S4–S12, S12–S21, S18–S21 (Sommer & Traut, 1975, 1976; Chang &

Flaks, 1972; Lutter, Zeichhardt, Kurland & Stöffler, 1972). In addition, both proteins S1 and S21 can be cross-linked to the 3′-end 16S RNA by the periodate/NaBH$_4$ procedure (Czernilofsky, Kurland & Stöffler, 1975) suggesting that they are also neighbours. Many cross-links have been reported among the proteins implicated in factor binding: S7–S13, S7–S9, S11–S13, S11–S18–S21, S12–S13, S12–S21, S13–S19 and S14–S19 (Shih & Craven, 1973; Bode, Lutter & Stöffler, 1974; Lutter, Bode, Kurland & Stöffler, 1974; Lutter *et al.*, 1975; Sun, Bollen, Kahan & Traut, 1974; Sommer & Traut, 1975, 1976). Furthermore, S7, S9, S13, S14 and S19 all bind to adjacent regions of 16S RNA (Brimacombe, Nierhaus, Garrett & Wittmann, 1976).

Before proceeding to correlate the results for site location with those from immune electron microscopy, some uncertainties in the latter approach should be pointed out. First, the two groups which have obtained the most information using this technique differ somewhat both in their respective models for ribosome structure, and in their placement of specific protein determinants (Tischendorf, Zeichhardt & Stöffler, 1975; Stöffler & Wittmann, 1977; Lake & Kahan, 1975; J. A. Lake & L. Kahan, unpublished results). Secondly, several of the 30S proteins of interest appear to have elongated conformations, with the possibility that two or several proteins may be near neighbours in more than one region of the subunit. This can create ambiguity for the assignment of ligand binding sites although future work should permit many of the present ambiguities to be resolved. Thus, for example, identification of covalently labelled peptides within an affinity-labelled protein, combined with an extension of the technique of immune electron microscopy using antibodies to ribosomal protein fragments, should allow a more definitive assignment of binding sites. The feasibility of this approach has been demonstrated in recent work of Yaguchi, Stöffler, Pongs, Lanka and co-workers (Stöffler & Wittmann, 1977). They have located the covalent incorporation of an AUG affinity label into the N-terminal region of protein S18 and have shown further which of the two S18 determinants found on the 30S particle includes this region.

Determinants for ten of the thirteen proteins listed in Table 9 are found in the limited, right temporal, region I of the Stöffler and Tischendorf model for the 30S particle, as indicated in Fig. 7a. This region includes determinants for all five mRNA site proteins, for eight out of ten initiation factor site proteins, and for three out of four fMet–tRNA site proteins. A corresponding, but more extended, region (I′) can be defined in the Lake and Kahan model (Fig. 7b) containing determinants

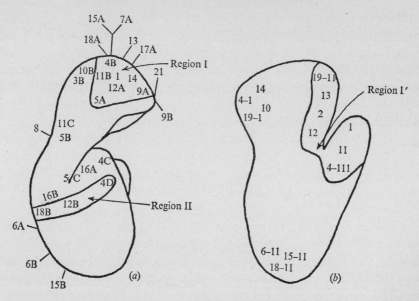

Fig. 7. Models for the 30S subunit: (*a*) Stöffler and Tischendorf, from Stöffler & Wittmann (1977); (*b*) Lake and Kahan, from J. A. Lake & L. Kahan (unpublished). The numbers correspond to those of the proteins; letters differentiate multiple antigenic sites where observed.

for eleven out of the thirteen proteins. In the orientation presented, seven of these determinants are shown, and those for four others, S7, S9, S18 and S21 are found in neighbouring areas which are not visible. Both regions I and I′ contain determinants for all five proteins implicated in mRNA binding, so the location of this site is clear. It may be noted that the Stöffler and Tischendorf model contains a second region (II) in which determinants for three of the proteins, S4, S12, and S18, are in close proximity. However, S1 and S21 are only found in region I, and it is the S18 determinant in region I which includes the site of AUG analogue affinity labelling.

For the initiation-factor binding sites, the Stöffler and Tischendorf model lacks determinants in region I for protein S2 and S19. Protein S2 is cross-linked only to IF2 and is a minor product, whereas protein S19 is a major cross-linked protein interacting with all three factor proteins and its absence would pose a real problem for location of the factor site. In recent unpublished work, Tischendorf and Stöffler now report a second S19 site, close to S14, and thus probably in region I (Stöffler & Wittmann, 1977). This would appear to resolve the problem and is in accord with unpublished results of Lake and Kahan, who have found two determinants for S19 of which one is in the region I′. Thus the initiation-factor binding sites can also be located in region I/I′.

The case for a single site for fMet–tRNA^{fMet} is less clear, since of the four proteins implicated, one, S3, is missing from both regions I and I', and a second, S14, is missing from I' as well. It may be that the 3'-terminus of tRNA is not fixed in a single location on the 30S particle. Region I/I' emerges from this discussion as *the* site of mRNA and initiation factor binding, as well as a probable site, perhaps of several, for the binding of the 3'-terminus of fMet–tRNA^{fMet}.

MECHANISM OF ELONGATION IN PROTEIN SYNTHESIS

THE ELONGATION CYCLE

Outline of elongation

A brief description of the cycle of elongation will be given first, and then individual steps will be discussed in greater detail under the headings of the particular components involved. The sequence of events is presented in Fig. 8. A ternary complex is formed between EF–Tu·GTP and aatRNA which then binds to the ribosome·mRNA·fMet–tRNA complex or, in further elongation steps, to the ribosome·mRNA·peptidyl–tRNA complex. A particular ternary complex is selected from the population of different species by the specific codon–anti-codon interaction. GTP is then hydrolysed, allowing EF–Tu·GDP to dissociate from the ribosome, which now contains fMet–tRNA^{fMet} or peptidyl–tRNA in the P site and aatRNA in the A site. The fMet– or peptidyl moiety is transferred to the aatRNA in the A site by the transferase activity of the large ribosomal subunit. Elongation factor (EF–G) and GTP, probably as a binary complex, now bind to the ribosomal complex. The uncharged tRNA in the P site is displaced by the newly formed peptidyl–tRNA and mRNA is translocated by one codon. One molecule of GTP is hydrolysed and EF–G and GDP dissociate from the ribosome, which is then ready to repeat the first step of the cycle. In a shuttle process catalysed by EF–Ts, the ternary complex aatRNA·EF–Tu·GTP is reformed from EF–Tu·GDP.

Elongation factor EF–Tu

The role of this protein factor is to catalyse the binding of aatRNA to the A site of the ribosome, a process involving interactions of EF–Tu with a considerable number of components. These have been studied in some detail and are now largely understood. EF–Tu is a single polypeptide chain of MW 44000 (Miller & Weissbach, 1970; Blumen-

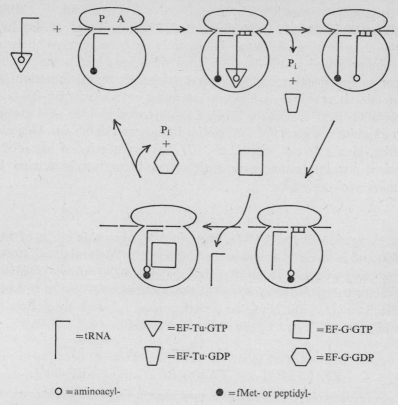

Fig. 8. Cycle of elongation.

thal, Landers & Weber, 1972) which represents about 5% of the soluble proteins of the cell. It is thus present in about the same molar quantity as tRNA, and about ten times that of EF–Ts (see below) and ribosomes (Furano, 1975). Highly purified EF–Tu is readily obtained (Miller & Weissbach, 1974; Furano, 1975) and crystallized. Free EF–Tu is extremely unstable, but is protected against thermal denaturation by GDP, GTP, GMPPCP or EF–Ts, each of which can form a complex with EF–Tu. GDP binds much more tightly than GTP ($K_d = 4.9 \times 10^{-9}$ M and 3.6×10^{-7} M, respectively; Arai, Kawakita & Kaziro, 1974a) and the binding is highly dependent on Mg^{2+} (K_d for GDP in the absence of Mg^{2+} is 2.2×10^{-6} M). The GTP analogue GMPPNP binds as strongly as GTP and much more strongly than GMPPCP ($K_d = 3.7 \times 10^{-7}$ M and 2.2×10^{-6} M, respectively). Replacement of GDP by GTP results in changes, probably rather small, in the conformation of the molecule (Printz & Miller, 1973; Crane & Miller, 1974; Arai et al., 1974b; Arai, Arai, Kawakita & Kaziro, 1975a), which increase the affinity of the binary complex for aatRNA by at least five orders of magnitude (Miller, Cashel & Weissbach, 1973; Shulman *et al.*, 1974; Arai *et al.*, 1974*a*). Even in the absence of aatRNA, an interaction has been observed between the binary complex, EF-Tu·GTP, and ribosomes which protects the factor against inactivation by *N*-ethylmaleimide. By this criterion, the interaction is specific to the triphosphate and is not seen with EF-Tu·GDP (Kawakita, Arai & Kaziro, 1974). EF-Tu contains three -SH groups, two of which are essential for activity in polypeptide chain elongation (Miller, Hachmann & Weissbach, 1971). The integrity of one -SH is required for interaction with aatRNA and another is needed for guanine nucleotide binding.

Elongation factor EF–Ts

The elongation factor, EF-Ts, is a single polypeptide chain of MW 19000 and is much more thermostable that EF-Tu (Arai *et al.*, 1974*a*). By reacting with EF-Tu·GDP or EF-Tu·GTP to form EF-Tu·EF-Ts, the factor catalyses the exchange of the nucleotides (Weissbach, Miller & Hachmann, 1970). Thus, the overall role of EF-Ts is to catalyse the reutilization of EF-Tu by the series of reactions:

$$EF-TU·GDP+EF-Ts \rightleftharpoons EF-Tu·EF-Ts+GDP$$
$$EF-Tu·EF-Ts+GTP \rightleftharpoons EF-Tu·GTP+EF-Ts$$
$$aatRNA+EF-Tu·GTP \rightleftharpoons aatRNA·EF-Tu·GTP$$
$$Rib+aatRNA·EF-Tu·GTP \rightleftharpoons Rib·aatRNA+EF-Tu·GDP+P_i$$

EF-Tu can thus be considered to participate in a cyclic reaction driven from equilibrium by GTP hydrolysis on the ribosome. EF-Ts contains one -SH group which is required for catalytic activity in GDP exchange. Reaction of the -SH with *N*-ethylmaleimide is inhibited when EF-Ts interacts with EF-Tu (Miller *et al.*, 1971).

Ternary complex aatRNA·EF–Tu·GTP

Charged tRNA binds rapidly and tightly to EF-Tu·GTP to form a ternary complex with a dissociation constant estimated to be 10^{-8} M (Miller *et al.*, 1973; Arai *et al.*, 1974*a*). Study of the requirement for ternary complex formation has been greatly facilitated by the observation that, unlike EF-Tu·GTP, it is not retained on Millipore filters. The complex can also be isolated by gel filtration (Gordon, 1968). To interact with EF-Tu·GTP the tRNA must be charged (Ravel, Shorey & Shive, 1967; Gordon, 1967; Skoultchi, Ono, Moon & Lengyel, 1968) but the amino group of the amino acid must not be acylated

(Ravel et al., 1967; Weissbach et al., 1970). Deamination of Phe–tRNA to give phenyl–lactyl–tRNA does not, however, abolish binding and the ternary complex can bind to ribosomes in the presence of poly(U) (Fahnestock, Weissbach & Rich, 1972). Inactivity of N-acetylated aatRNA is more likely to be related to steric hindrance than to loss of the charged amino group. If tRNA^{Phe} is oxidized with periodate and then reduced with borohydride, the resulting diol in the 3'-terminal adenosine can be charged but the molecule will not bind to EF–Tu·GTP (Chen & Ofengand, 1970). Recent studies using tRNA species terminating in 2'- or 3'-deoxyadenosine indicate, in contradiction to earlier reports (Ringer & Chladek, 1975; Sprinzl et al., 1975) that EF–Tu can bind to both positional isomers of aatRNA with no obvious preference for either (Hecht, Tan, Chinault & Arcari, 1977). Denatured Leu–tRNA$_3$^{Leu} from yeast will not form a ternary complex, although denatured Trp–tRNA from E. coli has been reported to do so (Ofengand, 1974). Some aspects of the tRNA tertiary structure thus appear to be necessary. On the other hand, the association of aatRNA and EF–Tu·GTP is not affected by selective modification of the anti-codon loop (Krauskopf, Chen & Ofengand, 1972; Ghosh & Ghosh, 1970; Thang, Springer, Thang & Grunberg-Manago, 1971), the anti-codon stem (Abelson et al., 1970), or the dihydro-uridine loop (Abelson et al., 1970) or stem (Krauskopf et al., 1972).

Met–tRNA^{fMet}, however, will not react with EF–Tu·GTP even when it is not formylated (Schulman & Her, 1973). The aspect of the tRNA structure that prevents the interaction is thought to be the absence of a Watson–Crick base-pair in position 1–73, since bisulphite modification of C1, which establishes U1·A73, allows binding (Schulman, Pelka & Sundari, 1974). This may not be a general requirement, however, because a modified tyrosine suppressor tRNA from a mutant which lacks the 5'-terminal base-pair (Smith & Celis, 1973) must be capable of at least some ternary complex formation for biological activity.

The GTPase activity associated with EF–Tu is normally dependent upon both charged tRNA and ribosomes, but in the presence of 20% methanol the requirement for tRNA is partially lost. This further suggests an interaction between the binary complex EF–Tu·GTP and the ribosome (Ballesta & Vazquez, 1972). The 50S but not the 30S subunit can substitute for 70S ribosomes in this reaction. GTPase activity is also seen when EF–Tu·GTP interacts with ribosomes containing pre-bound tRNA. Interestingly, the specificity with respect to tRNA is less restricted than in ternary complex formation. Thus, deacylated tRNA and N-acetyl–Phe–tRNA pre-bound to the A site can complement EF–Tu·GTP in this activity (Kawakita *et al.*, 1974). More striking, however, is the effect of the antibiotic kirromycin upon GTPase activity. As well as enhancing the ribosome-dependent activity, kirromycin induces a ribosome-independent GTPase, and thereby demonstrates that the catalytic centre is primarily located on the elongation factor itself (Chinali, Wolf & Parmeggiani, 1977) rather than on the ribosome, as previously suggested (Pongs, Nierhaus, Erdmann & Wittmann, 1974). In the presence of the antibiotic, aatRNA can interact with EF–Tu and GDP to form a complex sufficiently stable to be isolated by gel filtration. This observation can be seen to explain the effect of kirromycin on elongation, which it blocks by inhibiting the release of EF–Tu·GDP after GTP hydrolysis (Wolf, Chinali & Parmeggiani, 1977).

tRNA selection

The binding of aatRNA molecules to the ribosomal A site reflects primarily the codon–anti-codon interaction. Many factors may be expected to influence this interaction, including the way in which the codon is presented on the ribosomal surface (about which little is known) and the modification of bases in the anti-codon and in other positions in the anti-codon loop. This last factor has been easier to study but is far from being completely understood.

The association between oligonucleotides and tRNA has been used to study the relative stabilities of complexes involving the wobble pairs predicted by Crick (1966). In general, tRNA·oligonucleotide complexes containing these pairs (G·U, I·C, I·U and I·A) are much less stable than the corresponding complexes employing Watson–Crick base-pairs (Unger & Takemura, 1973). Whereas model systems employing oligo-nucleotides of known sequence have illuminated many aspects of base-pairing between complementary RNA strands, it has been shown recently that the prediction of binding energies (Gralla & Crothers, 1973*a*; Borer, Dengler, Tinoco & Uhlenbeck, 1974) cannot be applied successfully to the interactions between RNA loops. In studies of the association between complementary anti-codons, by the technique of Eisinger (1971), it has been shown that there is no systematic relation between the GC content and the stability of the base-paired complexes (H. J. Grojean, S. de Henau and D. M. Crothers, unpublished results). The effect of hypermodified purines on the 5'-side of the anti-codon, which appears to be a mechanism for strengthening adjacent A·U base-pairs (Högenauer, Turnowsky & Unger, 1972; Miller, Barrett & Ts'O, 1974; Grosjean, Söll & Crothers, 1976; Grosjean *et al.*, unpublished results) offers only a partial explanation for this observation. Clearly, standardization of the energy of cognate tRNA–codon interactions may be of great importance in distinguishing them from non-cognate interactions. The 'short' wobble pairs $U \times C$, $U \times C$ and their 2-thiolated derivatives (Crick, 1966) contribute significantly to helix stability in the complementary anti-codon system (Grosjean *et al.*, unpublished results). It is suggested that the ribosome may be able to discriminate against the rotations of the $C'5-C'4$ and $P-O'5$ bonds in the codon wobble nucleotide that are needed to accommodate these pairs. Ribosomal discrimination also appears necessary in weak interactions involving I and ψ. Apart from these peculiarities, a comparison of cognate and non-cognate interactions confirms that there are frequently mismatched complexes with lifetimes at least 1% of those of correctly matched complexes.

Whereas the coding rules must be followed strictly in order to avoid certain misreading errors, in other situations similar base mismatching may be unimportant. This is so in codon groups where all four codons, differing only in the third nucleotide, code for the same amino acid. At least during protein synthesis *in vitro*, it appears that each of the tRNA^{Val} iso-accepting species, with anti-codons UAC, GAC and IAC, are capable of reading all four Val codons (Mitra, Lustig, Akesson & Lagerkvist, 1977). A similar situation may prevail in the Arg codon group (J. Weissenbach, personal communication). Experiments *in vivo* show that during rapid growth or in an $su6^+$ strain, the utilization of tRNA^{Leu} iso-acceptors in *E. coli* is difficult to reconcile with the wobble rules (Holmes, Goldman, Minert & Hatfield, 1977). Thus in certain cases the genetic code appears to be strict only as regards the first two letters.

Codon specificity is not only influenced by the anti-codon sequence. Thus UGA–su$^+$ tRNA^{Trp} possesses the same anti-codon as wild-type tRNA^{Trp} and differs from it by only a single base in the dihydro-uridine stem (Hirsh, 1971; Hirsh & Gold, 1971). The suppressor species translates UGU as well as UGA and UGG, suggesting that it responds essentially to two-letter codons (R. H. Buckingham and C. G. Kurland, unpublished results). At present, it is not clear whether two-letter reading in groups of codons specifying a single amino acid is related to the particular properties of these codon–anti-codon interactions or to other features of the tRNA molecules which allow them to respond to a two-letter codon.

The fidelity of translation can be strongly affected by antibiotics and by mutations affecting ribosomes. Streptomycin and other aminoglycoside antibiotics can stimulate misreading (Davies, Gilbert & Gorini, 1964; Van Knippenberg, Grijm-Vos, Veldstra & Bosch, 1965; Gorini, 1974). Streptomycin affects reading of 5′-terminal and internal positions in the codon, one at a time, whereas neomycin can provoke in addition misreading of the 3′-codon position and of two bases at a time (Davies, Jones & Khorana, 1966). Two classes of mutants have been characterized that differ in their behaviour with respect to streptomycin. *StrA* mutants show lower levels of naturally occurring errors and lower efficiency of nonsense suppression whereas *ram* mutations have the opposite effect (Gorini, 1974). These mutations have been shown to affect ribosomal proteins S4, S5 and S12 (Wittmann & Wittman-Liebold, 1974). A kinetic analysis of missense and nonsense suppression in *StrA* and *ram* mutants suggests a rather simple relationship between the observed effects and molecular parameters (Ninio, 1974).

Very few errors are made during messenger decoding *in vivo*. Estimates of the order of 1 in 10^4 have been made for the mistranslation of isoleucine as valine, or of arginine as cysteine (Loftfield & Vanderjagt, 1972; Edelmann & Gallant, 1977). All that is known at present about the codon–anti-codon interaction suggests that in many cases a non-cognate tRNA will be available having a rate constant for dissociation, which is the major discriminating factor, of no more than one hundred times that of the cognate tRNA. However, many anti-codons which would be liable to mistranslate are perhaps not used in the cell and are functionally replaced by different anti-codons (Ninio, 1971, 1973). There may also be selection against codons liable to be mistranslated where this is functionally important.

How then does the ribosome achieve such precision? A number of attempts have been made to explain how the ribosome responds in a non-linear fashion to the association lifetime. Reaction schemes have been proposed by Hopfield (1974) and by Ninio (1975) which might be used in a variety of recognition processes and may be considered in relation to the particular problem of tRNA selection. Both schemes depend on there being two points at which the substrate may dissociate, thus allowing the discrimination available at dissociation to be used twice. In order to increase discrimination above that of a single-step process, it is essential that substrate entry should be predominantly by one of these gates and not by both. This is achieved by coupling the reaction cycle to an energy-consuming step so that it is driven continuously out of equilibrium. The schemes of Hopfield and of Ninio differ in the step at which the coupling is introduced. Fig. 9 shows one way, consistent with our current understanding of the elongation cycle, in

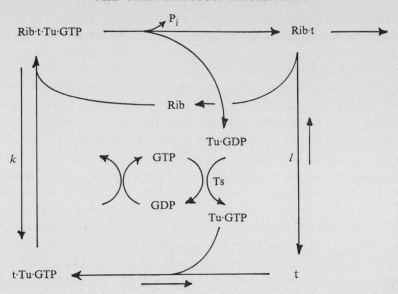

Fig. 9. A hypothesis for the elongation cycle with kinetic amplification of tRNA selection. t represents aatRNA.

which these principles may be applied to tRNA selection. The only rate constants which discriminate between cognate and non-cognate aatRNAs are k and l. The interaction between aatRNA·EF–Tu·GTP and the ribosome is not far from equilibrium so that good discrimination may be obtained at this stage. After GTP hydrolysis and EF–Tu·GDP dissociation, the model requires that there should be a possibility of dissociation of aatRNA as opposed to peptidyl transfer. In order that discrimination at this stage should be effective, a significant proportion of cognate tRNA molecules as well as a large proportion of non-cognate molecules would need to dissociate rather than act as acceptors of the peptidyl moiety. In order that aatRNA may not associate directly with the ribosome by this reaction, Hopfield (1974) supposes that the Rib·aatRNA complex after EF–Tu·GDP dissociation should be a 'high energy intermediate', direct formation of which is improbable. This may not be necessary if aatRNA molecules after dissociation are sequestered rapidly by a pool of EF–Tu·GTP, as the former binds very tightly to this complex (Fig. 9).

What evidence is there that the ribosome might function in this way? It may be significant that, during the early periods after ternary complex binding to the ribosome, there is much less aatRNA bound stably than there is GTP hydrolysed (Weissbach, Redfield & Brot, 1971). Perhaps after EF–Tu·GDP dissociation, aatRNA may also dissociate

until the binding is stabilized by the establishment of other aatRNA–
ribosome interactions such as the entry of the aminoacyl 3′-terminus
into the peptidyl-transferase centre. More recently, evidence has been
presented which suggests that the amount of amino acid participating
in dipeptide formation is a much smaller fraction of the GTP hydrolysed
when near-cognate ternary complexes bind to poly(U)-programmed
ribosomes than when cognate tRNA is used (Thompson & Stone, 1977).
A quite different model for tRNA selection has been proposed by
Kurland, Rigler, Ehrenberg & Bomberg (1975). This model operates
near equilibrium and postulates that sequential reactions, associated
with conformational changes in the bound tRNA molecule and the
establishment of further ribosomal binding intereactions, are characterized by equilibrium constants that discriminate between cognate and
non-cognate codon–anti-codon interactions. Thus the free energy
changes associated with the formation of additional interactions with
the ribosome are greater for non-cognate than for cognate tRNAs.
Experiments in the absence of ribosomes are considered to show only a
part of the difference between cognate and non-cognate tRNA–codon
binding that can be realized when ribosomes are present.

Possible alternatives to models like these are systems of selection
which, after the initial step of ternary complex–ribosome association,
operate well away from equilibrium and discriminate through forward
rate constants rather than equilibrium constants. For example, it is
conceivable that the activation energy for EF–Tu-dependent GTPase
activity might be influenced by the codon–anti-codon interaction. Until
the factors which influence the GTPase activity are better understood
such schemes must be considered purely speculatory.

Binding of ternary complex to ribosomes

When Phe–tRNA·EF–Tu·GTP is added to ribosomes carrying poly(U)
and N-acetyl–Phe–tRNA in the P site, Phe–tRNA is very rapidly
bound, more rapidly than when the separate constituents of the ternary
complex are added simultaneously (Ertel et al., 1968; Lucas-Lenard &
Haenni, 1968). This observation, and the fact that free aatRNA does
not compete with the ternary complex in binding at 0 °C, demonstrate
that the complex is an intermediate in aatRNA binding. Stable binding
is accompanied by hydrolysis of one molecule of GTP (Ono, Skoultchi,
Waterson & Lengyel, 1969; Ravel et al., 1969) and the release of P_i and
EF–Tu·GDP (Gordon, 1969; Shorey, Ravel, Garner & Shive, 1969;
Ono et al., 1969; Waterson, Beaud & Lengyel, 1970; Skoultchi, Ono,
Waterson & Lengyel, 1970). A site binding is specifically inhibited by

the antibiotic tetracycline (Weissbach *et al.*, 1971; Shorey *et al.*, 1969; Haenni & Lucas-Lenard, 1968; Lucas-Lenard, Tao & Haenni, 1969). Elongation can procede in the complete absence of EF–Tu and GTP hydrolysis but is much slower than in the factor-mediated process (Gavrilova *et al.*, 1976). When GTP is replaced by GMPPCP or GMPPNP, the ternary complex is bound to the ribosome, but peptidyl transfer does not occur until EF–Tu is removed from the ribosomal complex (Haenni & Lucas-Lenard, 1968; Skoultchi *et al.*, 1970; Shorey, Ravel & Shive, 1971). This occurs only slowly in the absence of GTP hydrolysis (Yokosawa *et al.*, 1973, 1975; Girbes, Vazquez & Modolell, 1976).

Nevertheless, the antibiotic puromycin, an analogue of aminoacyl-adenosine, can enter the peptidyl-transferase centre and accept *N*-acetyl-Phe from the donor tRNA in the P site even before removal of EF–Tu (Haenni & Lucas-Lenard, 1968; Modolell & Vazquez, 1973). It seems reasonable to suppose that EF–Tu maintains its tight binding to the aatRNA until GTP is hydrolysed, but establishes, in addition, an interaction with the 50S subunit which activates the GTPase centre. The relation of this to codon-specific selection is not altogether clear, although the lifetime of the codon–anti-codon association is presumably a major factor. Any ribosome–EF–Tu interaction prior to GTP hydrolysis must be sufficiently weak not to impede excessively the dissociation of incorrect codon–anti-codon complexes. Whether or not EF–Tu plays any role in positioning the tRNA during codon recognition is at present only conjectural.

When EF–Tu·GDP dissociates from the ribosomal complex, the aminoacyl–ACC terminus of the tRNA in the A site can enter the peptidyl-transferase centre. The extent of the rearrangement of the tRNA on the ribosome at this stage is unknown, though there have been suggestions that other parts of the tRNA molecule are involved in an interaction with ribosomal components. Some data suggest an involvement of the TψCG sequence in ribosomal binding (Ofengand & Henes, 1969; Richter, Erdmann & Sprinzl, 1973) perhaps dependent on the codon–anti-codon interaction (Schwarz, Menzel & Gassen, 1976). The tetranucleotide TψCG can replace uncharged intact tRNA in the codon-specific stringent factor-dependent synthesis of guanosine tetra- and pentaphosphate (Richter, Erdmann & Sprinzl, 1974), but it remains to be demonstrated that a direct interaction of this sequence takes place during elongation.

A comparison of EF–Tu-free and EF–Tu-dependent poly(U) translation shows that the factor greatly stimulates translation at low concentrations of Mg^{2+} (Gavrilova *et al.*, 1976). This stimulatory activity, and its ability to discriminate against uncharged tRNA and initiator tRNA, may be considered the main functions of EF–Tu.

Peptidyl-transferase

After dissociation of EF–Tu·GDP from the ribosomal complex, peptide bond formation takes place. The enzymic activity, peptidyl-transferase, which catalyses this reaction, is an integral part of the 50S ribosomal subunit. That puromycin is an effective acceptor substrate for the enzyme suggested a two-site model for ribosomal function. Thus, aatRNAs which are puromycin-reactive are said to be in the P site, while ribosome-bound but puromycin-unreactive substrates are said to be in the A site (acceptor site). The sites of the peptidyl-transferase which bind the CCA termini of substrates in the A and P sites have been named the A' and P' sites (Pestka, 1972; Harris & Symons, 1973). In the presence of methanol or certain other solvents the peptidyl-transferase activity becomes independent of 30S subunits and template (Monro, Cerna & Marcker, 1968; Monro, Staehelin, Celma & Vazquez, 1969). By using such conditions, the specificity of this enzymic centre in terms of both P' site and A' site ligands has been investigated in some detail. Fragments of aatRNA have been used to study the specificity of the donor site. The smallest fragment with donor activity is CCAaa, which interacts with puromycin only when formylated or acetylated (Monro *et al.*, 1968, 1969). The fragments are, however, less active than the tRNA derivatives, suggesting that another part of the tRNA molecule is involved in an interaction with the 50S subunit, probably outside the peptidyl-transferase centre. The integrity of the terminal ribose is important for donor activity in as far as neither N-acetyl–Phe–tRNA–C–C–3'–dA nor N-acetyl–Phe–tRNA$_{oxi-red}$ (made from tRNA^{Phe} lacking the C2'–C3' bond of the terminal ribose following periodate oxidation and borohydride reduction) are active (Chinali, Sprinzl, Parmeggiani & Cramer, 1974). It remains to be demonstrated that non-isomerizable analogues of 3'-O-peptidyl–tRNA are active donors, although present results suggest that this isomer is the effective donor in peptidyl transfer.

The A site appears to display considerable specificity for small MW substrates which carry the aminoacyl acceptor group on the 3'-OH group (Nathans & Neidle, 1963; Ringer & Chladek, 1974; Chladek, Ringer & Zemlicka, 1973; Pozdynakov *et al.*, 1972), although inhibition is shown by some 2'-O-aminoacyl analogues (Ringer, Quiggle & Chladek, 1975). However, both Phe–tRNA–C–C–3'–dA and the oxidized-reduced product of Phe–tRNA can act as acceptors but with much reduced rates of peptidyl transfer (Chinali *et al.*, 1974). Thus, it seems that both isomers can bind but that the 3'-*O*-aminoacyl isomer is the preferred acceptor. The enzyme can also catalyse transesterification. For example, α-hydroxypuromycin can act as an acceptor (Fahnestock, Neumann, Shashoua & Rich, 1970) and phenyl-lactyl–tRNA and lactyl–tRNA can form esters during protein synthesis *in vitro* (Fahnestock & Rich, 1971*a*, *b*). Even ethanol can substitute as an acceptor (Scolnick, Milman, Rosman & Caskey, 1970; Caskey, Beaudet, Scolnick & Rosman, 1971). Transfer to water (hydrolysis) will be considered in connection with the termination process.

Although the substrate requirements of peptidyl-transferase have by now been rather well defined, comparatively little is known about the enzymic mechanism itself. A formal comparison may be made with proteases which are capable of both peptide and ester bond formation, but there is little experimental evidence of the type of intermediates in the reaction on the ribosome.

Elongation factor EF–G

The complex series of reactions which make up the translocation process are catalysed by a single extra-ribosomal protein factor, EF–G. How the ribosome and EF–G promote messenger movement and the displacement of deacylated tRNA by peptidyl–tRNA is unknown, but there is some understanding of the interaction of EF–G with the ribosome and the role of GTP hydrolysis. EF–G is a single polypeptide chain, MW about 80000 (Parmeggiani, 1968; Leder, Skogerson & Nau, 1969; Rohrbach, Dempsey & Bodley, 1974). The biological activity of the factor is particularly sensitive to thiol reagents (Nishizuka & Lipmann, 1966*a*, *b*; Sutter & Moldave, 1966; Kaziro *et al.*, 1969; Kaziro, Inoue-Yokosawa & Kawakita, 1972) but some protection can be conferred by guanine nucleotides (Marsh, Chinali & Parmeggiani, 1975; Rohrbach & Bodley, 1976*b*; Baca, Rohrbach & Bodley, 1976). Gel filtration, equilibrium dialysis and fluorescence studies have also revealed an interaction with GDP and GTP (Arai *et al.*, 1975*b*; Baca *et al.*, 1976), and GDP binding displays an increased affinity in the presence of ribosomes (Baca *et al.*, 1976). Thus, as with EF–Tu, the site of binding of the GTP molecule necessary for EF–G function lies on the factor itself, but the binding affinity is much smaller ($K_a = 2.7 \times 10^4$ M^{-1}) (Baca *et al.*, 1976). The complex between factor and ribosome displays a GTPase activity independent of tRNA-and hence uncoupled from the function of EF–G in promoting translocation (Conway & Lipmann, 1964; Nishizuka & Lipmann, 1966b). Only the 50S subunit is necessary for this reaction, although addition of the 30S subunit stimulates it. Kinetic analysis of the uncoupled GTPase reaction shows an ordered sequence of association, with GTP binding to EF–G before an interaction with the ribosome occurs (Rohrbach & Bodley, 1976a; Baca et al., 1976). Direct binding of EF–G to the ribosome results in a dead-end complex which cannot then bind and hydrolyse GTP. The dissociation of the post-hydrolytic complex EF–G·GDP·Rib is also ordered, with GDP being the last component released from EF–G (Rohrbach & Bodley, 1976a). The complex may be stabilized by the antibiotic fusidic acid which must then be released before further dissociation can occur (Bodley, Zieve, Lin & Zieve, 1970a; Rohrbach & Bodley, 1976a). Thus fusidic acid allows only one round of GTP hydrolysis, associated with the formation of the complex (Bodley, Zieve & Lin, 1970b; Brot, Spears & Weissbach, 1971). A stable complex is also formed when GTP is replaced by the non-hydrolysable analogue GMPPCP (Brot et al., 1969, 1971). Both these stable complexes may be made using the 50S subunit alone.

Whether or not EF–G can bind to the ribosome depends upon what other ligands are bound already. Ribosomes carrying N-acetyl–Phe–Phe–tRNA or N-acetyl–Phe–tRNA in the A site and deacylated tRNA in the P site can interact with EF–G and GTP and undergo subsequent translocation (Lucas-Lenard & Lipmann, 1971; Modolell, Cabrer & Vazquez, 1973b; Watanabe, 1972) as discussed below. P-site-bound N-acetyl–Phe–tRNA, however, inhibits the interaction of ribosomes with EF–G as shown by the inability to form a stable complex in the presence of fusidic acid or GMPPCP, or to complement EF–G in the uncoupled GTPase reaction (Modolell, Cabrer & Vazquez, 1973a). All these activities may be restored by adding puromycin, which accepts the N-acetyl–Phe moiety and leaves deacylated tRNA in the P site. It has been clearly demonstrated that EF–G cannot bind to ribosomes that have EF–Tu still bound to them (Richter, 1972) and it has been concluded that the binding of these two factors is mutually exclusive (Cabrer, Vazquez & Modolell, 1972; Richman & Bodley, 1972; Miller, 1972; Richter, 1972; Ballesta & Vazquez, 1972; Modolell & Vazquez, 1973). It is also possible that EF–G cannot bind to ribosomes which contain deacylated tRNA^{Phe} in the P site and Phe–tRNA bound enzymically in the A site even after EF–Tu removal. Likewise, EF–G binding may be inhibited when Phe–tRAA has been bound non-enzymically (Modolell & Vazquez, 1973). These observations are less easy to interpret as such ribosomal complexes differ from those which arise normally as intermediates in the elongation cycle. The experimental data are generally consistent, however, with the hypothesis that the only intermediate in the elongation cycle to which EF–G·GTP will bind is the pre-translocation complex, and that EF–G leaves the ribosome before enzymic binding of aatRNA to the A site.

Translocation

Following peptidyl transfer, the ribosomal A site contains peptidyl–tRNA and the P site a deacylated donor tRNA. EF–G and GTP will bind to such a ribosomal complex and promote the displacement of the deacylated donor tRNA by the peptidyl–tRNA molecule (Haenni & Lucas-Lenard, 1968; Lucas-Lenard & Haenni, 1969; Erbe, Nau & Leder, 1969; Ishitsuka, Kuriki & Kaji, 1970) which then becomes reactive to puromycin. Deacylated tRNA in the P site cannot be released by EF–G and GTP if the A site is unoccupied (Lucas-Lenard & Haenni, 1969; Modolell et al., 1973a).

Catalytic amounts of EF–G suffice for this reaction in the presence of GTP, but not with the non-hydrolysable analogue GMPPCP or when fusidic acid is added. When, however, substrate amounts of EF–G are used a substantial part of the peptidyl–tRNA (or N-acetyl–Phe–tRNA) bound in the A site becomes puromycin reactive with either GMPPCP or GMPPNP provided that the analogues are added in sufficient concentration (Modolell et al., 1973b; Modolell, Girbes & Vazquez, 1975; Inoue-Yokosawa, Ishikawa & Kaziro, 1974). Thus, translocation itself does not need GTP hydrolysis. It is rather less clear whether or not removal of EF–G from the ribosomal complex is necessary for translocation to be completed. However, the observation that EF–Tu-dependent binding of Phe–tRNA is slow in the presence of EF–G and GMPPNP whereas translocation itself is rapid (Girbes, Vazquez & Modolell, 1976) suggests that it is only the former process that is impeded by the continued presence of EF–G on the ribosome (Cabrer et al., 1972; Miller, 1972; Richman & Bodley, 1972; Richter, 1972; Modolell et al., 1973a) and that the dissociation of EF–G is not necessary for translocation to be completed. Furthermore, fusidic acid is not inhibitory to N-acetyl–Phe–puromycin synthesis provided that substrate amounts of EF–G are used (Modolell et al., 1973b; Inoue-Yokosawa et al., 1974). Thus supposing that fusidic acid does not alter the mechanism of translocation, the decreased rate of dissociation of EF–G·GDP in the presence of fusidic acid seems not to affect the rate of translocation. As with EF–Tu function, the essential role of GTP hydrolysis appears to be one of accelerating the release of the factor. Although the most direct indication that one molecule of GTP must be hydrolysed for the removal of each factor per elongation cycle has come from experiments on the isolated steps, recent experiments *in vitro* on the complete translation system are consistent with this conclusion (Cabrer, San-Millan, Vazquez & Modolell, 1976).

LOCATION OF THE TRANSLATION COMPONENTS

Component sites

The methods used to locate translational components are essentially the same as those discussed previously for initiation components and we proceed immediately to a discussion of results, following the same outline used previously (see pp. 61–3).

tRNA

Previously cited experiments (see p. 65) showed that electrophilic N-acyl derivatives of Phe–tRNA^{Phe} can affinity-label isolated 30S particles. In similar experiments with 70S particles, covalent incorporation occurs almost exclusively into the 50S subunit. When the P site is occupied the principal proteins labelled are L2 and L27 (electrophilic derivatives) (Pellegrini, Oen, Eilat & Cantor, 1974; Czernilofsky, Collatz, Stöffler & Kuechler, 1974) and L11 and L18 (photolabile derivatives) (Hsiung & Cantor, 1974; Hsiung, Reines & Cantor, 1974). When the derivative of tRNA is shifted into the A site, in the presence of 30 mM Mg^{2+} and excess deacylated tRNA, the electrophilic derivative tested labels L16 in addition to L2 and L27 (Eilat *et al.*, 1974*a*). Furthermore, the labelling of L16 is blocked by added tetracycline, which is an A-site directed antibiotic, to a much greater extent than either L2 or L27. In addition Eilat *et al.* (1974*b*) found, using a series of electrophilic oligo-peptidyl–tRNA^{Phe} derivatives, that as the electrophilic group was placed farther and farther away from the 3'-terminus of the tRNA, other proteins were labelled, notably L24 and L32/33. Several indirect studies correlate well with these results. Thus, in reconstitution experiments, L16 has been shown to be essential for peptidyl-transferase activity (Moore *et al.*, 1975). L11 apparently stimulates this activity (Nierhaus & Montejo, 1973) and is also necessary for the binding of thiostrepton (another A-site directed antibiotic) (Highland *et al.*, 1975). In addition antibodies to L11 and L16 inhibit peptidyl transferase (Tate, Caskey & Stöffler, 1975). Furthermore, Fahnestock (1975) has found that when photo-oxidized L3 from *B. stearothermophilus*, corresponding to L2 in *E. coli*, is used in a *B. stearothermophilus* reconstitution system, peptidyl-transferase activity is lost, as is non-enzymic Phe–tRNA^{Phe} binding and EF–G-dependent GTPase.

Several indirect studies have also implicated 30S proteins as being important for EF–Tu dependent tRNA binding. Thus, Fab fragments directed against S3, S9, S10, S11, S19 and S21 are strongly inhibitory (Lelong et al., 1974); addition of S2, S3 and S14 to salt-washed ribosomes enhances such binding (Randall-Hazelbauer & Kurland, 1972).

Elongation factors

Proteins L7/12 appear to be important for the interaction of the three factors IF2, EF–Tu, and EF–G with the 70S ribosome. Thus, all three factors have been found to cross-link with L7/12 (Achary, Moore & Richards, 1973; Heimark, Hershey & Traut, 1976a; San-José, Kurland & Stöffler, 1976; Fabian, 1976) and antibodies to L7/L12 inhibit the binding of each of the three factors (Highland et al., 1973; Tate et al., 1975; Stöffler, 1974). Moreover, the interactions are at least partly competitive. Thus, bound IF2 inhibits both EF–G binding and EF–Tu-dependent Phe–tRNA^{Phe} binding (Benne & Voorma, 1972; Springer et al., 1971; Grunberg-Manago, Dondon & Graffe, 1972). In recent work, EF–Tu has also been found to cross-link to proteins L1, L5, L15, L20, L30, L33 (San-José et al., 1976) and to L23, L28 > L1, L3, L24 (Fabian, 1976).

GTP

All three factors, IF2, EF–Tu, and EF–G, have ribosome-associated GTPase activities. Recent evidence indicates that in each this activity is located on the factor itself. The clearest demonstration is for EF–Tu, which develops GTPase activity in the absence of ribosomes merely on interaction with the antibiotic kirromycin (Chinali et al., 1977). The evidence is also strong for EF–G, Using photolabile GTP derivatives, Girshovich, Pozdnyakov & Ovchinnikov (1976a) have used a photo-affinity-labelling approach to demonstrate that EF–G is the site of GTP fixation in the ribosome·EF–G complex. A recent experiment goes further in providing evidence that EF–G is also the GTPase centre in this complex. Thus, when EF–G to which a photolabile GTP derivative is covalently linked, is incubated with ribosomes to form the ternary complex, inorganic phosphate is released (Girshovich, Kurtskhalia, Pozdnyakov & Ovchinnikov, 1977). Evidence that GTP interacts directly with IF2 comes from experiments demonstrating GTP protection of IF2 against thiol substitution and thermal inactivation (see p. 53). A complex of 5S RNA and proteins L5, L18, and L25 has GTPase and ATPase activity, and is inhibited by both thiostrepton and fusidic acid (Horne & Erdmann, 1972; Roth & Nierhaus, 1974). The low specific activities found in these latter studies, coupled with the results cited above locating GTPase activity in the factors, make it questionable whether the hydrolase activity of the 5S RNA·protein complex is important in protein synthesis.

Peptidyl-transferase antibiotics

Two quite similar electrophilic derivatives of chloramphenicol have been used in affinity-labelling studies with ribosomes. Pongs, Bald & Erdmann (1973) found L16 as the major labelled protein and L24 as a secondary site, with either 70S or 50S particles, whereas Sonenberg, Wilchek & Zamir (1973) found L2 and L27 with 50S particles as the target. A difficulty with the latter experiment is that chloramphenicol present in excess was found to have no effect on the labelling reaction. In reconstitution experiments, L16 is found to be necessary for chloramphenicol binding (Nierhaus & Nierhaus, 1973) and antibodies to chloramphenicol attach to chloramphenicol·50S complexes in a region containing determinants for L16, L19 and L27 (Stöffler & Wittmann, 1977). Puromycin affinity-labels 70S ribosomes in a photo-induced process, with L23 as the major product (Cooperman, Jaynes, Brunswick & Luddy, 1975; Jaynes et al., 1977). In the presence of chloramphenicol, the labelling pattern is changed and S14 becomes the major affinity-labelled protein (P. G. Grant, E. N. Jaynes, Jr & B. S. Cooperman, unpublished result). This change may reflect the conformational alteration in the ribosome which is known to result from chloramphenicol binding (Miskin & Zamir, 1974; Langlois, Cantor, Vince & Pestka, 1977).

Aminoglycoside antibiotics

Several antibiotics that cause misreading and act on 30S subunits have been linked to specific 30S proteins. The greatest variety of data is available for streptomycin, although there is little overlap in the results obtained by different techniques. With 30S subunits as the target, a photolabile analogue of streptomycin labels S7 and, to a minor extent, S14 (Girshovich et al., 1976b). Interestingly, when 70S particles are the target both 30S and 50S proteins are labelled, although the identities of the proteins have not yet been established. Mutations modifying streptomycin effects on bacterial growth and protein biosynthesis have been located, affecting proteins S4, S5, and S12 (Wittmann & Wittmann-Liebold, 1974; Gorini, 1974). Antibodies especially to protein S19, as

Table 10. *Summary of the site location data in the elongation reaction*

Site / L protein	3'-terminus of tRNA	Peptidyl-transferase inhibitors	Factors	Present in region III
1			+	v
2	++	+		v
3			+	–
5	+		+	v
7/12			++	–
11	++			v
14	+			v
15	+		+	o
16	++	++		v
18	++			v
20			+	–
23		++	+	v
24	++	+	+	o
27	++	+		v
28			+	o
30			+	o
32/33	++		+	o

++ Strongly implicated; + implicated; v yes; – no; o not known.

well as to proteins S1, S10, S11, S12, S20, and S21 inhibit streptomycin binding (Lelong *et al.*, 1974). Mutants resistant to kasugamycin have adenosine residues in place of dimethyladenosine residues at positions 25 and 26 (counting from the 3'-end) of 16S RNA (Helser, Davies & Dahlberg, 1971, 1972). Proteins S5 and/or S12 are also altered in mutants resistant to spectinomycin and neamine (Piepersberg, Böck, Yaguchi & Wittmann, 1975; Dewilde *et al.*, 1975) and some kasuga-mycin-resistant mutants have an altered S2 (Okuyama, Yoshikawa & Tanaka, 1974; Yoshikawa, Okuyama & Tanaka, 1975).

Summary of location results

The proteins implicated in the binding sites for elongation components on the 50S ribosomal subunit are listed in Table 10. As expected, there is a great deal of overlap between the 3'-terminus of tRNA and chlor-amphenicol, the proteins labelled in common being particularly L16, as well as L2, L24, and L27. Protein L23, labelled by puromycin, is also found as one of the two major cross-linked proteins in one of the studies with EF–Tu (Fabian, 1976). There is also considerable over-lap between the proteins found cross-linked to EF–Tu and those associated with the 3'-terminus of tRNA (L5, L15, L24, L32/33). It is particularly interesting that L7/L12, despite its importance in factor binding, is apparently not directly involved at the peptidyl-transferase centre.

Five proteins, S4, S5, S7, S12, and S14 are most strongly implicated in aminoglycoside binding sites. Determinants are present for all five in region I of the 30S subunit (see page 67) and for all but S5 in region I'. A possible explanation for the apparent disagreement with respect to S5 is that Tischendorf and Stöffler have found three determinants for S5, whereas Lake and Kahan report only one. The 3'-end of 16S RNA is important for kasugamycin binding (see page 85), as it is for mRNA and initiation-factor binding (see page 64), and a recent RNA location study using immune electron microscopy (Politz & Glitz, 1977) has located the dimethyladenosines near the 3'-terminus of 16S RNA within region I'. There is a strong overlap between 30S proteins indirectly implicated in tRNA binding to a 70S particle and 30S proteins cross-linked to initiation factors (S2, S9, S11, S18, S19 and S21) (see pp. 63–4). Two other proteins, S3 and S14, implicated indirectly in tRNA binding to 70S particles, have also been located at the 3'-terminus of fMet–tRNA^{fMet} by affinity-labelling experiments (see pp. 64–5).

Correlation of site location and structural studies

There are few cross-linking data available for 50S proteins. However, a L2–L27–L32/33 trimer has been identified (Barritault, Expert-Bezancon & Milet, 1975), tying together three proteins implicated at the 3'-terminus of tRNA, as well as an L5–L23 pair (R. R. Traut, private communication), linking the puromycin labelled protein to the 3'-terminus of tRNA and to the EF–Tu binding site. The major structural evidence comes, as with the 30S subunit, from combined immunological and electron microscopy studies (Stöffler & Wittmann, 1977). Although these studies are incomplete, from what is already known it is clear that the great majority of the proteins implicated at sites discussed above have antibody determinants falling within a limited region (III) of the 50S subunit (Fig. 10). Determinants for twelve of the seventeen proteins listed in Table 10 have been located. Of these, nine fall within the region for chloramphenicol and for puromycin binding, including all of those known for the 3'-terminus of tRNA. Determinants for three of the proteins (L1, L5, L23) cross-linked to EF–Tu fall within region III, and two (L3, L20) are outside but nearby, as are some of the determinants for L7/L12. It thus seems very clear that peptidyl transfer occurs within region III, with the elongation factors positioned in adjacent or partially overlapping positions.

It has been suggested that the TψCG loop common to *E. coli* tRNAs interacts with a complementary region of 5S RNA (Erdmann, Sprinzl

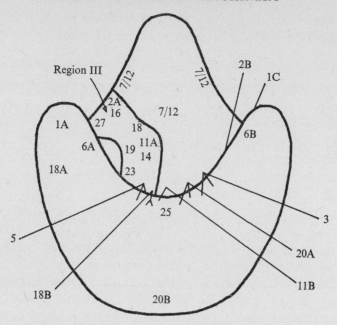

Fig. 10. The Stöffler and Tischendorf model for the 50S subunit. From Stöffler & Wittmann (1977).

& Pongs, 1973). 5S RNA forms a complex with proteins L5, L18, and L25 and proteins L2 and L6 are necessary for binding this complex to 23S (Monier, 1974). Three of the proteins, L2, L5, and L18, have determinants within region III, and the determinants for L6 and L25 are adjacent, supporting the concept of a 5S RNA·tRNA interaction.

Region III is essentially identical to the region indicated by Stöffler & Wittmann (1977) as constituting an overlap between the 30S and 50S subunits, largely on the basis of inhibition of subunit association by antibody fragments. This method also places proteins S9, S11, S12, S14 and S20 at the interface. The first four are also implicated in the binding of initiation components in region I (Table 10). Other evidence for the proximity of these two regions comes from the puromycin photo-induced affinity-labelling studies (see page 84), that showed a chloramphenicol-induced change in the principal site of labelling from L23 in region III to S14 in region I. These results strongly suggest that in the 70S ribosome one or both tRNA molecule(s) is (are) held by the anti-codon loop in region I on the 30S particle, while the 3'-terminus is held in region III on the 50S particle. The other ribosomal ligands affecting elongation – i.e. the protein factors and the antibiotic inhibitors – appear to bind within, or immediately adjacent to, these regions.

Given the conformational lability of the ribosome, allosteric effects should also be possible, but there is no strong evidence for them at present.

PEPTIDE CHAIN TERMINATION

Termination of the polypeptide chain occurs in response to the nonsense codons UAG, UAA and UGA and results in the release of the polypeptide by hydrolysis of the ester bond by which it is linked to tRNA, translocation being required before this event can take place (Capecchi & Klein, 1969; Tompkins, Scolnick & Caskey, 1970). The termination codons are recognized by soluble protein factors, which bind to the ribosome and appear to modify the specificity of the peptidyl-transferase centre in such a way that the polypeptide is transferred to H_2O rather than to the $-NH_2$ of an aatRNA in the A′ site. Release factors were isolated by using their ability to release fMet–tRNA from the complex fMet–tRNA·AUG·ribosome in the presence of the trinucleotide terminator codons (Caskey et al., 1968). Two release factors were found: RF1, which responds to UAG and UAA, and RF2, specific for UAA and UGA (Scolnick, Tompkins, Caskey & Nirenberg, 1968; Klein & Capecchi, 1971). RF1 and RF2 have MWs of 44000 and 47000, respectively, and are present to the extent of about one molecule of each factor per 50 ribosomes (Klein & Capecchi, 1971). The release factors bind at the ribosomal A site but binding is inhibited if EF–G is present on the ribosomes (Tompkins et al., 1970; Tate, Baudet & Caskey, 1973). Competition occurs between release factor and tRNA nonsense suppressors (Beaudet & Caskey, 1970; Ganoza & Tomkins, 1970). A third protein factor RF3 has been isolated which, under appropriate conditions, can stimulate the activity of RF1 and RF2 (Milman, Goldstein, Scolnick & Caskey, 1969). The role of this factor is not altogether clear; it can stimulate both the binding to ribosomes of RF1 and RF2 and their release, and it appears that guanine nucleotides may be important to its activity (Goldstein & Caskey, 1970). Whether or not a mechanism exists for stimulating the dissociation of the complex between ribosomes, mRNA and deacylated tRNA is still an unresolved problem.

PERSPECTIVES

The study of protein synthesis in bacteria over the past 20 years has generated impressive amounts of new knowledge which can be considered under the following general categories: definition of the components involved; determination of the pathway whereby the components interact; elucidation of the precise molecular mechanisms involved; and elucidation of control mechanisms which operate under various physiological conditions. We shall broadly examine each of these, both historically and critically.

The major components of protein synthesis are thought to be defined. Significant stimulation of various partial reactions has been obtained for each of the initiation, elongation and termination factors. However, reconstitution of an efficient, totally purified protein synthesis system has not yet been reported. This leaves open the possibility that additional factors affecting protein synthesis may yet be found. Also lacking is a rigorous demonstration that all of the components are necessary for optimum protein synthesis *in vivo*. The latter problem is particularly severe for IF1 and possibly for certain ribosomal proteins. The isolation of genetic mutants affecting the components of the system is an ideal way to demonstrate whether each is necessary. Studies on the genetic organization and expression of macromolecular components may also provide helpful information. Further progress along these lines of research can be expected in the near future.

The general pathway of protein synthesis is well understood, especially for the steps of elongation. The precise pathway whereby mRNA, fMet–tRNA and initiation factors bind to 30S subunits during initiation is not well elucidated, however, and a number of alternative pathways are possible. Further kinetic studies are required to determine whether a unique pathway for assembly of the 30S initiation complex exists, or whether several parallel pathways operate simultaneously. In addition, the termination events which occur directly following peptidyl–tRNA hydrolysis remain obscure. Furthermore, it is not yet known whether reinitiation on polycistronic mRNAs occurs by the same pathway as initiation at the first cistron.

The most impressive gains in recent years have been made in the area of mechanism and particularly in the role played by ribosomes. It is now clear that ribosomal RNA plays a critical role in mRNA binding and in ribosomal subunit association, while a ribosomal protein is responsible for peptidyl-transferase activity. In general, however, the intimate interaction of both RNA and protein contribute to ribosome function. A great deal of new knowledge about ribosomal structure has been gained, and much effort has gone into elucidating the sites on the ribosomes where the other components of protein synthesis react. However, structural studies, apart from RNA primary sequencing, have not yet reached the degree of precision which is needed to shed new light on precise molecular mechanisms. We expect further refinements in determining the topological arrangement of proteins and RNA in ribosomes and their complexes.

On the other hand, the soluble protein factors appear to carry the active sites for GTP hydrolysis. The role of GTP hydrolysis appears to involve the efficient ejection of protein factors from ribosomal complexes. This is necessary because the factors bind to overlapping sites, and this in part contributes to the chronological ordering of the various steps in the pathway of protein synthesis. The fact that most of these reactions occur in a rather restricted region of the ribosome raises the question of whether the other regions contribute to functions such as ribosome–membrane interactions. Recent kinetic proof-reading theories which explain the low error rate in protein synthesis suggest a further function for GTP hydrolysis. However, the role of GTP hydrolysis during translocation remains obscure, as does this molecular mechanism in general. Elucidation of the molecular mechanism of translocation remains one of the great challenges in the field of protein synthesis, as this step may resemble the basic elements of the general mechanism of unidirectional movement in biological systems. Greater precision in defining ribosomal structures and sites will be especially crucial here.

Mechanisms of control of protein synthesis have long been actively sought. Since the genetic approach has not yet been sufficiently fruitful, most studies involved the search for factors affecting in-vitro systems of protein synthesis. Such systems have been quite crude, and artifactual stimulations or inhibitions have been difficult to exclude. At present there is no well-characterized mechanism which modulates the selection of mRNAs for translation in uninfected bacterial cells. The development of the genetic approach will provide assistance in this search. The measurement of rates of biosynthesis and levels of the components of protein synthesis in cells grown under different physiological conditions may also shed light on how protein synthesis is controlled.

We greatly benefited from the excellent reviews of Dr Steitz and Dr Stöffler kindly provided before publication.

The financial support came from the following grants: Centre National de la Recherche Scientifique and Délégation Générale de la Recherche Scientifique et Technique, and Ligue Nationale Française contre le Cancer. B. S. Cooperman was supported by grants from NSF and INSERM and J. W. B. Hershey by EMBO and NATO fellowships and the Philippe Foundation. We wish to thank G. Anguérin for skilful assistance in the preparation of the manuscript.

REFERENCES

ABELSON, J. N., GEFTER, M. L., BARNETT, L., LANDY, A., RUSSELL, R. L. & SMITH, J. D. (1970). Mutant tyrosine transfer ribonucleic acids, *Journal of Molecular Biology*, **47**, 15–28.

ACHARY, A. S., MOORE, P. B. & RICHARDS, F. M. (1973). Cross-linking of elongation factor EF-G to the 50S ribosomal subunit of *Escherichia coli*. *Biochemistry*, **12**, 3108–14.

ADAMS, J. M. (1968). On the release of the formyl group from nascent protein. *Journal of Biological Chemistry*, **33**, 571–89.

ARAI, K., ARAI, T., KAWAKITA, M. & KAZIRO, Y. (1975a). Conformational transition of polypeptide chain elongation factor Tu. Studies with hydrophobic probes. *Journal of Biochemistry* (*Tokyo*), **77**, 1095–106.

ARAI, K., ARAI, T. & KAZIRO, Y. (1975b). Formation of a binary complex between elongation factor G and guanosine nucleotides. *Journal of Biochemistry* (*Tokyo*), **78**, 243–6.

ARAI, K., KAWAKITA, M. & KAZIRO, Y. (1974a). Studies on the polypeptide elongation factors from *Escherichia coli*. Properties of various complexes containing EF-Tu and EF-Ts. *Journal of Biochemistry* (*Tokyo*) **76**, 293–306.

ARAI, K. I., KAWAKITA, M., KAZIRO, Y., MAEDA, T. & OHNISHI, S. (1974b). Conformational transition in polypeptide elongation factor Tu as revealed by electron spin resonance. *Journal of Biological Chemistry*, **249**, 3311–13.

ARNOLD, H. H., SCHMIDT, W. & KEROTEN, H. (1975). Tetrahydrofolate dependent biosynthesis of m^{5}U in tRNAs of two grampositive organisms and initiation of protein synthesis by m^{5}U deficient unformylated initiator tRNAs. *FEBS Proceedings 10th Meeting* (*Paris*), Abstract no. 388.

BAAN, R. A., DUIJFJES, I. I., VAN LEERDAM, E., VAN KNIPPENBERG, P. H. & BOSCH, L. (1976). Specific in-situ cleavage of 16S ribosomal RNA of *Escherichia coli* interferes with the function of initiation factor IF1. *Proceedings of the National Academy of Sciences*, *USA*, **73**, 702–6.

BAAN, R. A., HILBERS, C. W., VAN CHARDORP, R., VAN LEERDAM, E., VAN KNIPPENBERG, P. H. & BOSCH, L. (1977). High resolution proton magnetic study of the secondary structure of the 3' terminal 49 nucleotide fragment of 16S rRNA from *Escherichia coli*. *Proceedings of the National Academy of Sciences*, *USA*, **74**, 1028–31.

BACA, O. G., ROHRBACH, M. S. & BODLEY, J. W. (1976). Equilibrium measurements of the interactions of guanine nucleotides with *Escherichia coli* elongation factor G and the ribosome. *Biochemistry*, **15**, 4570–4.

BALLESTA, J. P. G. & VAZQUEZ, D. (1972). Elongation factor T-dependent hydrolysis of guanosine triphosphate resistant to thiostrepton. *Proceedings of the National Academy of Sciences*, *USA*, **69**, 3058–62.

BARRELL, B. G. (1971). Fractionation and sequence analysis of radioactive nucleotides. In *Procedures in Nucleic Acid Research*, ed. G. L. Cantoni & D. R. Davies, vol. 2, pp. 751–79. New York: Harper and Row.

BARRELL, B. G. & CLARK, B. F. C., ed. (1974). *Handbook of Nucleic Acid Sequences*. pp. 32–5. Oxford: Joynson-Bruvvers.

BARRITAULT, D., EXPERT-BEZANCON, A. & MILET, M. (1975). Etude des relations de voisinage entre des proteines de la sous-unité 50S d'*Escherichia coli*. *Comptes rendus de l'Académie des Sciences de Paris*, **281**, 1043–6.

BEAUDET, A. L. & CASKEY, C. T. (1970). Release factor translation of RNA phage terminator codons. *Nature, London*, **227**, 38–40.

BENNE, R. & POUWELS, P. H. (1975). The role of IF3 in the translation of T7 and φ80 trp messenger RNA. *Molecular and General Genetics*, **139**, 311–19.

BENNE, R. & VOORMA, H. O. (1972). Entry site of formyl-methionyl-tRNA. *FEBS Letters*, **20**, 347–51.

BERNAL, S. D., BLUMBERG, B. M. & NAKAMOTO, T. (1974). Requirement of initiation factor 3 in the initiation of polypeptide synthesis with N-acetylphenylalanyl-tRNA. *Proceedings of the National Academy of Sciences, USA*, **71**, 774–8.

BEYREUTHER, K., ADLER, K., GEISLER, N. & KLEMM, A. (1973). The amino acid sequence of *lac* repressor. *Proceedings of the National Academy of Sciences, USA*, **70**, 3576–80.

BLUMENTHAL, T., LANDERS, T. A., & WEBER, K. (1972). Bacteriophage Qβ replicase contains the protein biosynthesis elongation factors EFTu and EFTs. *Proceedings of the National Academy of Sciences, USA*, **69**, 1313–17.

BODE, U., LUTTER, L. C. & STÖFFLER, G. (1974). Proteins S14 and S19 are near-neighbors in the *Escherichia coli* ribosome. *FEBS Letters*, **45**, 232–6.

BODLEY, J. W., ZIEVE, F. J., LIN, L. & ZIEVE, S. T. (1970a). Studies on translocation. Conditions necessary for the formation and detection of a stable ribosome–G factor–guanosine diphosphate complex in the presence of fusidic acid. *Journal of Biological Chemistry*, **245**, 5656–61.

BODLEY, J. W., ZIEVE, F. J. & LIN, L. (1970b). Studies on translocation. The hydrolysis of a single round of guanosine triphosphate in the presence of fusidic acid. *Journal of Biological Chemistry*, **245**, 5662–7.

BOEDTKER, H. & GESTELAND, R. F. (1975). Physical properties of RNA bacteriophages and their RNA. In *RNA phages*, ed. N. D. Zinder, pp. 1–26. Cold Spring Harbor Laboratory, USA.

BOLLEN, A., HERMARK, R. L., COZZONE, A., TRAUT, R. R. & HERSHEY, J. W. B. (1975). Cross-linking of initiation factor IF2 to *Escherichia coli* 30S ribosomal proteins with dimethylsuberimidate. *Journal of Biological Chemistry*, **250**, 4310–14.

BOON, T. (1971). Inactivation of ribosomes *in vitro* by colicin E₃. *Proceedings of the National Academy of Sciences, USA*, **68**, 2421–5.

BORER, P. N., DENGLER, B., TINOCO, I., Jr. & UHLENBECK, O. (1974). Stability of ribonucleic acid double stranded-helices. *Journal of Molecular Biology*, **86**, 843–53.

BOWMAN, C. M., DAHLBERG, J. E., IKEMURA, T., KONISKY, J. & NOMURA, M. (1971). Specific inactivation of 16S ribosomal RNA induced by colicin E *in vivo*. *Proceedings of the National Academy of Sciences, USA*, **68**, 964–8.

BRANLANT, C. & EBEL, J.-P. (1977). Studies on the primary structure of *Escherichia coli*: 23S RNA; Nucleotide sequence of the ribonuclease T1 digestion products containing more than one uridine residue. *Journal of Molecular Biology*, **111**, 215–56.

BRAUER, D. & WITTMANN-LIEBOLD, B. (1977). The primary structure of the initiation factor IF3 from *Escherichia coli*, *FEBS Letters* **79**, 269–75.

BRETSCHER, M. S. (1969). Direct translation of bacteriophage fd DNA in the absence of neomycin B. *Journal of Molecular Biology*, **42**, 595–8.

BRIMACOMBE, R., NIERHAUS, K. H., GARRETT, R. A. & WITTMANN, H. G. (1976). The ribosome of *Escherichia coli*. *Progress in Nucleic Acid Research*, **18**, 1–44.

BROT, N., SPEARS, C. & WEISSBACH, H. (1969). The formation of a complex containing ribosomes, transfer factor G and a guanosine nucleotide. *Biochemical and Biophysical Research Communications*, **34**, 843–8.

BROT, N., SPEARS, C. & WEISSBACH, H. (1971). The interaction of transfer factor G, ribosome, and guanosine nucleotide in the presence of fusidic acid. *Archives of Biochemistry and Biophysics*, **143**, 286–96.

CABRER, B., SAN-MILLAN, M. J., VAZQUEZ, D. & MODOLELL, J. (1976). Stoichiometry of polypeptide chain elongation. *Journal of Biological Chemistry*, **251**, 1718–22.

CABRER, B., VAZQUEZ, D. & MODOLELL, J. (1972). Inhibition by elongation factor EF-G of aminoacyl tRNA binding to ribosomes. *Proceedings of the National Academy of Sciences, USA*, **69**, 133–6.

CAPECCHI, M. R. & KLEIN, H. A. (1969). Characterization of three proteins involved in polypeptide chain termination. *Cold Spring Harbor Symposia on Quantitative Biology*, **34**, 469–77.

CASKEY, C. T., BEAUDET, A. L., SCOLNICK, E. M. & ROSMAN, M. (1971) Hydrolysis of fMet-tRNA by peptidyl transferase. *Proceedings of the National Academy of Sciences, USA*, **68**, 3163–7.

CASKEY, C. T., TOMPKINS, R., SCOLNICK, E., CARYK, T. & NIRENBERG, M. (1968). Sequential translation of trinucleotide codons for the initiation and termination of protein synthesis. *Science*, **162**, 135–8.

CHANG, F. N. & FLAKS, J. G. (1972). The specific cross-linking of two proteins from the *Escherichia-coli* 30S ribosomal subunit. *Journal of Biological Chemistry*, **68**, 177–80.

CHAPMAN, N. M. & NOLLER, H. F. (1977). Differential accessibility of specific sites in 16S RNA in 30S and 70S ribosomes. *Journal of Molecular Biology*, **109**, 131–49.

CHEN, C. M. & OFENGAND, J. (1970). Inactivation of the Tu-GTP recognition site in aminoacyl-tRNA by chemical modification of the tRNA. *Biochemical and Biophysical Research Communications*, **41**, 190–8.

CHINALI, G., SPRINZL, M., PARMEGGIANI, A. & CRAMER, F. (1974). Participation in protein biosynthesis of transfer ribonucleic acid bearing altered 3′-terminal ribosyl residues. *Biochemistry*, **13**, 3001–10.

CHINALI, G., WOLF, H. & PARMEGGIANI, A. (1977). Effect of kirromycin on elongation factor Tu. Location of the catalytic center for the ribosome. *European Journal of Biochemistry*, **75**, 55–65.

CHLADEK, S., RINGER, D. & ZEMLICKA, J. (1973). L-phenylalanine esters of openchain analogue of adenosine as substrates for ribosomal peptidyl transferase, *Biochemistry*, **12**, 5135–8.

CONWAY, T. & LIPMANN, F. (1964). Characterization of a ribosome linked guanosine triphosphatase in *Escherichia coli* extracts. *Proceedings of the National Academy of Sciences, USA*, **52**, 1462–9.

COOPERMAN, B. S. (1977). Affinity labelling of ribosomes. *Bioorganic Chemistry*, in press.

COOPERMAN, B. S., JAYNES, E. N., BRUNSWICK, D. J. & LUDDY, M. A. (1975). Photoincorporation of puromycin and *N*-(ethyl-2- diazomalonyl)puromycin into *Escherichia coli* ribosomes. *Proceedings of the National Academy of Sciences, USA*, **72**, 2974–8.

CRANE, L. J. & MILLER, D. L. (1974). Guanosine triphosphate and guanosine diphosphate as conformation-determining molecules. *Biochemistry*, **13**, 933–9.

CRÉPIN, M., LELONG, J C. & GROS, F. (1973). Early steps in the formation of a translation initiation complex on newly transcribed mRNA. *Research methods in reproductive endovirology*, 6th Karolinska symposium, pp. 33–53.

CRICK, F. H. C. (1966). Codon–anticodon pairing: the wobble hypothesis. *Journal of Molecular Biology*, **19**, 548–55.

CZERNILOFSKY, A. P., COLLATZ, E. E., STÖFFLER, G. & KUECHLER, E. (1974). Proteins at the tRNA binding sites of *Escherichia coli* ribosomes. *Proceedings of the National Academy of Sciences, USA*, **71**, 230–4.

CZERNILOFSKY, A. P., KURLAND, C. G. & STÖFFLER, G. (1975). 30S ribosomal proteins associated with the 3′-terminus of 16S RNA. *FEBS Letters*, **58**, 281–4.

DAHLBERG, A. E., LUND, E., KJELDEGAARD, N. O., BOWMAN, C. M. & NOMURA, M. (1973). Colicin E₃ induced cleavage of 16S ribosomal ribonucleic acid: blocking effects of certain antibiotics. *Biochemistry*, **12**, 948–50.

DANCHIN, A. (1973). Does formylation of initiator tRNA act as regulatory signal in *Escherichia coli? FEBS Letters*, **34**, 327–32.

DAVIES, J., GILBERT, W. & GORINI, L. (1964). Streptomycin suppression and the code. *Proceedings of the National Academy of Sciences, USA*, **51**, 883–90.

DAVIES, J., JONES, D. S. & KHORANA, H. G. (1966). A further study of misreading of codons induced by streptomycin and neomycin using ribopolynucleotides containing two nucleotides in alternating sequences as templates. *Journal of Molecular Biology*, **18**, 48–57.

DEBEY, P., HUI BON HOA, G., DOUZOU, P., GODEFROY-COLBURN, Th., GRAFFE, M. & GRUNBERG-MANAGO, M. (1975). Ribosomal subunit interactions as studied by light-scattering. Evidence of different classes of ribosome preparations. *Biochemistry*, **14**, 1553–9.

DE GRAAF, F. K., PLANTA, R. J. & STOUTHAMER, A. H. (1971). Effect of a bacteriocin produced by *Enterobacter cloacae* on protein biosynthesis. *Biochimica et Biophysica Acta*, **240**, 122–36.

DELK, A. S. & RABINOWITZ, J. C. (1974). Partial nucleotide sequence of a prokaryote initiator tRNA that functions in its non-formylated form. *Nature, London*, **252**, 106–9.

DELK, A. S. & RABINOWITZ, J. C. (1975). Biosynthesis of ribosylthymine in the transfer RNA of *Streptoccocus faecalis*: a folate dependent methylation involving S-adenosyl methionine. *Proceedings of the National Academy of Sciences, USA*, **72**, 528–30.

DEUSSER, E. (1972). Heterogeneity of ribosomal populations in *Escherichia coli* cells grown in different media. *Molecular and General Genetics*, **119**, 249–58.

DE WACHTER, R., MERREGAERT, J., VANDENBERGHE, A., CONTRERAS, R. & FIERS, W. (1971). Studies on the bacteriophage MS₂. The untranslated 5′ terminal nucleotide sequence preceding the first cistron. *European Journal of Biochemistry*, **22**, 400–14.

DEWILDE, M., CABEZON, T., VILLAREL, R., HERZOG, A. & BOLLEN, A. (1975). Cooperative control of translational fidelity by ribosomal proteins in *Escherichia coli. Molecular and General Genetics*, **142**, 19–33.

DONDON, J., GODEFROY-COLBURN, TH., GRAFFE, M. & GRUNBERG-MANAGO, M. (1974). IF-3 requirement for initiation complex formation with synthetic messengers in *Escherichia coli* system. *FEBS Letters*, **45**, 82–7.

DUBNOFF, J. S., LOCKWOOD, A. H. & MAITRA, U. (1972). Studies on the role of guanosine triphosphate in polypeptide chain initiation in *Escherichia coli. Journal of Biological Chemistry*, **247**, 2884–94.

DUBNOFF, J. S. & MAITRA, U. (1971). Isolation and properties of polypeptide chain initiation factor FII from *Escherichia coli*: evidence for a dual function. *Proceedings of the National Academy of Sciences, USA*, **68**, 318–23.

DUBNOFF, J. S. & MAITRA, U. (1972). Characterization of the ribosome-dependent guanosine triphosphatase activity of polypeptide chain initiation factor IF2. *Journal of Biological Chemistry*, **247**, 2876–3.

EDELMANN, P. & GALLANT, J. (1977). Mistranslation in *Escherichia coli. Cell* **10**, 131–7.

EILAT, D., PELLEGRINI, M., OEN, H., DE GROOT, N., LAPIDOT, Y. & CANTOR, C. R. (1974a). Affinity labelling the acceptor site of the peptidyl transferase centre of the *Escherichia coli* ribosomes. *Nature, London*, **250**, 514–16.

EILAT, D., PELLEGRINI, M., OEN, H., LAPIDOT, Y. & CANTOR, C. R. (1974b). A chemical mapping technique for exploring the location of proteins along the ribosome-bound peptide chain. *Journal of Molecular Biology*, **88**, 831–40.

EISENSTADT, J. M. & BRAWERMANN, G. (1966). A factor from *Escherichia coli* concerned with the stimulation of cell-free polypeptide synthesis by exogenous ribonucleic acid. I. Evidence for the occurrence of a stimulation factor. *Biochemistry*, **5**, 2777–83.

EISINGER, J. (1971). Complex formation between transfer RNAs with complementary anticodons. *Biochemical and Biophysical Research Communications*, **43**, 854–61.

ERBE, R. W., NAU, M. M. & LEDER, P. (1969). Translation and translocation of defined RNA messengers. *Journal of Molecular Biology*, **38**, 441–60.

ERDMANN, V. A., SPRINZL, M. & PONGS, O. (1973). The involvement of 5S RNA in the binding of tRNA to ribosomes. *Biochemical and Biophysical Research Communications*, **54**, 942–8.

ERTEL, R., BROT, N., REDFIELD, B., ALLENDE, J. E. & WEISSBACH, H. (1968). Binding of guanosine 5′-triphosphate by soluble factors required for polypeptide synthesis. *Proceedings of the National Academy of Sciences, USA*, **59**, 861–8.

FABIAN, U. (1976). Identification of proteins located in the neighbourhood of the binding site for the elongation factor EF-Tu on *Escherichia coli* ribosomes. *FEBS Letters*, **71**, 256–60.

FAHNESTOCK, S. R. (1975). Evidence of the involvement of a 50S ribosomal protein in several active sites. *Biochemistry*, **14**, 5321–7.

FAHNESTOCK, S., NEUMANN, H., SHASHOUA, V. & RICH, A. (1970). Ribosome-catalyzed ester formation. *Biochemistry*, **9**, 2477–83.

FAHNESTOCK, S. & RICH, A. (1971a). Synthesis by ribosomes of viral coat protein containing ester linkages. *Nature New Biology*, **229**, 8–10.

FAHNESTOCK, S. & RICH, A. (1971b). Ribosome-catalysed polyester formation. *Science*, **173**, 340–3.

FAHNESTOCK, S., WEISSBACH, H. & RICH, A. (1972). Formation of a ternary complex of phenyllactyl-tRNA with transfer factor Tu and GTP. *Biochimica et Biophysica Acta*, **269**, 62–6.

FAKUNDING, J. L. & HERSHEY, J. W. B. (1973). The interaction of radioactive initiation factor IF-2 with ribosomes during initiation of protein synthesis. *Journal of Biological Chemistry*, **248**, 4206–12.

FIERS, W., CONTRERAS, R., DUERINCK, F., HAEGEMAN, G. ISERENTANT, D., MERREGAERT, J., MIN JOU, W., MOLEMANS, F., RAEYMAEKERS, A., VAN DEN BERGHE, A., VOLCKAERT, G. & YSEBAERT, M. (1976). Complete nucleotide sequence of bacteriophage MS2 RNA: primary and secondary structure of the replicase gene. *Nature, London*, **260**, 500–7.

FIERS, W., CONTRERAS, R., DUERINCK, F., HAEGEMAN, G., MERREGAERT, J., MIN JOU, W., RAEYMAEKERS, A., VOLCKAERT, G., YSEBAERT, M., VAN DE KERCKHOVE, J., NOLF, F. & VAN MONTAGU, M. (1975). A-protein gene of bacteriophage MS2. *Nature, London*, **256**, 273–8.

FILES, J. G., WEBER, K., COULONDRE, C. & MILLER, J. H. (1975). Identification of the UUG codon as a translational initiation codon *in vivo*. *Journal of Molecular Biology*, **95**, 327–30.

FILES, J. G., WEBER, K. & MILLER, J. H. (1974). Translational reinitiation: reinitiation of *lac* repressor fragments at three internal sites early in the *lac i* gene of *Escherichia coli*. *Proceedings of the National Academy of Sciences, USA*, **71**, 667–70.

FISER, I., MARGARITELLA, P. & KUECHLER, E. (1975a) Photo affinity reaction between polyuridylic acid and protein S1 on the *Escherichia coli* ribosome. *FEBS Letters*, **52**, 281–3.

FISER, I., SCHEIT, K. H., STÖFFLER, G. & KUECHLER, E. (1974). Identification of protein S1 at the messenger RNA binding site of the *Escherichia coli* ribosome. *Biochemical and Biophysical Research Communications*, **60**, 1112–18.

FISER, I., SCHEIT, K. H., STÖFFLER, G. & KUECHLER, E. (1975b). Proteins at the mRNA binding site of the *Escherichia coli* ribosome. *FEBS Letters*, **56**, 226–9.

FURANO, A. (1975). Content of elongation factor Tu in *Escherichia coli*. *Proceedings of the National Academy of Sciences, USA*, **12**, 4780–4.

GANEM, D., MILLER, J. H., FILES, J. G., PLATT, T. & WEBER, K. (1973). Reinitiation of a *lac* repressor fragment at a codon other than AUG. *Proceedings of the National Academy of Sciences, USA*, **70**, 3165–9.

GANOZA, M. C. & TOMKINS, J. L. N. (1970). Popypeptide chain termination *in vitro*: competition for nonsense codons between a purified release factor and suppressor tRNA. *Biochemical and Biophysical Research Communications*, **40**, 1455–67.

GASSEN, H. G. (1977). Translation of synthetic and natural polynucleotides. *Poznan, symposium*, Blazejewko, ed. A. B. Legocki, Agricultural University, Poznan, Poland, in press.

GAVRILOVA, L. P., KOSTIASHKINA, O. E., KOTELIANSKY, V. E., RUTKEVITCH, N. M. & SPIRIN, A. S. (1976). Factor-free ('non enzymic') and factor dependent systems of translation of polyuridylic acid by *Escherichia coli* ribosomes. *Journal of Molecular Biology* **101**, 537–52.

GHOSH, K. & GHOSH, H. P. (1970). Role of modified nucleoside adjacent to the 3'-end of anticodon in codon–anticodon interaction. *Biochemical and Biophysical Research Communications*, **40**, 135–43.

GIEGÉ, R., EBEL, J. P., SPRINGER, M. & GRUNBERG-MANAGO, M. (1973). Initiation of protein synthesis with mischarged tRNA^fMet^ from *Escherichia coli*. *FEBS Letters*, **37**, 166–9.

GINZBURG, I. & ZAMIR, A. (1975). Characterization of different conformational forms of 30S ribosomal subunits in isolated and associated states: possible correlations between structure and function. *Journal of Molecular Biology*, **93**, 465–76.

GIRBES, D., VAZQUEZ, D. & MODOLELL, J. (1976). Polypeptide-chain elongation promoted by Guanyl yl-5'-imidodiphosphate. *European Journal of Biochemistry*, **67**, 257–65.

GIRSHOVICH, A. S., BOCHKAREVA, E. S., KRAMAROV, V. A. & OVCHINNIKOV, YU. A. (1974a). *Escherichia coli*, 30S and 50S ribosomal subparticle components in the localization region of the tRNA acceptor terminus. *FEBS Letters*, **45**, 213–17.

GIRSHOVICH, A. S., BOCHKAREVA, E. S. & OVCHINNIKOV, YU. A. (1976b). Identification of components of the streptomycin-binding centre of *Escherichia coli* MRE 600 ribosomes by photo-affinity labelling. *Molecular General Genetics*, **144**, 205–12.

GIRSCHOVICH, A. S., BOCHKAREVA, E. S. & POZDNYAKOV, V. A. (1974b). Affinity labelling of functional centres of *Escherichia coli* ribosomes. *Acta Biologica Medica Germanica*, **33**, 639–48.

GIRSHOVICH, A. S., KURTSKHALIA, T. V., POZDNYAKOV, V. A. & OVCHINNIKOV, YU. A. (1977). Localization of the GTP-binding centre on the elongation factor G1. *FEBS Letters*, **80**, 161–3.

GIRSHOVICH, A. S., POZDNYAKOV, V. A. & OVCHINNIKOV, YU. A. (1976a). Localization of the GTP-binding site in the ribosome-Elongation factor G–GTP complex. *European Journal of Biochemistry*, **69**, 321–8.

GODEFROY-COLBURN, T., WOLFE, A. D., DONDON, J., GRUNBERG-MANAGO, M., DESSEN, P. & PANTALONI, D. (1975). Light-scattering studies showing the effect of initiation factors on the reversible dissociation of *Escherichia coli* ribosomes. *Journal of Molecular Biology*, **94**, 461–78.

GOLDBERG, M. L. & STEITZ, J. A. (1974). Cistron specificity of 30S ribosomes heterologously reconstituted with components from *Escherichia coli* and *Bacillus stearothermophilus*. *Biochemistry*, **13**, 2123–8.

GOLSTEIN, J. L. & CASKEY, C. T. (1970). Peptide chain termination: effect of protein S on ribosomal binding of release factors. *Proceedings of the National Academy of Sciences, USA*, **67**, 537–43.

GORDON, J. (1967). Interaction of guanosine 5'-triphosphate with a supernatant fraction from *Escherichia coli* and aminoacyl-sRNA. *Proceedings of the National Academy of Sciences, USA*, **58**, 1574–8.

GORDON, J. (1968). A stepwise reaction yielding a complex between a supernatant fraction from *Escherichia coli*, guanosine-5'-triphosphate and aminoacyl-sRNA. *Proceedings of the National Academy of Sciences, USA*, **59**, 179–83.

GORDON, J. (1969). Hydrolysis of guanosine 5'-triphosphate associated with binding of aminoacyl transfer ribonucleic acid to ribosomes. *Journal of Biological Chemistry*, **244**, 5680–6.

GORINI, L. (1974). Streptomycin and misreading of the genetic code. In *Ribosomes*, ed. M. Nomura, A. Tissières & P. Lengyel, pp. 791–803. Cold Spring Harbour laboratory, USA.

GRALLA, J. & CROTHERS, D. M. (1973a). The free energy of imperfect nucleic acid helices. II Small hairpin loops, *Journal of Molecular Biology*, **73**, 497–511.

GRALLA, J. & CROTHERS, D. M. (1973b). Free energy of imperfect nucleic acid helices. *Journal of Molecular Biology*, **78**, 301–19.

GROSJEAN, H., SÖLL, D. & CROTHERS, D. M. (1976). Studies of the complex between transfer RNAs with complementary anticodons. *Journal of Molecular Biology*, **103**, 499–519.

GRUNBERG-MANAGO, M., CLARK, B. F. C., REVEL, M., RUDLAND, P. S. & DONDON, J. (1969). Stability of different ribosomal complexes with initiator tRNA and synthetic mRNA. *Journal of Molecular Biology*, **40**, 33–44.

GRUNBERG-MANAGO, M., DONDON, J. & GRAFFE, M. (1972). Inhibition by thiostrepton of the IF-2-dependent ribosomal GTPase. *FEBS Letters*, **22**, 217–21.

GRUNBERG-MANAGO, M. & GROS, F. (1977). Initiation mechanisms of protein synthesis. In *Progress in Nucleic Acid Research and Molecular Biology*, **20**, ed. W. Cohn, 209–84. New York: Academic Press.

GUALERZI, C., PON, C. L. & KAJI, A. (1971). Initiation factor dependent release of aminoacyl-tRNA from complexes of 30S ribosomal subunits, synthetic polynucleotide and aminoacyl-tRNA. *Biochemical and Biophysical Research Communications*, **45**, 1312–19.

GUALERZI, C., RISULEO, G. & PON, C. L. (1977). Initial rate kinetic analysis of the mechanism of initiation complex formation and the role of initiation factor IF-3. *Biochemistry*, **16**, 1684–9.

HAENNI, A. L. & LUCAS-LENARD, J. (1968). Stepwise synthesis of a tripeptide. *Proceedings of the National Academy of Sciences, USA*, **61**, 1363–9.

HARDY, S. J. S. (1975). The stoichiometry of the ribosomal proteins of *Escherichia coli*. *Molecular and General Genetics*, **140**, 253–74.

HARRIS, R. J. & SYMONS, R. H. (1973). A detailed model of the active centre of *Escherichia coli* peptidyl transferase. *Bioorganic Chemistry*, **2**, 286–92.

HARVEY, R. J. (1973). Growth and initiation of protein synthesis in *Escherichia coli* in the presence of trimethoprim. *Journal of Bacteriology*, **114**, 309–22.

HAWLEY, R. A., SLOBIN, L. I. & WAHBA, A. J. (1974). The mechanism of action of initiation factor 3 in protein synthesis. *Biochemical and Biophysical Research Communications*, **61**, 544–60.

HECHT, S. M., TAN, K. H., CHINAULT, A. C. & ARCARI, P. (1977). Isomeric amino-acyl-tRNA are both bound by elongation factor Tu. *Proceedings of the National Academy of Sciences, USA*, **74**, 437–441.

HEIMARK, R. L., HERSHEY, J. W. B. & TRAUT, R. R. (1976a). Cross-linking of initiation factor IF2 to proteins L7/L12 in 70S ribosomes of *Escherichia coli*. *Journal of Biological Chemistry*, **251**, 7779–84.

HEIMARK, R. L., KAHAN, L., JOHNSTON, K., HERSHEY, J. W. B. & TRAUT, R. R. (1976b). Cross-linking of initiation factor IF3 to proteins of the *Escherichia coli* 30S subunits. *Journal of Molecular Biology*, **105**, 219–20.

HELD, W. A., GETTE, W. R. & NOMURA, M. (1974). Role of 16S ribosomal ribonucleic acid and the 30S ribosomal protein S12 in the initiation of natural messenger ribonucleic acid translation. *Biochemistry*, **13**, 2115–22.

HELSER, T. L., DAVIES, J. E. & DAHLBERG, J. E. (1971). Change in methylation of 16S ribosomal RNA associated with mutation to kasugamycin resistance in *Escherichia coli*. *Nature New Biology*, **233**, 12–14.

HELSER, T. L., DAVIES, J. E. & DAHLBERG, J. E. (1972). Mechanism of kasugamycin resistance in *Escherichia coli*. *Nature New Biology*, **235**, 6–9.

HERSHEY, J. W. B., DEWEY, K. F. & THACH, R. E. (1969). Purification and properties of initiation factor f-1. *Nature, London*, **222**, 944–7.

HERSHEY, J. W. B., YANOV, J., JOHNSON, K. & FAKUNDING, J. L. (1977). Purification and characterization of protein synthesis initiation factors IF1, IF2 and IF3 from *Escherichia coli*. *Archives of Biochemistry and Biophysics*, **128**, 626–36.

HIGHLAND, J. H., BODLEY, J. W., GORDON, J., HASENBANK, R. & STÖFFLER, G. (1973). Identity of the ribosomal proteins involved in the interaction with elongation factor G. *Proceedings of the National Academy of Sciences, USA*, **70**, 147–50.

HIGHLAND, J. H., OCHONER, E., GORDON, J., BODLEY, J., HASENBANK, R. & STÖFFLER, G. (1975). Identification of a ribosomal protein necessary for thiostrepton binding to *Esherichia coli* ribosomes. *Journal of Biological Chemistry*, **250**, 1141–5.

HIRSH, D. (1971). Tryptophan transfer RNA as the UGA suppressor. *Journal of Molecular Biology*, **58**, 439–58.

HIRSH, D. & GOLD, L. (1971). Translation of the UGA triplet *in vitro* by tryptophan transfer RNA's. *Journal of Molecular Biology*, **58**, 459–68.

HÖGENAUER, G., TURNOWSKY, F. & UNGER, F. M. (1972). Codon–anti-codon interaction of methionine specific tRNAs. *Biochemical and Biophysical Research Communications*, **46**, 2100–7.

HOLMES, W. M., GOLDMAN, E., MINERT, T. A. & HATFIELD, G. W. (1977). Differential utilization of leucyl-tRNAs by *Escherichia coli*. *Proceedings of the National Academy of Sciences, USA*, **74**, 1393–7.

HOPFIELD, J. J. (1974). Kinetic proof reading: a new mechanism for reducing errors in biosynthetic processes requiring high specificity. *Proceedings of the National Academy of Sciences, USA*, **71**, 4135–9.

HORNE, J. R. & ERDMANN, V. A. (1972). Isolation and characterization of 5S RNA–Protein complexes from *Bacillus stearothermophilus* and *Escherichia coli* ribosomes. *Molecular and General Genetics*, **119**, 337–44.

HSIUNG, N. & CANTOR, C. R. (1974). A new simpler photo affinity analogue of peptidyl tRNA. *Nucleic Acids Research*, **1**, 1753–62.

HSIUNG, N., REINES, S. A. & CANTOR, C. R. (1974). Investigation of ribosomal peptidyl transferase centre using a photo affinity label. *Journal of Molecular Biology*, **88**, 841–55.

HUI BON HOA, G., GRAFFE, M. & GRUNBERG-MANAGO, M. (1977). Thermodynamic studies of the reversible association of *Escherichia coli* ribosomal subunits. *Biochemistry*, **16**, 2800–5.

INOUE-YOKOSAWA, N., ISHIKAWA, C. & KAZIRO, Y. (1974). The role of guanosine triphosphate in translocation reaction catalyzed by elongation factor G. *Journal of Biological Chemistry*, **249**, 4321–5.

ISHITSUKA, H., KURIKI, Y. & KAJI, A. (1970). Release of transfer ribonucleic acid from ribosomes. A G factor and guanosine triphosphate-dependent reaction. *Journal of Biological Chemistry*, **245**, 3346–51.

ISONO, S. & ISONO, K. (1975). Role of ribosomal protein S1 in protein synthesis: effects of its addition to *Bacillus stearothermophilus* cell-free system. *European Journal of Biochemistry*, **56**, 15–22.

IWASAKI, K., SABOL, S., WAHBA, A. J. & OCHOA, S. (1968). Translation of the genetic message. VII. Role of initiation factors in formation of the chain initiation complex with *Escherichia coli* ribosomes. *Archives of Biochemistry and Biophysics*, **125**, 542–7.

JAYNES, E. N., JR, GRANT, P. G., WIEDER, R., GIANGRANDE, G. & COOPERMAN, B. S. (1977). Photoinduced affinity labelling of the *Escherichia coli* ribosome puromycin binding site *Biochemistry*, in press.

KAUFMAN, Y. & ZAMIR, A. (1972). Aminoacyl-transfer RNA binding to active 30S subunits: effect of 50S subunits and a new role for elongation factor T. *Journal of Molecular Biology*, **69**, 357–72.

KAWAKITA, M., ARAI, K. & KAZIRO, Y. (1974). Interaction of EF-Tu, GTP, and ribosome; role of ribosome-bound tRNA in GTPase reaction. *Journal of Biochemistry (Tokyo)*, **76**, 801–9.

KAY, A. C., GRAFFE, M. & GRUNBERG-MANAGO, M. (1976). Purification and properties of two initiation factors from *B. stearothermophilus*. *Biochimie*, **58**, 183–99.

KAZIRO, Y., INOUE-YOKOSAWA, N. & KAWAKITA, M. (1972). Studies on polypeptide elongation factor from *Escherichia coli*. Crystalline factor G. *Journal of Biochemistry (Tokyo)*, **72**, 853–63.

KAZIRO, Y., INOUE-YOKOSAWA, N., KURIK, Y., MIZUMOTO, K., TANAKA, M. & KAWAKITA, M. (1969). Purification properties of factor G. *Cold Spring Harbor Symposia on Quantitative Biology*, **34**, 385–93.

KLEIN, H. A. & CAPECHI, M. R. (1971). Polypeptide chain termination: purification of release factors R_1 and R_2 from *Escherichia coli*. *Journal of Biological Chemistry*, **246**, 1055–61.

KLIBER, J. S., HUI BON HOA, G., DOUZOU, P. & GRUNBERG-MANAGO, M. (1976). Implications of electrostatic potentials on ribosomal proteins. *Nucleic Acids Research*, **3**, 3423–38.

KOLAKOFSKY, D., DEWEY, K. F., HERSHEY, J. W. B. & THACH, R. E. (1968). Guanosine 5' triphosphatase activity of initiation factor f2. *Proceedings of the National Academy of Sciences, USA*, **61**, 1066–70.

KRAUSKOPF, M., CHEN, C. M. & OFENGAND, J. (1972). Interaction of fragmented and cross-linked *Escherichia coli* valine transfer ribonucleic acid with Tu factor–guanosine triphosphate complex. *Journal of Biological Chemistry*, **247**, 842–50.

KRAUSS, S. W. & LEDER, P. (1975). Regulation of initiation and elongation factor levels in *Escherichia coli* as assessed by a quantitative immunoassay. *Journal of Biological Chemistry*, **250**, 3752–8.

KURLAND, C. G. (1977). Aspects of ribosome structure and function. In *Molecular mechanisms of protein synthesis*, ed. H. Weissbach & S. Pestka, pp. 81–116. New York: Academic Press.

KURLAND, C. G., RIGLER, R., EHRENBERG, M. & BOMBERG, C. (1975). Allosteric mechanism for codon-dependent tRNA selection on ribosomes. *Proceedings of the National Academy of Sciences, USA*, **72**, 4248–51.

LAKE, J. A. (1976). Ribosome structure determined by electron microscopy of *Escherichia coli* small subunits, large subunits and monomeric ribosomes. *Journal of Molecular Biology*, **105**, 131–59.

LAKE, J. A. & KAHAN, L. (1975). Ribosomal proteins S5, S11, S13 and S19 localized by electron microscopy of antibody-labelled subunits. *Journal of Molecular Biology*, **99**, 631–44.

LANGBERG, S., KAHAN, L., TRAUT, R. R. & HERSHEY, J. W. B. (1977). Identification of 30S ribosomal proteins near the binding site of initiation factor 1F1. *Journal of Biological Chemistry*, in press.

LANGLOIS, R., CANTOR, C. R., VINCE, R. & PESTKA, S. (1977). Interaction between the erythromycin and chloramphenicol binding sites on the *Escherichia coli* ribosomes. *Biochemistry*, **16**, 2349–56.

LEDER, P., SKOGERSON, L. E. & NAU, M. M. (1969). Translocation of mRNA codons, the preparation and characteristics of a homogeneous enzyme. *Proceedings of the National Academy of Sciences, USA*, **62**, 457–60.

LEE-HUANG, S. & OCHOA, S. (1971). Messenger discriminating species of initiation factor F3. *Nature New Biology*, **234**, 236–9.

LEE-HUANG, S. & OCHOA, S. (1973). Purification and properties of two messenger-discriminating species of *Escherichia coli* initiation factor 3. *Archives of Biochemistry and Biophysics*, **156**, 84–96.

LEE-HUANG, S., SILLERO, M. A. G. & OCHOA, S. (1971). Isolation and properties of crystalline initiation factor F1 from *Escherichia coli* ribosomes. *European Journal of Biochemistry*, **18**, 536–43.

LEFFLER, S. & SZER, W. (1974a). Purification and properties of initiation factor IF3 from *Caulobacter crescentus*. *Journal of Biological Chemistry*, **249**, 1458–64.

LEFFLER, S. & SZER, W. (1974b). Polypeptide chain initiation in *Caulobacter crescentus* without initiation factor IF1. *Journal of Biological Chemistry*, **249**, 1465–8.

LELONG, C., GROS, D., GROS, F., BOLLEN, A., MASCHLER, R. & STÖFFLER, G. (1974). Function of individual 30S subunit proteins of *Escherichia coli*. Effect of specific immuno globulin fragments (Fab) on activities of ribosomal decoding sites. *Proceedings of the National Academy of Sciences, USA*, **71**, 248–52.

LELONG, J. C., GRUNBERG-MANAGO, M., DONDON, J., GROS, D. & GROS, F. (1970). Interaction between guanosine derivatives and factors involved in the initiation of protein synthesis. *Nature, London*, **226**, 505–10.

LODISH, H. F. (1970). Specificity in bacterial protein synthesis: role of initiation factors and ribosomal subunits. *Nature, London*, **226**, 705–7.

LODISH, H. F. (1971). Thermal melting of bacteriophage f2 RNA and initiation of synthesis of the maturation protein. *Journal of Molecular Biology*, **56**, 627–32.

LODISH, H. F. & ROBERTSON, H. D. (1969). Regulation of *in vitro* translation of bacteriophage f2 RNA. *Cold Spring Harbor Symposia on Quantitative Biology*, **34**, 655–73.

LOFTFIELD, R. B. & VANDERJAGT, D. (1972). The frequency of errors in protein biosynthesis. *Biochemical Journal*, **128**, 1353–6.

LUCAS-LENARD, J. & HAENNI, A.-L. (1968). Requirement of guanosine 5′-triphosphate for ribosomal binding of aminoacyl-sRNA. *Proceedings of the National Academy of Sciences, USA*, **59**, 554–60.

LUCAS-LENARD, J. & HAENNI, A.-L. (1969). Release of transfer RNA during peptide chain elongation. *Proceedings of the National Academy of Sciences, USA*, **63**, 93–7.

LUCAS-LENARD, J. & LIPMANN, F. (1967). Initiation of polyphenylalanine synthesis by N-acetyl-phenylalanyl sRNA. *Proceedings of the National Academy of Sciences, USA*, **57**, 1050–7.

LUCAS-LENNARD, J. & LIPMANN, F. (1971). Protein biosynthesis. *Annual Review of Biochemistry*, **40**, 409–48.

LUCAS-LENARD, J., TAO, P. & HAENNI, A.-L. (1969). Further studies on bacterial polypeptide elongation. *Cold Spring Harbor Symposia on Quantitative Biology*, **34**, 455–62.

LUHRMANN, R., GASSEN, H. G. & STÖFFLER, G. (1976). Identification of the 30S ribosomal proteins at the decoding site by affinity labelling with a reactive oligonucleotide. *European Journal of Biochemistry*, **66**, 1–9.

LUTTER, L. C., BODE, U., KURLAND, C. G. & STÖFFLER, G. (1974). Ribosomal protein neighbourhoods. *Molecular and General Genetics*, **129**, 167–76.

LUTTER, L. C., KURLAND, C. G. & STÖFFLER, G. (1975). Protein neighbourhoods in the 30S ribosomal subunit of *Escherichia coli*. *FEBS Letters*, **54**, 144–50.

LUTTER, L. C., ZEICHHARDT, H., KURLAND, C. G. & STÖFFLER, G. (1972). Ribosomal protein neighbourhoods. *Molecular and General Genetics*, **119**, 357–66.

MARCKER, K. A. (1965). The formation of *N*-formyl-methionyl-sRNA. *Journal of Molecular Biology*, **14**, 63–70.

MARSH, R. C., CHINALI, G. & PARMEGGIANI, A. (1975). Function of sulphydryl groups in ribosome–elongation factor *G* reactions. Assignment of guanine nucleotide binding site to elongation factor G. *Journal of Biological Chemistry*, **250**, 8344–52.

MAZUMDER, R., CHAE, Y. B. & OCHOA, S. (1969). Polypeptide chain initiation: properties of initiation factor F2. *Federation Proceedings*, **28**, 597.

McLAUGHLIN, C. S., DONDON, J., GRUNBERG-MANAGO, M., MICHELSON, A. M. & SAUNDERS, G. (1968). Stability of messenger RNA–transfer RNA–ribosome complex. *Journal of Molecular Biology*, **32**, 521–42.

MEIER, D., LEE-HUANG, S. & OCHOA, S. (1973). Factor requirements for initiation complex formation with natural and synthetic messengers in *Escherichia coli* systems. *Journal of Biological Chemistry*, **248**, 8613–15.

MELSER, T. L., DAVIES, J. E. & DAHLBERG, J. E. (1971). Change in methylation of 16S ribosomal RNA associated with mutation to kasugamycin resistance in *Escherichia coli*. *Nature New Biology*, **232**, 12–14.

MILLER, D. L. (1972). Elongation factors EFT$_u$ and EF$_g$ interact at related sites on ribosomes. *Proceedings of the National Academy of Sciences, USA*, **69**, 752–5.

MILLER, P. S., BARRETT, J. C. & Ts'O, P. O. P. (1974). Synthesis of oligodeoxyribonucleotide ethyl phosphotriesters and specific complex formation with transfer ribonucleic acid. *Biochemistry*, **13**, 4887–96.

MILLER, D. L., CASHEL, M. & WEISSBACH, H. (1973). The interaction of Guanosine 5′ diphosphate 2′(3′) diphosphate with the bacterial elongation factor Tu. *Archives of Biochemistry and Biophysics*, **154**, 675–82.

MILLER, D. L., HACHMANN, J. & WEISSBACH, H. (1971). The reaction of the sulphydryls groups on the elongation factors Tu and Ts. *Archives of Biochemistry and Biophysics*, **144**, 115–121.

MILLER, M. J. & WAHBA, A. J. (1973). Chain initiation factor 2: purification and properties of two species from *Escherichia coli*. *Journal of Biological Chemistry*, **248**, 1084–90.

MILLER, D. L. & WEISSBACH, H. (1970). Studies on the purification and properties of factor Tu from *Escherichia coli*. *Archives of Biochemistry and Biophysics*, **141**, 26–37.

MILLER, D. L. & WEISSBACH, H. (1974). Elongation factor Tu and the aminoacyl tRNA-EFTu GTP complex. In *Methods in Enzymology*, **30**, ed. L. Grossman & K. Moldave, pp. 219–32. New York: Academic Press.

MILMAN, G., GOLDSTEIN, J., SCOLNICK, E. & CASKEY, T. (1969). Peptide chain termination. Stimulation of *in vitro* termination. *Proceedings of the National Academy of Sciences, USA*, **63**, 183–90.

MISKIN, R. & ZAMIR, A. (1974). Enhancement of peptidyl transferase activity by antibiotics acting on the 50S ribosomal subunit. *Journal of Molecular Biology*, **87**, 121–34.

MITRA, S. K., LUSTIG, F., AKESSON, B. & LAGERKVIST, U. (1977). Codon–anticodon recognition in the valine codon family. *Journal of Biological Chemistry*, **252**, 471–8.

MODOLELL, J., CABRER, B. & VAZQUEZ, D. (1973a). The interaction of elongation factor G with N-acetylphenylalanyl–Transfer RNA–ribosome complexes. *Proceedings of the National Academy of Sciences, USA*, **70**, 3561–5.

MODOLELL, J., CABRER, B. & VAZQUEZ, D. (1973b). The stoichiometry of ribosomal translocation. *Journal of Biological Chemistry*, **248**, 8356–60.

MODOLELL, J., GIRBES, T. & VAZQUEZ, D. (1975). Ribosomal translocation promoted by guanylylimido diphosphate and guanylyl-methylene diphosphate. *FEBS Letters*, **60**, 109–113.

MODOLELL, J. & VAZQUEZ, D. (1973). Inhibition by amino acyl tRNA of elongation factor G-dependent binding of guanosine nucleotide to ribosomes. *Journal of Biological Chemistry*, **248**, 488–93.

MONIER, R. (1974). 5S RNA. In *Ribosomes* ed. N. Nomura, A. Tissières & P. Lengyel, pp. 141–68. Cold Spring Harbor Laboratory, USA.

MONRO, R. E., CERNA, J. & MARCKER, K. A. (1968). Ribosome–catalyzed peptidyl transfer: substrate specificity at the P-site. *Proceedings of the National Academy of Sciences, USA*, **61**, 1042–9.

MONRO, R. E., STAEHELIN, T., CELMA, M. L. & VAZQUEZ, D. (1969). The peptidyl transferase activity of ribosomes. *Cold Spring Harbor Symposia on Quantitative Biology*, **34**, 357–68.

MOORE, V. G., ATCHINSON, R. E., THOMAS, G., MORGAN, M. & NOLLER, H. F. (1975). Identification of a ribosomal protein essential to peptidyl transferase activity. *Proceedings of the National Academy of Sciences, USA*, **72**, 844–8.

NATHANS, D. & NEIDLE, A. (1963). Structural requirements for puromycin inhibition of protein synthesis. *Nature, London*, **197**, 1076–7.

NIERHAUS, K. H. & MONTEJO, V. (1973). A protein involved in the peptidyl transferase activity of *Escherichia coli* ribosome. *Proceedings of the National Academy of Sciences, USA*, **70**, 1931–5.

NIERHAUS, D. & NIERHAUS, K. H. (1973). Identification of the chloramphenicol binding protein in *Escherichia coli* ribosome by partial reconstitution. *Proceedings of the National Academy of Sciences, USA*, **70**, 2224–8.

NINIO, J. (1971). Codon-anticodon recognition: The missing triplet hypothesis. *Journal of Molecular Biology*, **56**, 63–82.

NINIO, J. (1973). Recognition in nucleic acids and the anticodon families. In *Progress in Nucleic acid Research and Molecular Biology*, ed. J. N. Davidson & W. E. Cohn, vol. 13, pp. 301–37. New York and London: Academic Press.

NINIO, J. (1974). A semi quantitative treatment of missense and nonsense suppression in the str A and ram ribosomal mutants of *Escherichia coli*. Evaluation of some molecular parameters of translation *in vivo*. *Journal of Molecular Biology*, **84**, 297–313.

NINIO, J. (1975). Kinetic amplification of enzyme discrimination. *Biochimie*, **57**, 587–95.

NISHIZUKA, Y. & LIPMANN, F. (1966a). The interrelationship between guanosine triphosphatase and amino acid polymerization. *Archives of Biochemistry and Biophysics*, **116**, 344–51.

NISHIZUKA, Y. & LIPMANN, F. (1966b). Comparison of guanosine triphosphate split and polypeptide synthesis with a purified *Escherichia coli* system. *Proceedings of the National Academy of Sciences, USA*, **55**, 212–19.

NOLL, M. & NOLL, H. (1972). Mechanism and control of initiation in the translation of R17 RNA. *Nature New Biology*, **238**, 225–8.

NOLL, M. & NOLL, H. (1976). Structure dynamics of bacterial ribosomes. V. Magnesium-dependent dissociation of tight couples into subunits. *Journal of Molecular Biology*, **105**, 111–27.

NOLL, H., NOLL, M., HAPKE, B. & VAN DIEIJEN, G. (1973). Mechanism of subunit interaction as a key to the understanding of ribosome function. In *Regulation of transcription and translation in eukaryotes*, 24th Mosbach-Baden Colloquium, ed. E. F. K. Bautz, P. Karlson & H. Kersten, p. 257–311. Berlin and New York: Springer-Verlag.

NOMURA, M., MIZUSHIMA, S., OZAKI, M., TRAUB, P. & LOWRY, C. V. (1969). Structure and function of ribosomes and their molecular components. *Cold Spring Harbor Symposia on Quantitative Biology*, **34**, 49–61.

OFENGAND, J. (1974). Assay for aa-tRNA recognition by the EF-Tu–GTP complex of *Escherichia coli*. In *Methods in Enzymology*, **29**, ed. L. Grossmann & K. Moldave, pp. 661–7. New York: Academic Press.

OFENGAND, J. (1977). tRNA and amino acyl-tRNA synthetases. In *Molecular mechanisms of protein synthesis*, ed. H. Weissbach & S. Pestka, pp. 8–79. New York: Academic press.

OFENGAND, J. & HENES, C. (1969). The function of pseudouridylic acid in transfer ribonucleic acid. *Journal of Biological Chemistry*, **244**, 6241–53.

OKUYAMA, A., YOSHIKAWA, M. & TANAKA, N. (1974). Alteration of ribosomal protein S_2 in kasugamycin-resistant mutant derived from *Escherichia coli* AB 312. *Biochemical and Biophysical Research Communications*, **60**, 1163–9.

ONO, Y., SKOULTCHI, A., WATERSON, J. & LENGYEL, P. (1969). Stoichiometry of aminoacyl-transfer RNA binding and GTP cleavage during chain elongation and translocation. *Nature, London*, **223**, 697–701.

PARMEGGIANI, A. (1968). Crystalline transfer factor from *Escherichia coli*. *Biochemical and Biophysical Research Communications*, **30**, 613–19.

PELLEGRINI, M., OEN, H., EILAT, D. & CANTOR, C. R. (1974). The mechanism of covalent reaction of bromoacetyl–phenylalanyl transfer RNA with the peptidyl transfer RNA binding site of the *Escherichia coli* ribosome. *Journal of Molecular Biology*, **88**, 809–29.

PESTKA, S. (1972). Studies on transfer ribonucleic acid–ribosome complexes. *Journal of Biological Chemistry*, **247**, 4669–78.

PETERSEN, H. U., DANCHIN, A. & GRUNBERG-MANAGO, M. (1976*a*). Toward an understanding of the formylation of initiator tRNA methionine in prokaryotic protein synthesis. I. *In vitro* studies of the 30S and 70S tRNA ribosomal complex. *Biochemistry*, **15**, 1357–61.

PETERSEN, H. U., DANCHIN, A. & GRUNBERG-MANAGO, M. (1976*b*). Toward an understanding of the formylation of initiator tRNA methionine in prokaryotic protein synthesis. II. A two state model for the 70S ribosome. *Biochemistry*, **15**, 1362–9.

PIEPERSBERG, W., BÖCK, A., YAGUCHI, M. & WITTMANN, H. G. (1975). Genetic position and amino acid replacements of several mutations in ribosomal protein S5 from *Escherichia coli*. *Molecular and General Genetics*, **143**, 43–52.

PLATT, T., WEBER, K., GANEM, D. & MILLER, J. H. (1972). Translational restarts: AUG reinitiation of a *lac* repressor fragment. *Proceedings of the Nationa Academy of Sciences, USA*, **69**, 897–901.

PLATT, T. & YANOFSKY, C. (1975). An intercistronic region and ribosome-binding site in bacterial messenger RNA. *Proceedings of the National Academy of Sciences, USA*, **72**, 2399–403.

POLITZ, S. M. & GILITZ, D. G. (1977). Ribosome structure-localization of N^6,N^6-dimethyladenosine by electron microscopy of a ribosome–antibody complex. *Proceedings of the National Academy of Sciences, USA*, **74**, 1468–72.

PONGS, O., BALD, R. & ERDMANN, V. A. (1973). Identification of chloramphenicol binding protein in *Escherichia coli* ribosome by affinity labelling. *Proceedings of the National Academy of Sciences, USA*, **70**, 2229–33.

PONGS, O. & LANKA, E. (1975a). Affinity labelling of ribosomes. II Synthesis of a chemically reactive analog of the initiation codon. Its reaction with ribosomes of *Escherichia coli. Hoppe-Seyler's Zeitschrift für physiologische Chemie*, **356**, 449–58.

PONGS, O. & LANKA, E. (1975b). Affinity labelling of the ribosomal decoding site with an AUG-substrate analog. *Proceedings of the National Academy of Sciences, USA*, **72**, 1505–9.

PONGS, O., LANKA, E. & BALD, R. (1975a). Proteins involved in the ribosomal decoding site. *10th FEBS Meeting Abstract* (*Paris*), no. 447.

PONGS, O., NIERHAUS, K. H., ERDMANN, V. A. & WITTMANN, H. G. (1974). Active sites in *Escherichia coli* ribosomes, *FEBS Letters*, **40**, S28–37.

PONGS, O. & ROSSNER, E. (1975). Synthesis of a chemically reactive analog of the nonsense codon UGA. Its reaction with ribosomes of *Escherichia coli. Hoppe-Seyler'Zeitschrift für physiologische Chemie*, **356**, 1297–304.

PONGS, O. & ROSSNER, E. (1976). Comparison of the reactions of chemically reactive analogs of UGA and of AUG with ribosomes of *Escherichia coli. Nucleic Acids Research*, **3**, 1625–33.

PONGS, O., STÖFFLER, G. & BALD, R. W. (1976). Location of protein S_1 of *Escherichia coli* ribosomes at the 'A'-site of the codon binding site-affinity labelling studies with a 3'-modified A-U-G analog. *Nucleic Acids Research*, **3**, 1635–6.

PONGS, O., STÖFFLER, G. & LANKA, E. J. (1975b). The codon binding site of the *Escherichia coli* ribosome as studied with a chemically reactive AUG analog. *Journal of Molecular Biology*, **99**, 301–15.

POZDNYAKOV, V. A., MITIN, Yu. Y., KUKHANOVA, M. K., NIKOLAEVA, L. V., KRAYEVSKY, A. A. & GOTTIKH, B. P. (1972). On the mechanism of peptide bond synthesis in the ribosome. 3'-O-Phenylalanyl-2'-O-methyladenosine as a peptide acceptor. *FEBS Letters*, **24**, 177–80.

PRINTZ, M. P. & MILLER, D. L. (1973). Evidence for conformational changes in elongation factor Tu induced by GTP and GDP. *Biochemical and Biophysical Research Communications*, **53**, 149–56.

RANDALL-HAZELBAUER, L. L. & KURLAND, C. G. (1972). Identification of three 30S proteins contributing to the ribosomal A-site. *Molecular and General Genetics*, **115**, 234–42.

RAVEL, J. M., SHOREY, R. L., GARNER, C. W., DAWKINS, R. C. & SHIVE, W. (1969). The role of an aminoacyl-tRNA–GTP-protein complex in polypeptide synthesis. *Cold Spring Harbor Symposia on Quantitative Biology*, **34**, 321–30.

RAVEL, J. M., SHOREY, R. L. & SHIVE, W. (1967). Evidence for a guanine nucleotide aminoacyl tRNA complex as an intermediate in the enzymatic transfer of aa-tRNA to ribosomes. *Biochemical and Biophysical Research Communications*, **29**, 68–73.

REVEL, M., AVIV, H., GRONER, Y. & POLLACK, Y. (1970). Fractionation of translation initiation factor B (F3) into cistron specific species. *FEBS Letters*, **9**, 213–17.

REVEL, M. & GREENSHPAN, H. (1970). Specificity in the binding of *Escherichia coli* ribosomes to natural messenger RNA. *European Journal of Biochemistry*, **16**, 117–22.

RICHMAN, N. & BODLEY, J. W. (1972). Studies on translocation XII. Ribosomes cannot interact simultaneously with elongation factors EF-Tu and EF-G. *Proceedings of the National Academy of Sciences, USA*, **69**, 686–9.

RICHTER, D. (1972). Inability of *Escherichia coli* ribosomes to interact simultaneously with the bacterial elongation factors EF-Tu and EF-G. *Biochemical and Biophysical Research Communications*, **46**, 1850–6.

RICHTER, D., ERDMANN, V. A. & SPRINZL, M. (1973). Specific recognition of GTψC loop (loop IV) of tRNA by 50S ribosomal subunits from *Escherichia coli*. *Nature New Biology*, **246**, 132–5.

RICHTER, D., ERDMANN, V. A. & SPRINZL, M. (1974). A new transfer RNA fragment reaction: TpψpCpGp bound to a ribosome–messenger RNA complex induces the synthesis of guanosine tetra- and pentaphosphates. *Proceedings of the National Academy of Sciences, USA*, **71**, 3226–9.

RINGER, D. & CHLADEK, S. (1974). Inhibition of the peptidyl transferase A-site function by 2′-*O*-aminoacyloligonucleotides. *Biochemical and Biophysical Research Communcations*, **56**, 760–6.

RINGER, D. & CHLADEK, S. (1975). Interaction of elongation factor Tu with 2′(3′)-*O*-aminoacyloligonucleotides derived from the 3′ terminus of aminoacyl tRNA. *Proceedings of the National Academy of Sciences, USA*, **72**, 2950–4.

RINGER, D., QUIGGLE, K. & CHLADEK, S. (1975). Recognition of the 3′ terminus of 2′-*O*-aminoacyl transfer ribonucleic acid by the acceptor site of ribosomal peptidyltransferase. *Biochemistry*, **14**, 514–20.

ROHRBACH, M. S. & BODLEY, J. W. (1976a). Steady state kinetic analysis of the mechanism of guanosine triphosphate hydrolysis catalyzed by *Escherichia coli* elongation factor G and the ribosome. *Biochemistry*, **15**, 4565–8.

ROHRBACH, M. S. & BODLEY, J. W. (1976b). Selective chemical modification of *Escherichia coli* elongation factor G. *Journal of Biological Chemistry*, **251**, 930–3.

ROHRBACH, M. S., DEMPSEY, M. E. & BODLEY, J. W. (1974). Preparation of homogeneous elongation factor G and examination of the mechanism of guanosine triphosphate hydrolysis. *Journal of Biological Chemistry*, **249**, 5094–101.

ROTH, H. E. & NIERHAUS, K. H. (1975). Structural and functional studies of ribonucleoprotein. *Journal of Molecular Biology*, **94**, 111–21.

RUDLAND, P. S., WHYBROW, W. A. & CLARK, B. F. C. (1971). Recognition of bacterial initiator tRNA by an initiation factor. *Nature New Biology*, **231**, 76–8.

RUDLAND, P. S., WHYBROW, W. A., MARCKER, K. A. & CLARK, B. F. C. (1969). Recognition of bacterial tRNA by initiation factors. *Nature, London*, **222**, 750–3.

SABOL, S. & OCHOA, S. (1971). Ribosomal binding of labelled initiation factor F3. *Nature New Biology*, **234**, 233–6.

SAN-JOSÉ, C., KURLAND, C. G. & STÖFFLER, G. (1976). The protein neighbourhood of ribosome-bound elongation factor Tu. *FEBS Letters*, **71**, 133–7.

SANTER, M. & SHANE, S. (1977). The area of 16S RNA at or near the interface between 30S and 50S ribosomes of *Escherichia coli*. *Journal of Bacteriology*, **130**, 900–10.

SCHENKMAN, M. L., WARD, D. C. & MOORE, P. B. (1974). Covalent attachment of a messenger RNA to the *Escherichia coli* ribosome. *Biochimica et Biophysica Acta*, **353**, 503–8.

SCHIFF, N., MILLER, M. J. & WAHBA, A. J. (1974). Purification and properties of chain initiation factor 3 from T4-infected and uninfected *Escherichia coli* MRE 600: stimulation of translation of synthetic and natural messengers. *Journal of Biological Chemistry*, **249**, 3797–802.

SCHULMAN, L. H. & HER, M. O. (1973). Recognition of altered *Escherichia coli* formylmethionine transfer RNA by bacterial T factor. *Biochemical and Biophysical Research Communications* **51**, 275–82.

SCHULMAN, L. H. & PELKA, H. (1975). The structural basis for the resistance of *Escherichia coli* formylmethionyl transfer ribonucleic acid to cleavage by *Escherichia coli* peptidyl transfer ribonucleic acid hydrolase. *Journal of Biological Chemistry*, **250**, 542–7.

SCHULMAN, L. H., PELKA, H. & SUNDARI, R. M. (1974). Structural requirements for recognition of *Escherichia coli* initiator and non-initiator transfer ribonucleic acids by bacterial T factor. *Journal of Biological Chemistry*, **249**, 7102–10.

SCHWARTZ, U., MENZEL, H. M. & GASSEN, H. G. (1976). Codon-dependent rearrangement of the three-dimensional structure of phenylalanine tRNA, exposing the T-ψ-C-G sequence for binding to the 50S ribosomal subunit. *Biochemistry*, **15**, 2484–90.

SCOLNICK, E., MILMAN, G., ROSMAN, N. & CASKEY, T. (1970). Transesterification by peptidyl transferase. *Nature, London*, **225**, 152–4.

SCOLNICK, E. M., TOMPKINS, R., CASKEY, T. & NIRENBERG, M. (1968). Release factors differing in specificity for terminator codons. *Proceedings of the National Academy of Sciences, USA*, **61**, 768–74.

SENIOR, B. W. & HOLLAND, I. B. (1971). Effect of colicin-E₃ upon the 30S ribosomal subunit of *Escherichia coli*. *Proceedings of the National Academy of Sciences, USA*, **68**, 959–63.

SHIH, C. T. & CRAVEN, G. R. (1973). Identification of neighbour relationship among proteins in the 30S ribosome. *Journal of Molecular Biology*, **78**, 651–63.

SHINE, J. & DALGARNO, L. (1974). The 3′-terminal sequence of *Escherichia coli* 16S ribosomal RNA complementary to nonsense triplets and ribosome binding sites. *Proceedings of the National Academy of Sciences, USA*, **71**, 1342–6.

SHINE, J. & DALGARNO, L. (1975). Determinant of cistron specificity in bacterial ribosomes. *Nature, London*, **254**, 34–8.

SHOREY, R. L., RAVEL, J. M., GARNER, C. W. & SHIVE, W. (1969). Formation and properties of the aminoacyl transfer ribonucleic acid–guanosine triphosphate–protein complex. *Journal of Biological Chemistry*, **244**, 4555–64.

SHOREY, R. L., RAVEL, J. M. & SHIVE, W. (1971). The effect of guanylyl 5′ methylene diphosphonate on binding of aminoacyl transfer ribonucleic acid to ribosomes. *Archives of Biochemistry and Biophysics*, **146**, 110–17.

SHULMAN, R. G., HILBERS, C. W. & MILLER, D. L. (1974). Nuclear magnetic resonance studies of protein–RNA-interactions. I. The elongation factor Tu-GTP–aminoacyl tRNA complex. *Journal of Molecular Biology*, **90**, 601–7.

SKOULTCHI, A., ONO, Y., MOON, H. M. & LENGYEL, P. (1968). On three complementary amino acid polymerization factors from *B. stereothermophilus*: separation of a complex containing two of the factors, guanosine 5′-triphosphate and aminoacyl-transfer RNA. *Proceedings of the National Academy of Sciences, USA*, **60**, 675–82.

SKOULTCHI, A., ONO, Y., WATERSON, J. & LENGYEL, P. (1970). Peptide chain elongation: indications for the binding of an aminoacid polymerization factor, guanosine 5′-triphosphate–aminoacyl-transfer ribonucleic acid complex to the messenger–ribosome complex. *Biochemistry*, **9**, 508–14.

SMITH, J. D. & CELIS, J. E. (1973). Mutant tyrosine transfer RNA that can be charged with glutamine. *Nature New Biology*, **243**, 66–71.

SOMMER, A. & TRAUT, R. R. (1975). Identification by diagonal gel electrophoresis of nine neighbouring protein pairs in the *Escherichia coli* 30S ribosome cross-linked with methyl-4-mercaptobutyrimidate. *Journal of Molecular Biology*, **97**, 471–81.

SOMMER, A. & TRAUT, R. R. (1976). Identification of neighboring protein pairs in the

Escherichia coli 30S ribosomal subunit by cross-linking with methyl-4-mercapto-butyrimidate. *Journal of Molecular Biology*, **105**, 995–1015.

SONENBERG, N., WILCHEK, M. & ZAMIR, A. (1973). Mapping of *Escherichia coli* ribosomal components involved in peptidyl transferase activity. *Proceedings of the National Academy of Sciences, USA*, **70**, 1423–6.

SPRAGUE, K. U., STEITZ, J. A., GRENLEY, R. M. & STOCKING, C. E. (1977). 3′-terminal sequences of 16S rRNA do not explain translational specificity differences between *Escherichia coli* and B. stearothermophilus ribosomes. *Nature, London*, **267**, 462–4.

SPRINGER, M., DONDON, J., GRAFFE, M., GRUNBERG-MANAGO, M., LELONG, J. C. & GROS, F. (1971). Role of translation factors in the interaction between aminoacyl tRNAs and their ribosomal decoding sites. *Biochimie*, **53**, 1047–57.

SPRINGER, M., GRAFFE, M. & GRUNBERG-MANAGO, M. (1977a). Characterization of an *Escherichia coli* mutant with a thermolabile initiator factor IF3 activity. *Molecular and General Genetics*, **151**, 17–26.

SPRINGER, M., GRAFFE, M. & HENNECKE, H. (1977b). Specialized transducing phage for the initiation factor IF3 gene in *Escherichia coli*. *Proceedings of the National Academy of Sciences, USA*, **74**, 3970–4.

SPRINZL, M., CHINALI, G., PARMEGGIANI, A., SCHEIT, K.-H., MAELICKE, A., STERNBACH, H., van der Haar, F. & Cramer, F. (1975). Modified tRNAs in the investigation of the mechanism of protein biosynthesis. In *Structure and Conformation of Nucleic Acids and Protein–Nucleic Acid Interactions*, ed. M. Sundalingham, & S. T. Rao, pp. 293–301. Baltimore: University Park Press.

STEEGE, D. A. (1977). The 5′-terminal nucleotide sequence of the *Escherichia coli* lactose repressor messenger RNA: features of translational initiation and reinitiation sites. *Proceedings of the National Academy of Sciences, USA*, in press.

STEITZ, J. A. (1968). Isolation of the A protein from bacteriophage R17. *Journal of Molecular Biology*, **33**, 937–45.

STEITZ, J. A. (1969). Polypeptide chain initiation: nucleotide sequences of the three ribosomal binding sites in bacteriophage R17 RNA. *Nature, London*, **224**, 957–64.

STEITZ, J. A. (1973). Specific recognition of non-initiator regions in RNA bacteriophage messengers by ribosomes of *Bacillus stearothermophilus*. *Journal of Molecular Biology*, **73**, 1–16.

STEITZ, J. A. (1977). Genetic signals and nucleotide sequences in messenger RNA. In *Biological Regulations and Development*, ed. R. Goldberger, New York: Plenum, in press.

STEITZ, J. A. & JAKES, K. (1975). How ribosomes select initiator regions in mRNA: base-pair formation between the 3′ terminus of 16S rRNA and the mRNA during initiation of protein synthesis in *Escherichia coli*. *Proceedings of the National Academy of Sciences, USA*, **72**, 4734–8.

STEITZ, J. A., WAHBA, A. J., LAUGHREA, M. & MOORE, P. B. (1977). Differential requirements for polypeptide chain initiation complex formation at the three bacteriophage R17 initiator regions. *Nucleic Acids Research*, **4**, 1–15.

STÖFFLER, G. (1974). Structure and function of the *Escherichia coli* ribosome. Immunochemical analysis. In *Ribosomes*, ed. M. Nomura, A Tissières & P. Lengyel, pp. 615–67. Cold Spring Harbour laboratory, USA.

STÖFFLER, G. & WITTMANN, H. G. (1977). Primary structure and three-dimensional arrangement of proteins within the *Escherichia coli* ribosome. In *Molecular Mechanisms of Protein Synthesis*, ed. H. Weissbach and S. Pestka, pp. 117–202. New York: Academic Press.

STRINGER, E. A., SARKAR, P. & MAITRA, U. (1977). Function of initiation factor 1 in the binding and release of initiation factor 2 from ribosomal initiation complexes in *Escherichia coli*. *Journal of Biological Chemistry*, **252**, 1739–44.

SUTTER, R. P. & MOLDAVE, K. (1966). The interaction of aminoacyl transferase II and ribosomes. *Journal of Biological Chemistry*, **241**, 1698–704.

SUTTLE, D. P., HARALSON, M. A. & RAVEL, J. M. (1973). Initiation factor 3 requirement for the formation of initiation complexes with synthetic oligonucleotides. *Biochemical and Biophysical Research Communications*, **51**, 376–82.

SUN, T. T., BOLLEN, A., KAHAN, L. & TRAUT, R. R. (1974). Topography of ribosomal proteins of the *Escherichia coli* 30S subunit as studied with the reversible cross-linking reagent methyl 4-mercaptobutyrimidate. *Biochemistry*, **13**, 2334–40.

SURYANARAYANA, T. & SUBRAMANIAN, A. R. (1977). Separation of two forms of IF-3 in *Escherichia coli* by two-dimensional gel electrophoresis *FEBS letters*, **79**, 264–8.

SZER, W., HERMOSO, J. M. & BOUBLIK, M. (1976). Destabilization of the secondary structure of RNA by ribosomal protein S1 from *Escherichia coli*. *Biochemical and Biophysical Research Communications*, **70**, 957–64.

TAKANAMI, M., YAN, Y. & JUKES, T. H. (1965). Studies on the site of ribosomal binding of f2 bacteriophage RNA. *Journal of Molecular Biology*, **12**, 761–73.

TATE, W. P., BEAUDET, A. L. & CASKEY, C. T. (1973). Influence of guanine nucleotides and elongation factors on interaction of release factors with the ribosome. *Proceedings of the National Academy of Sciences, USA*, **70**, 2350–5.

TATE, W. P., CASKEY, C. T. & STÖFFLER, G. (1975). Inhibition of peptide chain termination by antibodies specific of ribosomal proteins. *Journal of Molecular Biology*, **93**, 375–89.

THANG, M. N., SPRINGER, M., THANG, D. C. & GRUNBERG-MANAGO, M. (1971). Recognition by T factor of a tRNA^{Phe} yeast molecule recombined from 3′ and 5′ halves; and its non messenger dependent binding to ribosomes. *FEBS Letters*, **17**, 221–71.

THOMPSON, R. C. & STONE, P. J. (1977). Proofreading of the codon:anticodon interaction on ribosomes. *Proceedings of the National Academy of Sciences, USA*, **74**, 198–202.

TISCHENDORF, G. W., ZEICHHARDT, H. & STÖFFLER, G. (1975). Architecture of the *Escherichia coli* ribosome as determined by immune election microscopy. *Proceedings of the National Academy of Sciences, USA*, **72**, 4820–4.

TOMPKINS, R., SCOLNICK, E. & CASKEY, C. T. (1970). Peptide chain termination. The ribosomal and release factor requirements for peptide release. *Proceedings of the National Academy of Sciences, USA*, **65**, 702–8.

TRAUB, P. & NOMURA, M. (1968). Structure and function of *Escherichia coli* ribosomes. Reconstitution of functionally active 30S ribosomal particles from RNA and protein. *Proceedings of the National Academy of Sciences, USA*, **59**, 777–84.

UNGER, F. M. & TAKEMURA, S. (1973). A comparison between inosine- and guanosine-containing anticodons in ribosome-free codon–anticodon binding. *Biochemical and Biophysical Research Communications*, **52**, 1141–7.

VAN DER HOFSTAD, G. A. J. M., FOEKENS, J. A., BOSCH, L. & VOORMA, H. O. (1977). The involvement of a complex between formyl-methionyl-tRNA and initiation factor IF-2 in prokaryotic initiation. *European Journal of Biochemistry*, **77**, 69–75.

VAN DIGGELEN, O. P. & BOSCH, L. (1973). The association of ribosomal subunits of *Escherichia coli*. *European Journal of Biochemistry*, **39**, 499–510.

VAN DIGGELEN, O. P., OOSTROM, H. & BOSCH, L. (1973). The association of ribosomal subunits of *Escherichia coli*. *European Journal of Biochemistry*, **39**, 511–23.

VAN DUIN, J. & KURLAND, C. G. (1970). Functional heterogeneity of the 30S ribosomal subunit of *Escherichia coli*. *Molecular and General Genetics*, **109**, 169–76.

VAN DUIN, J., KURLAND, C. G., DONDON, J. & GRUNBERG-MANAGO, M. (1975). Near neighbours of IF3 bound to 30S ribosomal subunits. *FEBS Letters*, **59**, 287–90.

VAN DUIN, J., KURLAND, C. G., DONDON, J., GRUNBERG-MANAGO, M., BRANLANT, C. & EBEL, J. P. (1976). New aspects of the IF3 ribosome interaction. *FEBS Letters*, **62**, 111–14.

VAN KNIPPENBERG, P. H., GRIJM-FOS, M., VELDSTRA, H. & BOSCH, L. (1965). Effects of streptomycin on the translation of turnip yellow mosaic virus RNA *in vitro*. *Biochemical and Biophysical Research Communications*, **20**, 4–9.

VAN KNIPPENBERG, P. H., VAN DUIN, J. & LENTZ, H. (1973). Stoichiometry of initiation factor IF3 in exponentially growing *Escherichia coli* MRE 600. *FEBS Letters*, **34**, 95–8.

VERMEER, C., DE KIEVIT, R. J., VAN ALPHEN, W. J. & BOSCH, L. (1973*a*). Recycling of the initiation IF-3 on 30S ribosomal subunits of *Escherichia coli*. *FEBS Letters*, **31**, 273–6.

VERMEER, C., VAN ALPHEN, W. J., KNIPPENBERG, P. & BOSCH, L. (1973*b*). Initiation factor-dependent binding of MS_2 RNA to 30S ribosomes and the recycling of IF-3. *European Journal of Biochemistry*, **40**, 295–308.

VOYNOW, P. and KURLAND, C. G. (1971). Stoichiometry of the 30S ribosomal proteins of *Escherichia coli*. *Biochemistry*, **10**, 517–24.

WAHBA, A. J., IWASAKI, K., MILLER, M. J., SABOL, S., SILLERO, M. A. G. & VAZQUEZ, C. (1969). Initiation of protein synthesis in *Escherichia coli*, II. Role of the initiation factors in polypeptide synthesis. *Cold Spring Harbour Symposia on Quantitative Biology*, **34**, 291–9.

WATANABE, S. (1972). Interaction of siomycin with the acceptor site of *Escherichia coli* ribosomes. *Journal of Molecular Biology*, **67**, 443–57.

WATERSON, J., BEAUD, E. & LENGYEL, P. (1970). The S_1 factor in peptide chain elongation. *Nature, London*, **227**, 34–8.

WEBER, H. J. (1972). Stoichiometric measurements of 30S and 50S ribosomal proteins from *Escherichia coli*. *Molecular and General Genetics*, **119**, 233–48.

WEINER, A. M. & WEBER, H. (1971). Natural read-through at the UGA termination signal of Q7 coat protein cistron. *Nature New Biology*, **234**, 206–9.

WEISSBACH, H., MILLER, D. L. & HACHMANN, J. (1970). Studies on the role of factor T_S in polypeptide synthesis. *Archives of Biochemistry and Biophysics*, **137**, 262–9.

WEISSBACH, H., REDFIELD, B. & BROT, N. (1971). Aminoacyl-tRNA–Tu GTP interaction with ribosomes. *Archives of Biochemistry and Biophysics*, **145**, 676–84.

WEISSMANN, C., BILLETER, M. A., GOODMAN, H. M., HINDLEY, J. & WEBER, H. (1973). Structure and function of phage RNA. *Annual Reviews of Biochemistry*, **42**, 303–28.

WEISSMANN, C., TANIGUCHI, T., DOMINGO, E., SABO, D. FLAVELL, R. A. (1977). Site-directed mutagenesis as a tool in genetics. *Proceedings of the 9th Miami Winter symposium*, in press.

WHITE, B. N. & BAYLEY, S. T. (1972). Methionine transfer RNAs from the extreme halophile *Halobacterium cutirubrum*. *Biochimica et Biophysica Acta*, **272**, 583–7.

WISHNIA, A., BOUSSERT, A., GRAFFE, M., DESSEN, P. & GRUNBERG-MANAGO, M. (1975). Kinetics of the reversible association of ribosomal subunits: stopped-flow studies of the rate law and of the effect of Mg^{2+}. *Journal of Molecular Biology*, **93**, 499–515.

WITTMANN, H. G. & WITTMANN-LIEBOLD, B. (1974). Chemical structure of bacterial ribosomal proteins. In *Ribosomes*, ed. M. Nomura, A. Tissières & P. Lengyel, pp. 115–40. Cold Spring Harbor Laboratory, USA.

WOLF, H., CHINALI, G. & PARMEGGIANI, A. (1977). Mechanism of the inhibition of protein synthesis by kirromycin. Role of elongation factor Tu and ribosomes. *European Journal of Biochemistry*, **75**, 67–75.

YOKOSAWA, H., INOUE-YOKOSAWA, N., AKAI, K., KAWAKITA, M. & KAZIRO, Y. (1973). The role of guanosine triphosphate hydrolysis in elongation factor Tu-promoted binding of aminoacyl transfer ribonucleic acid to ribosomes. *Journal of Biological Chemistry*, **248**, 375–7.

YOKOSAWA, H., KAWAKITA, M., ARAI, K., INOUE-YOKOSAWA, N. & KAZIRO, Y. (1975). Binding of amino acyl tRNA to ribosomes promoted by elongation factor Tu. Further studies on the role of GTP-hydrolysis. *Journal of Biochemistry, Tokyo*, **77**, 719–28.

YOSHIKAWA, M., OKUYAMA, A. & TANAKA, N. (1975). Third kasugamycin resistance locus, $K_{sg}C$ affecting ribosomal protein S2 in *Escherichia coli* K-12. *Journal of Bacteriology*, **122**, 796–7.

YUAN, R. C., STEITZ, J. A. & CROTHERS, D. M. (1976). Direct evidence for secondary structure in the colicin E3 released 3′-terminal 16S RNA fragment. *Federation Proceedings*, **35**, 1351.

ZAMIR, A., MISKIN, R. & ELSON, D. (1971). Inactivation and reactivation of ribosomal subunits:aminoacyl-transfer RNA binding activity of the 30S subunit of *Escherichia coli*. *Journal of Molecular Biology*, **60**, 347–64.

ZAMIR, A., MISKIN, R., VOGEL, Z. & ELSON, D. (1973). Inactivation and reactivation of *Escherichia coli* ribosomes. In *Methods in Enzymology*, **30**, ed. K. Moldave & L. Grossman, pp. 406–26. New York: Academic Press.

STRUCTURE-FUNCTION RELATIONSHIPS OF THE GRAM-NEGATIVE BACTERIAL CELL ENVELOPE

VOLKMAR BRAUN

Lehrstuhl für Mikrobiologie II
Universität Tübingen, West Germany

INTRODUCTION

The Gram-negative cell is surrounded by two membranes, an outer membrane, also called cell wall, and an inner membrane designated cytoplasmic membrane. The cytoplasmic membrane harbours the enzyme systems of the respiratory chain, of oxidative phosphorylation, of active transport, of biosynthetic pathways which lead to major structural membrane components such as phospholipids, murein (mucopeptide, peptidoglycan) lipopolysaccharides and capsular polysaccharides. The enzymic composition of the cytoplasmic membrane is adapted to the growth conditions. For example, it differs greatly when cells are grown aerobically or anaerobically, or when the terminal electron acceptor is either oxygen or nitrate (Spencer & Guest, 1974). Most active transport systems are inducible. Newly inserted proteins can make up 1% of the total membrane protein. The composition of the membrane is not fixed in a stoichiometric way; whole enzyme systems can be suppressed and new ones can be inserted either in addition to existing ones or by replacing them. The structural composition, and with that the functional properties of the whole cell envelope, is to an astonishing degree variable according to physiological needs. It is also amenable to experimental manipulations by genetic means. On the one hand these properties offer great advantages in studies of structure–function relationships; on the other, they allow the cell to evade experimental restrictions set to study a particular function. Pleiotropic effects are often the result of a single mutation or of a certain physiological condition which can hamper severely the analysis of cause and effect.

In spite of the great structural and physiological diversity of Gram-negative bacteria, the general properties of the cytoplasmic membrane are common to all of them. The outer membrane also shows many common features. As far as it has been studied the chemical structure of the murein is the same (Rogers, 1970; Schleifer & Kandler, 1972;

Ghuysen & Shockman, 1973) and the outer membrane contains phospholipids, glycolipids and proteins in amounts and arranged as in most biological membranes. It is, therefore, justifiable to consider that a few species reveal the basic features which then will apply, with variations, to the others. *Escherichia coli* and *Salmonella* are the most extensively studied Gram-negative organisms and for this reason they will be treated preferentially in this review.

Three most recently discovered aspects of the Gram-negative cell envelope will mainly be discussed: the structural flexibility of the outer membrane, the multifunctional properties of membrane proteins and the functional interrelationships between the outer membrane and the cytoplasmic membrane.

COMPOSITION OF THE OUTER MEMBRANE OF
E. COLI AND SALMONELLA

The number of molecules of the major components in a 1 μm^2 section of the outer membrane is of the order of 10^5 for lipopolysaccharide (Mühlhardt, Menzel, Golecki & Speth, 1974; Smit, Kamio & Nikaido, 1975), 10^5 for proteins (Braun & Rehn, 1969; Rosenbusch, 1974; Inouye, Shaw & Shen, 1972; Henning, Höhn & Sonntag, 1973) and 10^6 for phospholipids (Smit *et al.*, 1975). The fatty acids in the lipid A portion of lipopolysaccharide cover 30–40% of the outermost layer of the outer membrane bilayer where they are exclusively located (Mühlradt *et al.*, 1974; Mühlradt, 1976). It is interesting that the number of fatty acids contributed by the lipopolysaccharide (3.4×10^6) equals, within the limits of detection, those provided by the phospholipid molecules (2.9×10^6). The phosphatidylethanolamine head groups are not hydrolysed by externally added phospholipase C of *Bacillus cereus* and they do not become covalently coupled to cyanogen bromide-activated dextran, suggesting that they are not exposed at the cell surface (Kamio & Nikaido, 1976). It is possible that the phospholipids make up the lipid phase of only the inner leaflet of the outer membrane whereas the lipid phase of the outer layer is formed by the fatty acids of the lipopolysaccharide. Such a peculiar distribution of the phospholipids in the outer membrane is also revealed by the following observations. The lateral diffusion constant for the lipopolysaccharide is five orders of magnitude lower than that for the phospholipids (Mühlradt *et al.*, 1974), showing that they are not immersed as single molecules in a highly liquid lipid phase. On the other hand, the phospholipid phase of the outer membrane shows transition points like those of the cytoplasmic membrane (Overath *et al.*, 1975; Nikaido, Takeuchi, Ohnishi & Nakae, 1977). These results make it highly likely that the phospholipids and lipopolysaccharides are segregated into separate domains in the outer membrane, but only 25–40% of the hydrocarbon chains of the outer membrane phospholipids, compared to 60–80% of the phospholipids of the cytoplasmic membrane, take part in the temperature-dependent phase transition (Overath *et al.*, 1975). This observation can be reconciled by assuming that the lipid covalently fixed to the lipoprotein (Braun, 1975) immerses itself into the inner leaflet of the outer membrane and binds there a portion of the phospholipid molecules, leading to their immobilization. Nearly 10^6 fatty acid chains in a 1 μm^2 section of the outer membrane are contributed by the lipoprotein.

Considerable progress has been made in the experimentally difficult field of the characterization of outer membrane proteins. These are integral membrane proteins strongly bound to each other, to phospholipids, lipopolysaccharide and some to murein. The possibility of separating the outer membrane and the cytoplasmic membrane (Osborn, Gander, Parisi & Carson, 1972) has greatly facilitated the identification of the relatively few outer membrane proteins. The resolution by electrophoresis on polyacrylamide gels in the presence of dodecyl sulphate has been improved to a degree that most bands contain only one protein species. With the use of ionic and nonionic detergents outer membrane proteins have been solubilized and purified on a preparative scale.

The nomenclature adopted by different research groups for the proteins has not yet been unified so Fig. 1 shows how they compare. The two proteins b and c (Ia, Ib, σ–8, σ–9) occur in *E. coli* K12 strains, but their relative amounts vary with the growth medium (Lugtenberg, Peters, Bernheimer & Berendsen, 1976). In *E. coli* B, protein b is missing. Protein a was first isolated and characterized from *E. coli* B and called matrix protein, because it was found to be arranged in a regular array with hexagonal symmetry and a periodicity of 7.5 nm on the outer face of the murein (Rosenbusch, 1974). Conditions have been devised to reconstitute the two *E. coli* K12 proteins with the isolated murein (Hasegawa, Yamada & Mizushima, 1976). No differences in the amino acid composition could be found between these two proteins so both are either coded by two very similar genes or they are subject to post-transcriptional modification. Mutants spontaneously resistant to phage TuI have either lost one protein or both (Schmitges & Henning, 1976). Protein d (II*, 3a) serves as binding site for phages (K3, TuII*), is absent in conjugation-deficient, recipient cells of the F-type (Skurray,

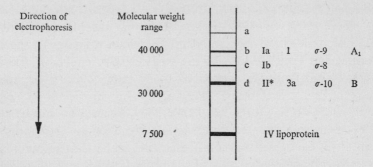

Fig. 1. 'Major' outer membrane proteins of *E. coli* K12 as revealed by gel electrophoresis in polyacrylamide gels in the presence of sodium dodecylsulphate using the system of Lugtenberg *et al.* (1975). The designation of the proteins can be taken from the above reference and from Henning & Haller (1975), Bragg & Hou (1971), Schnaitman (1974), Hasegawa *et al.* (1976). The estimation of the apparent molecular weight depends on the systems used.

Hancock & Reeves, 1974; Manning, Puspurs & Reeves, 1976) and is lacking in bacteriocin JF246-tolerant (*tolG*) mutants (Chai & Foulds, 1974). The isolated protein together with lipopolysaccharide specifically inhibits conjugation (Schweizer & Henning, 1977). Multiple functions are related with this protein which call for its surface exposure.

The major outer membrane protein with the fastest electrophoretic mobility is the lipoprotein the chemical structure of which is known (Braun, 1975). One third of the molecule is covalently bound to the underlying murein, while the rest occurs in the free form in the outer membrane (Inouye, 1975). The primary structure of the lipoprotein will be presented later in connection with its biosynthesis. Its conformation is apparently highly helical (Braun, Rotering, Ohms & Hagenmaier, 1976c) and unusually stable in sodium dodecyl sulphate (SDS). The matrix protein in contrast is rich in β-structured polypeptide, and also stable in SDS solution at room temperature (Rosenbusch, 1974; Nakamura & Mizushima, 1976) but changes its conformation to a predominantly α-helical structure upon heating in SDS. Lipoprotein and matrix protein were both treated with detergents during isolation and thus their 'native' conformation is not known, but since they have rather different ones in the 'renatured' state, one may conclude that their conformation in the membrane is also very different. Evidence that the conformation of the 'renatured' lipoprotein is not grossly different from that of the native material comes from serological data. Antibodies prepared against lipoprotein in living cells reacted strongly with the isolated lipoprotein and vice versa (Braun *et al.*, 1976a). It is likely that membrane proteins differ in their conformation as markedly as soluble proteins do and that there are no specific conformational demands for proteins to become integrated into membranes. A few hydrophobic amino acids or short hydrophobic sequences exposed on the proteins' surfaces suffice to fix proteins into membranes (see e.g. MacGregor & Schnaitman, 1974).

Extensive protein–protein interaction was demonstrated by the use of short cross-linking reagents which fixed all major proteins to each other (Haller & Henning, 1974). By the use of lower concentrations of cross-linking reagents, protein I was specifically paired, indicating a close interaction of polypeptides of the same kind (Palva & Randall, 1976). About 30 additional proteins, present in much lower amounts than the major proteins, could be linked to each other when deep rough mutants of the lipopolysaccharide were oxidized with $CuSO_4$–o-phenanthroline or ferricyanide or ferricyanide-ferrocene (Haller, Hoehn & Henning, 1975). These observations, and the large amount of protein in the outer membrane, indicate a big contribution of protein–protein binding to the overall organization of the outer membrane. This conclusion is supported by the electron microscopic finding that wild type *E. coli* strains are usually not cleaved through the outer membrane when freeze-etched but rather through the cytoplasmic membrane. Cleavage of the outer membrane would be expected if it were predominantly composed of a lipid bilayer structure (Bayer, Koplow & Goldfine, 1975). This apparently is not the case (Nanninga, 1970) (but see the following section about protein deficient mutants).

All of the major proteins of the outer membrane of *E. coli* serve as phage adsorption sites. Mutants lacking an outer membrane protein can conveniently be isolated by screening for phage resistant strains. Lipoprotein is an exception since no phage has been found which uses it as receptor. Studies with lipoprotein-specific antibodies have also shown that lipoprotein becomes accessible only in rough mutants (Rb_2), suggesting that lipoprotein is buried below the cell surface structures in wild type strains (Braun *et al.*, 1976*a*). Another approach has been used to localize outer membrane proteins in *S. typhimurium*. Three out of four major proteins reacted with cyanogen bromide-activated dextran (Kamio & Nikaido, 1977); lipoprotein was not considered in this study. One protein (33K) with an apparent molecular weight of 33000 was not coupled to dextran. Interestingly also, no phage has been found, so far, which binds to this protein (Nurminen *et al.*, 1976). All of the major outer membrane proteins with the likely exception of lipoprotein and the 33K protein of *S. typhimurium* have at least portions of the polypeptide chain exposed at the cell surface. The matrix protein (I, b, c, Fig. 1) apparently spans the whole thickness of the outer membrane starting from murein and reaching up to the cell surface.

Variations in the protein composition of the outer membrane

The main objective of such studies is to discover how far the composition of a membrane can be varied while still retaining its structural integrity and basic functions. The composition of the outer membrane has been followed under various growth conditions, or mutants have been isolated which have lost one or several proteins. Alternatively the influence of the variation of one component on other components has been studied, or genes specifying additional membrane proteins have either been induced, or transferred, into similar, or rather different, strains. The outcome of these studies is to find that membranes can accommodate substantial amounts of additional proteins; they can withstand drastic losses of proteins and cells regulate their membrane composition according to their physiological needs, as they do in adapting their cytoplasmic functions to environmental conditions.

Mutants lacking outer membrane proteins

Deep rough mutants of *S. typhimurium* LT2 (Rd, Re) which contain only a fraction of the total length of the saccharide chain of the lipopolysaccharide, were found to lack 60% of the protein from the outer membrane (Ames, Spudich & Nikaido, 1974). Heptose-less lipopolysaccharide mutants of *E. coli* showed a similar reduction in the major outer membrane proteins (Koplow & Goldfine, 1974). The number of lipopolysaccharide molecules remained the same as in the wild type but the phospholipid molecules per unit cell surface area increased by 70%. This had two interesting consequences. The phospholipid could no longer be accommodated in the inner layer of the outer membrane, as discussed before, and in fact the phospholipid was now accessible to added phospholipase C and cyanogen bromide-activated dextran (Kamio & Nikaido, 1976). Freeze-etching now cleaved within the outer membrane, showing a predominantly hydrophobic plane probably made up by a phospholipid bilayer (Smit *et al.*, 1975; Bayer *et al.*, 1975). The density of the particles, seen mainly in the outer half of the membrane, was decreased to the same extent as the reduction in proteins, supporting the notion that the particles are made up largely of proteins. It is unknown how the reduction in three major outer membrane proteins (except the lipoprotein which is present in normal amounts) is related to the lipopolysaccharide mutations. Other heptose-less *E. coli* mutants have been found to be deficient in only one protein (e.g. protein b,

Fig. 1) (van Alphen, Lugtenberg & Berendsen, 1976). Loss of each of the major outer membrane proteins alone (the lipoprotein was not considered) did not cause a reduction of the number of intramembranous particles, but in conjunction with heptose-deficient lipopolysaccharide, many fewer particles were observed (Verkleij, Lugtenberg & Ververgaert, 1976). *E. coli* mutants which lack two major outer membrane proteins, I and II* (see Fig. 1), had a phospholipid-to-protein ratio increased by a factor of 2.3, but normal amounts of lipopolysaccharide and of lipoprotein (Schweizer, Schwarz, Sonntag & Henning, 1976). Only in mutants lacking both proteins I and II* was the number of particles reduced in the outer half of the outer membrane. When both proteins were missing, it was no longer possible to cross-link the residual proteins so that they would form a container of the size and the shape of the cells, resistant to boiling in a 4% SDS solution. The authors conclude that 'the loss of two major proteins and the concomitant increase of phospholipid concentration has changed the architecture of the outer membrane from a highly oriented structure, with a large fraction of protein–protein interaction, to one predominantly exhibiting planar lipid bilayer characteristics. *E. coli* thus can assemble rather different outer membranes, a fact excluding that outer membrane formation constitutes a highly ordered or strictly sequential assembly-line process.'

Before Y. Hirota found a mutant lacking the lipoprotein entirely (Hirota, Suzuki, Nishimura & Yasuda, 1977), all the protein-deficient mutants described had contained lipoprotein in normal amounts. By crossing two lipoprotein-containing strains, an F' was obtained which carried a genetic defect in the gene for lipoprotein synthesis. It maps at 37.5 min in a region in which a mutation, leading to a lipoprotein lacking the diglyceride residue at the N-terminal cysteine residue, is located (Suzuki *et al.*, 1976). This mutant in addition carries an amino acid exchange, arginine→cysteine, at residue number 56 (H. Rotering (this laboratory) & M. Inouye, private communication). The mutant lacking the lipoprotein is hypersensitive to ethylenediamine tetraacetate and cationic dyes and somewhat sensitive to detergents (see Table 1). These physiological characteristics suggest that lipoprotein plays a role in maintaining the integrity of the outer membrane but not in the vital processes of growth and division. In ultra-thin sections of lipoprotein-deficient cells, the outer membrane was frequently detached from the cytoplasmic membrane (Y. Hirota, private communication). On the basis of the trypsin-induced release of lipoprotein from the murein which leads to the detachment of outer membrane and cytoplasmic

Table 1. *Some properties of* E. coli *mutants lacking major outer membrane proteins* (*Schweizer* et al., *1976; Manning* et al., *1976; Hirota* et al., *1977; U. Henning, private communication*)

Protein				Behaviour of cells
Ia	Ib	II*	Lp	
+	+	+	−	EDTA (1 mM) hypersensitive, leakage of periplasmic enzymes
+	+	−	+	Conjugation deficient recipient cell
+	+	−	−	Require 30 mM MgCl₂ in nutrient broth (Difco) for growth, still partial lysis
−	−	+	+	EDTA (0.5 M) sensitive
−	−	+	−	Like the mutant lacking lipoprotein alone
−	−	−	+	Phospholipase suicide, killed upon freezing in 30% glycerol

membrane, we proposed that lipoprotein fixes the murein to the outer membrane (Braun & Rehn, 1969). The observations made *in vivo* on the lipoprotein-deficient strain support this conclusion.

Among mutants of *S. typhimurium* which leak periplasmic proteins into the medium, *lykD* mutants form a class in which the outer membrane failed to invaginate during septum formation despite normal ingrowth of the cytoplasmic membrane and murein. The outer membrane instead formed a large bleb at the division site. The amount of lipoprotein covalently bound to murein was reduced to less than half the amount in wild type cells with a corresponding increase of the free form (Weigand, Vinci & Rothfield, 1976). Again the failure of these mutants to pull the outer membrane into the growing septum argues for a fixation of the outer membrane to the murein via the lipoprotein.

Additional *E. coli* mutants have been isolated with reduced amounts of lipoprotein. They showed abnormal cellular morphology and temperature sensitivity in growth and division (Wu & Lin, 1976). Such a mutant was also studied by Torti & Park (1976), who found a strict correlation between the reduced amount of lipoprotein in a temperature-sensitive strain having only a single mutation, the lack of growth and abnormal division at the non-permissive temperature. The amount of lipoprotein in cells could also be reduced by the antibiotic bicyclomycin which leads to elongated or spheroplast-like cells (Tanaka *et al.*, 1976). The spectrum of organisms affected by bicyclomycin corresponds with the known occurrence of lipoprotein. Under conditions where synthesis of cytoplasmic proteins was not affected, synthesis of envelope proteins was inhibited by 50%, free lipoprotein by 60% and murein-bound lipoprotein by 95%. In contrast the mutant of Hirota *et al.* (1977) lacks lipoprotein entirely and grows normally, thus a correlation between lipoprotein deficiency and physiological defects becomes questionable. It is still possible, however, that in strains with different genetic information, lipoprotein deficiency leads to variable effects.

Outer membranes containing additional proteins

E. coli lysogenized by the lambdoid phage PA–2 produced an additional major outer membrane protein, called 2, with an electrophoretic mobility between proteins 1 and 3a (Fig. 1) (Schnaitman, Smith & de Salsas, 1975). Protein 2 replaced protein 1. Protein 2 was subject to catabolite repression when glucose was added to the growth medium and concomitantly, protein 1 rose to some extent. Such a partial mutual exclusion of outer membrane proteins has been observed in several instances but apparently there exists no strict correlation. After phage infection many proteins have been found associated with the cell envelope. With the small RNA and DNA phages (for example, phage fr) the outer membrane could accommodate the phage envelope protein of the incoming phage, and also the newly synthesized protein was found in the cytoplasmic membrane before phage assembly took place (see literature cited in Wickner, 1976). The cells remained viable at 30 °C and kept on growing (Hoffmann-Berling & Mazé, 1964) whereas at 37 °C the cells lysed. Membrane aspects of virus–cell interactions have been reviewed by Braun & Hantke (1974) where further details and a list of references can be found.

The plasticity of membranes with regard to the protein composition has been demonstrated in a spectacular way by transferring the *E. coli* genes specifying protein receptors of the phages λ and T1, into *Proteus mirabilis* where they were expressed (Wohlhieter, Gemski & Baron, 1975). In another example, the *ompA* gene of *E. coli*, the structural gene specifying the outer membrane protein II* (Fig. 1), has been transferred into *S. typhimurium* and *P. mirabilis* (Datta, Krämer & Henning, 1976). Protein II* appeared in the outer membrane of the recipient cells and phage adsorbing to the protein was inactivated. The cells were killed but no phage developed in the foreign cells. The outer membrane proteins of *S. typhimurium* and *Proteus* are clearly different from those of *E. coli*. The fact that the *E. coli* protein can be incorporated in addition to the foreign host proteins and into a foreign environment shows that there are no very specific requirements for primary structure of proteins allowing incorporation into outer membranes. The degree of the expression of the *E. coli* protein in the foreign hosts differed from that in *E. coli* itself, and that of the host outer membrane proteins was also influenced. The data obtained did not suffice to draw conclusions with regard to regulatory mechanisms either on the level of transcription/translation, translocation through the cytoplasmic membrane or deposition in the outer membrane. Merodiploids of the *ompA* gene in *E. coli* showed no gene dosage effect; on the other hand a missense protein was much more strongly expressed than the wild type allele in a merodiploid. Variation in the relative and absolute amounts of major outer membrane proteins have been shown under several growth conditions, but the relations of cause and effect remain to be established. The situation is clearer with some minor outer membrane proteins for which beneficial functions for the cell have been found and which are expressed according to the cell's requirements (see later).

Another example of substantial amounts of an additional protein has been found in *E. coli* K12 carrying a derepressed, F-like R-factor. It contains 7×10^4 copies per cell of a protein similar to the subunit of the sex-pilus specified by the R-factor (Beard & Connolly, 1975). Also, induction of the *mal* and *malB* gene regions by maltose leads not only to synthesis of the intracytoplasmic maltose-catabolizing enzymes but also to the appearance of about 10^5 copies per cell of the phage λ receptor protein in the outer membrane (Schwartz, 1976; Braun & Krieger-Brauer, 1977), about 10^4 copies of maltose binding protein in the periplasm (Kellermann & Szmelcman, 1974), and two additional functions related to active transport (Hofnung, 1974). An additional function in the transport of maltose has been found for the λ receptor protein. Observations of this kind will now be summarized under the aspect of multiple functions of membrane proteins.

MULTIPLE FUNCTIONS OF OUTER MEMBRANE PROTEINS. INTERACTIONS OF OUTER MEMBRANE AND CYTOPLASMIC MEMBRANE FUNCTIONS

The statement that each membrane protein with a known function like a cytochrome, and which is present in quantity, also assumes a structural role in the three-dimensional architecture of the membrane may sound trivial. Each protein more or less determines the overall organization of the organelle, and in this sense it is already multifunctional. However, it has been found that 'minor' proteins of the outer membrane can serve simultaneously as receptors for more than one phage, for a colicin, and for the translocation of certain substrates across the outer membrane. The latter aspect has been reviewed most recently (Braun & Hantke, 1977) and only some general features will be discussed here together with important new findings. Transport systems

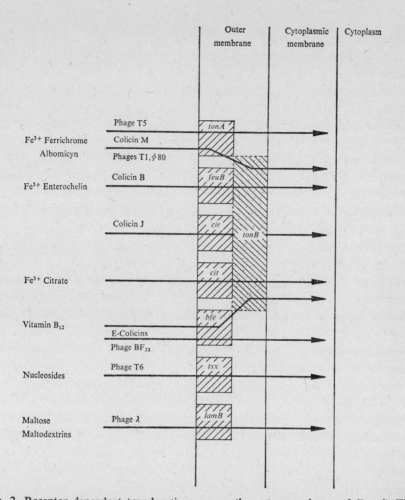

Fig. 2. Receptor dependent translocations across the outer membrane of *E. coli*. The drawing attempts to portray likely functional relationships of outer membrane proteins, designated according to their gene symbols, to translocations of the macromolecular phage nucleic acids and colicin proteins and low molecular substrates across the outer membrane. The proteins *tonA* (Braun, Schaller & Wolff, 1973; Braun & Wolff, 1973; Hantke & Braun, 1975), *feuB* and *cit* (Hancock, Hantke & Braun, 1976; Pugsley & Reeves, 1977), *cir* (Konisky & Liu, 1974) *bfe* (Sabet & Schnaitman, 1973), *tsx* (Hantke, 1976) and *lamB* (Randall-Hazelbauer & Schwartz, 1973) have been identified and located in the outer membrane. Competition experiments showed common binding sites (Wayne & Neilands, 1975; Wayne, Frick & Neilands, 1976; Bradbeer, Woodrow & Khalifah, 1976). The arrows which indicate the direction of the translocation go through an area designated *tonB* in cases where the *tonB* function is required. The *tonB* function has at present not been located and its mode of action is also unknown. Additional genes conferring colicin tolerance have been identified which may or may not influence the uptake of the substrates. Periplasmic binding proteins are involved in the uptake of maltose and with less certainty also for vitamin B12. Genes which probably specify permeases for the specific translocation steps across the cytoplasmic membrane are also known. For the sake of simplicity not all these transport functions have been included in the drawing. The single arrows drawn for one receptor protein should not be taken to imply that there also exist common components in the cytoplasmic membrane for the transfer of nucleic acids, colicins, and substrates; data collected so far indicate specific ways beyond the common receptor for each component to be taken up.

involving outer membrane proteins known to date are presented schematically in Fig. 2.

The outer membrane of Gram-negative bacteria forms a permeability barrier to hydrophilic substances which is the major reason for the pronounced resistance of this class of bacteria to many antibiotics. For *E. coli* and *S. typhimurium* it has been shown that the hindrance of permeation becomes effective for substances with molecular weights above 500–600 daltons (Payne & Gilvarg, 1968; Nakae & Nikaido, 1975). Nakae (1976*a, b*) was able to reconstitute an outer membrane vesicle of *E. coli* containing one protein, protein I (Fig. 1), phospholipids and lipopolysaccharide with an exclusion limit (pore size) for the passage of sugars similar to that found for the intact cell. The equivalent vesicle of *S. typhimurium* contained 3 major outer membrane proteins, designated 36K, 35K and 34K (K = 1000). The 33K protein and the lipoprotein, both probably not exposed at the cell surface (see earlier discussion) were absent from these vesicles. The one protein in the membrane of *E. coli* and the 3 analogous proteins in *S. typhimurium*, apparently form a water-filled channel through which hydrophilic substances, up to a certain size limit, penetrate the outer membrane. For substances above this critical size, such as the ferric iron complexes of citrate, enterochelin (enterobactin), ferrichrome, vitamin B12, malto-dextrins, there exist additional highly specific channels formed by proteins originally detected as phage and colicin receptors (Fig. 2). Phage- and colicin-resistant strains lacking the receptor proteins were unable to take up the iron complexes (MW above 700), vitamin B12 (MW 1351), and maltotetrose (MW 665).

The uptake of the nucleosides, maltose and maltotriose, with molecu-lar weights below the exclusion limit requires the receptor proteins only when these substances are supplied at very low, micromolar, concentra-tions in the medium. When the latter were supplied as growth factors at millimolar concentrations, receptor deficient mutants grew as fast as the wild type strains. The growth rate of *lamB* missense mutants on maltotriose (MW 504), was greatly reduced, however, in mutants containing normal amounts of λ receptor protein with an amino acid substitution (Szmelcman & Hofnung, 1975; Braun and Krieger-Brauer, 1977). Mutations in the λ receptor protein increased the K_m for maltose transport by a factor of 100–500 (Szmelcman, Schwartz, Silhavy & Boos, 1976). The maximal rate of transport at saturating maltose concentrations was found to be the same as in the wild type strain. This suggests that in λ receptor-deficient strains, maltose dif-fusion through the outer membrane becomes the rate-limiting step for transport at low maltose concentrations. The apparent K_m for transport is that for binding to the maltose binding-protein. In strains lacking the receptor the concentration of maltose in the periplasm is apparently much lower than the maltose concentration present in the medium. Translocation across the cytoplasmic membrane to concentrate maltose in the cytoplasm is the energy requiring step in maltose transport. Energization of transport generally occurs in the cytoplasmic membrane either via the electron transport chain and/or via ATP hydrolysis.

The molecular events at the receptor protein are not known. The iron complexes and vitamin B12 prevent binding of the phages and colicins to the receptor, whereas maltose and nucleosides do not. Again with maltose and nucleosides, no binding to the isolated receptor could be demonstrated. However, the λ receptor apparently translocates only maltose and maltodextrins, so some specific interaction between the protein and the substrate must occur to account for the selectivity. That the receptor protein reduces the uptake rate for both maltose and nucleosides points to specialized pores for small substrates too. Conceivably the outer membrane may be penetrated by an array of more or less substrate specific pores formed by 'minor' proteins and perhaps by general pores formed by the major protein I in *E. coli* or its equivalents in *S. typhimurium* (but see later regulation of 'minor' proteins).

Phage binding triggers at least two events based on conformational changes in protein ultrastructure: the release of the nucleic acids from the phage capsid and the transfer of the nucleic acid through two barriers, the outer and the cytoplasmic membranes.

Upon binding of colicins to their receptor proteins, ways must be opened for them to reach their targets, which are located in the membrane or the cytoplasm (for references see the recent review of Hardy, 1975).

Since very different compounds bind to the same receptor protein (Fig. 2), a specificity problem arises. At the beginning of these studies it was found that when the receptor function had been lost for one component it had disappeared for all components. It was soon realized that more than 90% of the mutants lacked the receptor protein in the outer membrane, irrespective of the system studied. In a few of the systems, the events at the receptor could be dissected to some extent. For example, missense mutants isolated as λ-resistant but still sensitive to a host range mutant, λh, showed reduced initial rates of maltose transport ranging from 6%–91% of the wild type (Szmelcman & Hofnung,

1975). Mutants resistant to λh but containing normal or reduced amounts of λ receptor protein in the outer membrane were additionally impaired in maltose transport but the residual transport activity ranging from 13% to 47% of the initial λ-resistant strain was not correlated with the amount of receptor protein present (Braun & Krieger-Brauer, 1977). The data suggest that mutations conferring absolute resistance against phage λ or λh only reduce maltose transport. Thus there is a functional overlap on the receptor protein, but it is not exactly the same site on the polypeptide chain that is involved. The specificity of the receptor-dependent maltose translocation across the outer membrane is apparently not so narrow as the binding of λ or λh. This is not an unexpected result if the receptor protein forms a pore through the outer membrane, the selectivity of which may not rely as strictly on a few critical amino acid residues, as does the phage binding site.

Phages adsorb at 0 °C to cell-bound receptors without transfer of the nucleic acid into the cell. An *E. coli* mutant called *pel* adsorbs phage λ which then fails to inject its DNA (Scandella & Arber, 1974). With the phages T1 and φ80, an additional function called *tonB* is required for irreversible adsorption and entry of the DNA into the cell.

The *tonB* function, not specified biochemically until now, plays a central role in processes following the initial binding to the receptor. The *tonB* dependent systems are depicted in Fig. 2. Its function has been studied most extensively for the irreversible binding of the phages T1 and φ80 (Hancock & Braun, 1976) and the transport of ferric iron as enterochelin complex (Frost & Rosenberg, 1975; Hancock, Hantke & Braun, 1977). Phage T1 binds reversibly to *E. coli* cells lacking the *tonB* function or to cells the energy metabolism of which is either poisoned or non-functional due to mutation. The required energy can either be provided by the electron transport chain or by ATP hydrolysis via the ATPase in the cytoplasmic membrane. On these grounds it was suggested that the *tonB* function may mediate the energy state of the cytoplasmic membrane to the receptor dependent infection process (Hancock & Braun, 1976). An important advance was the discovery of Frost & Rosenberg (1975) that the requirement for the *tonB* function in iron-enterochelin uptake could be bypassed when the intracellular synthesis of enterochelin was kept low. This was achieved in a mutant whose enterochelin synthesis was dependent on the supply of the biosynthetic precursor dihydroxybenzoate. They reasoned that the iron is carried into the outer membrane by dihydroxybenzoate or other complexing agents like carbonic acids and is then picked up by enterochelin in the periplasm and in turn translocated across the cyto-plasmic membrane. It was then shown that the receptor protein (*feuB*, Fig. 2) is bypassed, too. It is even more interesting that the type of energy required to take up iron changed with the bypass of the outer membrane. Ferric-enterochelin transport is strongly inhibited by dinitro-phenol (DNP) (Pugsley & Reeves, 1976), an uncoupler of oxidative phosphorylation thought to facilitate proton movement across the cytoplasmic membrane and thus to dissipate the energized membrane state. When ferric-enterochelin was supplied in the medium and had to pass through the outer membrane, DNP inhibited iron uptake by 85%, whereas under bypass conditions DNP inhibited only by 5–14% (Hancock, Hantke & Braun, 1977). The translocation of ferric-en-terochelin across the cytoplasmic membrane is under both normal and bypass conditions presumably the same. The electrochemical potential across the cytoplasmic membrane disturbed by DNP is therefore only required for the protein receptor and *tonB* dependent translocation step through the outer membrane. Possibly the perme-ability of the presumptive pore is controlled by the energy state of the cytoplasmic membrane and the *tonB* function serves as coupling device between the cytoplasmic membrane and the outer membrane. The recently discovered involvement of the *tonB* function in the second, energy dependent step of vitamin B12 transport (Bassfort, Bradbeer, Kadner & Schnaitman, 1976) supports the view that *tonB* mediates an energy function between the two membranes. Vitamin B12 transport, like ferric-enterochelin transport, requires an outer membrane receptor protein and an energized state of the cytoplasmic membrane (Bradbeer & Woodrow, 1976).

In the vitamin B12 uptake system it was detected with the use of a temperature sensitive *tonB* synthesis, that the function of *tonB* declines rapidly after cessation of its synthesis. The functional instability of the *tonB* product was not only observed in the rapid decrease of vitamin B12 transport but also in the loss of the cells' sensitivity to the *tonB* depend-ent killing actions of the colicins B, D and Ia (Bassford, Schnaitman & Kadner, 1977).

The functional state of a receptor protein and the routes taken by different substrates after binding to the same receptor have recently been studied. After cessation of synthesis of the vitamin B12 receptor, organisms first lost sensitivity to colicins E2 and E3 and subsequently to bacteriophage BF23. The decline in sensitivity occurred much faster than could be accounted for by receptor dilution due either to growth or to asymmetric distribution after cell separation (Bassford *et al.*, 1977).

The receptor, however, remained fully active in vitamin B12 uptake. In addition, binding of colicin E3 remained, so that the organisms developed a kind of colicin tolerance upon termination of receptor synthesis. It appears that newly synthesized receptor molecules are able to interact only for a short period with the systems responsible for subsequent steps of uptake of colicin E3 or phage BF23 DNA. They are apparently capable of continuously interacting with the systems responsible for B12 uptake, provided that the *tonB* gene product, which may be the immediate reaction partner for the receptor protein, is synthesized. It is also interesting that only the vitamin B12 uptake calls for the *tonB* product and that phage BF23 and colicin E infection is independent of its function (Fig. 2). In addition colicin E3 tolerant cells are fully sensitive to phage BF23 and active in vitamin B12 transport, demonstrating again the very different routes for compounds binding to the same receptor protein. In contrast compounds which bind to the receptors for iron complexes all require the *tonB* function except phage T5 (Fig. 2). In these systems, common elements are present in the reaction pathways of the various compounds beyond the binding to the same receptor.

Mutifunctional properties of membrane proteins have so far been described mainly for receptor proteins. It should also be mentioned that some periplasmic binding proteins (Boos, 1974) e.g. those for maltose, galactose and ribose, not only channel their substrates via the active transport systems but also via the chemotactic system (Hazelbauer, 1975; see references therein). The enzymes II of the phosphotransferase transport system for glucose, mannose, D-fructose and D-mannitol apparently also serve as chemoreceptors for these sugars (Adler & Epstein, 1974). The adenosine triphosphatase not only functions in oxidative phosphorylation and membrane energization via ATP hydrolysis but also as chemoreceptor for divalent metal ions in bacterial chemotaxis (Zukin & Koshland, 1976). Enzymes I and II of the phosphotransferase systems not only function in sugar transport but also in catabolite repression (Peterkofsky & Gazdar, 1975; Harwood, Gazdar, Prasad & Peterkofsky, 1976; Castro, Feucht, Morse & Saier, 1976). It is of great interest that a very important regulation mechanism acting on many operons in the cytoplasm is directed by two membrane associated enzyme systems, the adenylate cyclase and the phosphotransferase enzymes. The proteins involved must be able to interact with different components.

BIOGENESIS OF THE OUTER MEMBRANE

Aspects of the transport of low and high molecular weight compounds, from the medium into organisms, have been discussed in the previous section. The movement of molecules in the opposite direction, from the cytoplasm through the cytoplasmic membrane into the outer membrane, is not less problematic and is also not understood any better. How do proteins, lipopolysaccharides, and capsular polysaccharides overcome the barrier of the cytoplasmic membrane and why are they deposited exclusively in the outer membrane? Are there specific enzymes, translocases, which transfer macromolecules from the cytoplasmic membrane to the outer membrane? Alternatively is the affinity of outer membrane components for the existent outer membrane much stronger for physical reasons? Are there specific affinities between certain components in the outer membrane which determine where the constituents will finally end up? The problems in connection with outer membrane assembly cannot adequately be discussed here in the space allotted but some recent findings will be presented.

As discussed before, the outer membrane becomes impermeable for substances as small as tetrasaccharides or pentapeptides but specific 'pores' exist for some compounds of higher molecular weight. Proteins forming these pores also serve as receptors for phages and colicins and allow macromolecular nucleic acids and colicin proteins to penetrate through the outer membrane and the cytoplasmic membrane. It has long been known that cells become temporarily leaky upon phage infection and that phage coded functions contribute to the sealing of the membranes as long as the phage develops intracellularly (Braun & Hantke, 1974). Impermeability and permeability must always be well-balanced properties of any membrane. Secretion of macromolecules such as those forming the outer membrane has to occur while maintaining the low and high molecular content of the cytoplasm. Lipopolysaccharide seems to appear first at discrete sites on the cell surface and then to spread over the whole surface (Mühlradt, 1976). The excretion areas coincide with adhesion sites between the outer and cytoplasmic membrane. The phages T1–T7 and ϕX174 preferentially adsorb to areas at the cell surface, where adhesions between the cytoplasmic membrane and the wall are visible after plasmolysis (Bayer, 1975). Perhaps the export and import of macromolecules occurs at such adhesion sites. When S. anatum cells were infected with the phage Σ^{15}, radioactively labelled phage protein was first found associated with the outer membrane and afterwards with a fraction consisting of outer and cytoplasmic membrane (Tomita, Iwashita & Kanegasaki, 1976). The phage protein was found with the membrane mixture under conditions where the phage not only adsorbed to, but actually infected the cell. It is tempting to assume that the mixture of both membranes consisted of the adhesion sites.

When considering membrane biogenesis it is interesting to note that transport systems can be induced. For iron transport systems, transport with enterochelin or citrate is induced under iron deprivation of cells (Rosenberg & Young, 1974). The iron-related outer membrane proteins (see Fig. 2) appear under iron-limiting conditions (Braun, Hancock, Hantke & Hartmann, 1976b). Synthesis of the λ receptor protein is also induced to a large extent when maltose is supplied in the medium. In fully induced cells the number of protein copies for the iron related receptor proteins in the outer membrane can be estimated, from the density of the staining of the band in gels, to be in the range of 10^4 per cell. Similarly the λ receptor protein is present as about 10^5 copies per cell. The λ receptor preferentially appears in the vicinity of the forming septum during the last quarter of the cell's life cycle (Ryter, Shuman & Schwartz, 1975). Similar observations have been made for the incorporation of newly synthesized murein subunits into the cell wall of a growing cell (Schwarz et al., 1975). Cell cycle dependent synthesis of membrane proteins around the time of septum formation has been shown for cyto-chrome b1, the L-α-glycerophosphate transport system (Ohki, 1972), the galactose binding protein (Shen & Boos, 1973) and the insertion of free lipoprotein into the outer membrane (James & Gudas, 1976). The meaning of the incorporation of some membrane components during a short period of the life cycle is unknown. However, it shows that the assembly of the membrane is a discontinuous process, not only with regard to the supply of the membrane components but also with regard to the insertion sites in the existing membrane. Some constituents are not inserted randomly but rather at specific locations and afterwards distributed over the whole membrane.

A more detailed discussion of protein secretion from the cytoplasm across the cytoplasmic membrane into the outer membrane will be confined to the lipoprotein. In-vitro translation of the isolated mRNA has provided a lipoprotein with 20 additional amino acids at the N-terminal end (Fig. 3a). (Inouye et al., 1977). The additional sequence contains two hydrophobic peptides which may fix the growing peptide chain during synthesis to the cytoplasmic membrane. The long-discussed question of whether there exist membrane-bound ribosomes in E. coli can be answered in the sense that when membrane proteins are

Met—Lys—Ala—Thr—Lys—Leu—Val—Leu—Gly—Ala—Val—Ile—Leu—Gly—Ser—Thr—Leu—Leu—Ala—Gly—

Cys—Ser—Ser—Asn—Ala—Lys—Ile—Asp—Glu—Leu—Ser—Ser—Asp—Val—Gln—Thr—Leu—Asn—Ala—Lys—

Val—Asp—Glu—Leu—Ser—Asn—Asp—Val—Asn—Ala—Met—Arg—Ser—Asp—Val—Gln—Ala—Ala—Lys—

Asp—Asp—Ala—Ala—Arg—Ala—Asn—Gln—Arg—Leu—Asp—Asn—Met—Ala—Thr—Lys—Tyr—Arg—Lys

(a)

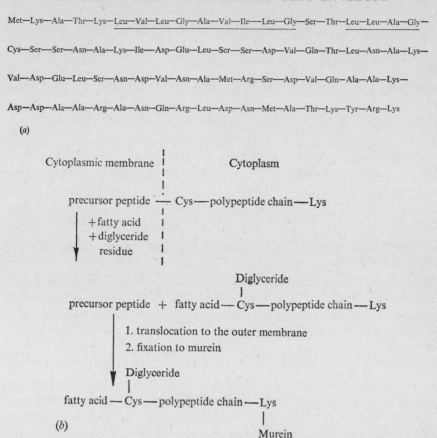

(b)

Fig. 3(a). Amino acid sequence of the pro-lipoprotein. The additional peptide of the pro-form is in the first line with the hydrophobic sequences underlined. (b) Hypothetical scheme of the biogenesis of lipoprotein based on the known structure and location of the final product (Braun, 1975) and the finding of pro-lipoprotein (Inouye *et al.*, 1977).

synthesized and the first hydrophobic N-terminal end grows into the membrane, ribosomes are fixed to the membrane as long as protein synthesis goes on. Such an additional sequence of amino acids at the N-terminal end has been found for many nascent secretory proteins in eukaryotes and they have been called signal peptides to account for the fact that they direct excretion by a certain pathway across membranes (Campbell & Blobel, 1976). In the lipoprotein the additional peptide is later replaced by a fatty acid. In addition, a diglyceride residue will be attached to the mercapto group of the N-terminal cysteine residue to form the final product found in the outer membrane (Fig. 3b) (Hantke & Braun, 1973). One third of the lipoprotein is covalently bound to the murein, two thirds remain free in the outer membrane. Linkage to murein is therefore probably not involved in the translocation across the cytoplasmic membrane into the outer membrane, but the reactions replacing the peptide of the prolipoprotein by the two lipid residues may well be part of the translocation process. Evidence against this assumption comes from a mutant containing a lipoprotein which lacks the diglyceride residue but which is still found in the outer membrane (Suzuki et al., 1976). However our results show that this mutant still contains one fatty acid (H. Rotering, this laboratory).

An additional peptide, 25 amino acids long, distinguishes the penicillinase of B. licheniformis which remains bound to the cell, from the same enzyme which is released into the surrounding medium (Yamamoto & Lampen, 1976). The same principles have apparently been developed in both prokaryotes and eukaryotes to ensure that proteins synthesized in the cytoplasm find their final destination either within membranes or outside the synthesizing cell. It can be foreseen that research on pro- and eukaryotic systems will be mutually fruitful in unravelling the specificity of translocation and that the use of mutants in prokaryotes will deliver the biochemical data necessary to understand the extremely selective insertion of proteins and other macromolecules into membranes.

CONCLUDING REMARKS

The structural versatility of the two membranes of E. coli and Salmonella encountered during the studies of the last few years has been unexpectedly great. In considering this phenomenon, one has to take into account the extensive variations in the membrane composition which go on as the membrane maintains its highly selective permeability and its structural role. In the most extreme case described, the protein content of a mutant's outer membrane is reduced by 60%, a loss which is balanced by an increase of 70% in the phospholipid content. Proteins are distributed exclusively in either the outer membrane or the cytoplasmic membrane, showing an extreme specificity for one or the other. Despite this fact, structural genes transferred to a foreign cell, e.g. from E. coli into Proteus, are expressed, and the proteins are incorporated into the foreign outer membrane environment. The lipoprotein of a mutant lacking the N-terminal diglyceride residue is nevertheless translocated from the cytoplasm across the cytoplasmic membrane into the outer membrane. The important detection of the pro-form containing an additional hydrophobic peptide at the N-terminal end opens the way for further studies on the translocation process which are significant not only for *E. coli* but also for our understanding of how cytoplasmic-ally synthesized proteins find their way into organelles such as mito-chondria, chloroplasts, lysosomes, the nucleus etc. of eukaryotic cells.

The regulation of membrane biogenesis is also an open field. Con-ceivably the presence or absence of one membrane component directs the occurrence of other constituents which are perhaps structurally but not functionally interlinked. The question arises of how information about structural requirements, if there is any, is transferred from the outer membrane to the cytoplasmic membrane or to the cytoplasm. What kind of feed back system exists? Does regulation act on the level of transcription, translation, translocation, or deposition in the mem-brane? Do components which cannot be incorporated in the outer membrane pile up in the cytoplasmic membrane or the cytoplasm, and do they inhibit their own synthesis?

The involvement of membrane proteins in different functions seems to occur quite frequently. The detection of specific 'pores' formed by receptor proteins for phages and colicins has interesting consequences with respect to the permeability properties of the outer membrane. The enhanced resistance of *E. coli* and other Gram-negative bacteria to antibiotics is frequently due to the outer membrane permeability barrier. With growing knowledge of the structural basis of outer mem-brane properties, new concepts can be foreseen that may be used to overcome the permeability barrier for drugs (Braun, 1977). Another application may arise from improved understanding of the adherence of bacteria to tissues, which is often the prerequisite of pathogenicity.

The fact that a defined chromosomal region produces a membrane protein, identified as single band in an electrophoretic gel system, does not guarantee the uniformity of the protein. Proteins which are specified by one gene and later modified or which are the products of very similar genes which may or may not be co-regulated can appear as a single band. Therefore mutations leading to the disappearance of one protein band have to be studied carefully to unravel the cause. The danger of possible pitfalls has recently been demonstrated by Boyd & Holland (1977) who showed that an outer membrane protein, designated protein D, once proposed as having a role in the control of DNA replication and the synthesis of which was thought to be restricted to a brief period during the cell cycle, is in fact rapidly induced by iron shortage and is protein *feuB* of Fig. 2. If the other examples of multifunctional proper-ties of single proteins described pass vigorous examination, it will be interesting to study the molecular mechanisms and to see whether they derive from a limited coding capacity of the bacterial genome or because a membrane has more functions to fulfill than it has space to incorporate all the required proteins.

The author's work was supported by the Deutsche Forschungsgemeinschaft (SFB 76). I wish to thank E. Oldmixon for reading the manuscript.

REFERENCES

ADLER, J. & EPSTEIN, W. (1974). Phosphotransferase-system enzymes as chemoreceptors for certain sugars in *Escherichia coli* chemotaxis. *Proceedings of the National Academy of Sciences, USA*, **71**, 2895–9.

VAN ALPHEN, W., LUGTENBERG, B. & BERENDSEN, W. (1976). Heptose-deficient mutants of *Escherichia coli* K12 deficient in up to three major outer membrane proteins. *Molecular and General Genetics*, **147**, 263–9.

AMES, G. F.-L., SPUDICH, E. N. & NIKAIDO, H. (1974). Protein composition of the outer membrane of *Salmonella typhimurium*: effect of lipopolysaccharide mutations. *Journal of Bacteriology*, **117**, 406–16.

BASSFORD, P. J., BRADBEER, C., KADNER, R. J. & SCHNAITMAN, C. A. (1976). Transport of vitamin B12 in *tonB* mutants of *Escherichia coli*. *Journal of Bacteriology*, **128**, 242–7.

BASSFORD, P. J., SCHNAITMAN, C. A. & KADNER, R. J. (1977). Functional stability of the *bfe* and *tonB* gene products in *Escherichia coli*. *Journal of Bacteriology*, **130**, 750–8.

BAYER, M. (1975). Role of adhesion zones in bacterial cell-surface function and biogenesis. In *Membrane Biogenesis, Mitochondria, Chloroplasts, and Bacteria*, ed. A. Tzagoloff, pp. 393–427. New York and London: Plenum.

BAYER, M. E., KOPLOW, J. & GOLDFINE, H. (1975). Alterations in envelope structure of heptose-deficient mutants of *Escherichia coli* as revealed by freeze-etching. *Proceedings of the National Academy of Sciences, USA*, **72**, 5145–9.

BEARD, J. P. & CONNOLLY, J. C. (1975). Detection of a protein, similar to the sex pilus subunit, in the outer membrane of *Escherichia coli* cells carrying a derepressed F-like R factor. *Journal of Bacteriology*, **122**, 59–65.

BOOS, W. (1974). Bacterial transport. *Annual Review of Biochemistry*, **43**, 123–46.

BOYD, A. & HOLLAND, I. B. (1977). Protein D, an iron transport protein induced by filtration of cultures of *Escherichia coli*. *FEBS Letters*, **76**, 20–4.

BRADBEER, C. & WOODROW, M. L. (1976). Transport of vitamin B12 in *Escherichia coli*: energy dependence. *Journal of Bacteriology*, **128**, 99–104.

BRADBEER, C., WOODROW, M. L. & KHALIFAH, L. I. (1976). Transport of vitamin B12 in *Escherichia coli*: common receptor system for vitamin B12 and bacteriophage BF23 on the outer membrane of the cell envelope. *Journal of Bacteriology*, **125**, 1032–9.

BRAGG, P. D. & HOU, C. (1971). Purification of three proteins from the outer membrane of the envelope of *Escherichia coli*. *FEBS Letters*, **15**, 142–4.

BRAUN, V. (1975). Covalent lipoprotein from the outer membrane of *Escherichia coli*. *Biochimica et Biophysica Acta* (*Reviews in Biomembranes*), **415**, 335–77.

BRAUN, V. (1977). Membranpermeation and Antibiotika-Resistenz bei Bakterien. *Die Naturwissenschaften*, **64**, 126–32.

BRAUN, V., BOSCH, V., KLUMPP, E. R., NEFF, I., MAYER, A. & SCHLECHT, S. (1976*a*). Antigenic determinants of murein lipoprotein and its exposure at the surface of enterobacteriaceae. *European Journal of Biochemistry*, **62**, 555–66.

BRAUN, V., HANCOCK, R. E. W., HANTKE, K. & HARTMANN, A. (1976*b*). Functional organization of the outer membrane of *Escherichia coli*. Phage and colicin receptors as components of iron uptake systems. *Journal of Supramolecular Structure*, **5**, 37–58.

BRAUN, V. & HANTKE, K. (1974). Biochemistry of bacterial cell envelopes. *Annual Review of Biochemistry*, **43**, 89–121.

BRAUN, V. & HANTKE, K. (1977). Bacterial receptors for phages and colicins as constituents of specific transport systems. In *Microbial Interactions*, ed. J. L. Reissig, in press. London: Chapman and Hall.

BRAUN, V. & KRIEGER-BRAUER, H. J. (1977). Interrelationship of the phage lambda receptor protein and maltose transport in mutants of *Escherichia coli* K12. *Biochimica et Biophysica Acta*, **469**, 89–98.

BRAUN, V. & REHN, K. (1969). Chemical characterization, spatial distribution and function of a lipoprotein (mureinlipoprotein) of the *Escherichia coli* cell wall. The specific effect of trypsin on the membrane structure. *European Journal of Biochemistry*, **10**, 426–38.

BRAUN, V., ROTERING, H., OHMS, J.-P. & HAGENMAIER, H. (1976c). Conformational studies on murein-lipoprotein from the outer membrane of *Escherichia coli*. *European Journal of Biochemistry*, **70**, 601–10.

BRAUN, V., SCHALLER, K. & WOLFF, H. (1973). A common receptor-protein for phage T5 and colicin M in the outer membrane of *Escherichia coli* B. *Biochimica et Biophysica Acta*, **323**, 87–97.

BRAUN, V. & WOLFF, H. (1973). Characterization of the receptor protein for phage T5 and colicin M in the outer membrane of *Escherichia coli* B. *FEBS Letters*, **34**, 77–80.

CAMPBELL, P. N. & BLOBEL, G. (1976). The role of organelles in the chemical modification of the primary translation products of secretory proteins. *FEBS Letters*, **72**, 215–24.

CASTRO, L., FEUCHT, B. U., MORSE, M. L. & SAIER JR, M. H. (1976). Regulation of carbohydrate permeases and adenylate cyclase in *Escherichia coli*. *Journal of Biological Chemistry*, **251**, 5522–7.

CHAI, T. & FOULDS, J. (1974). Demonstration of a missing outer membrane protein in *tolG* mutants of *Escherichia coli*. *Journal of Molecular Biology*, **85**, 465–74.

DATTA, D. B., KRÄMER, C. & HENNING, U. (1976). Diploidy for a structural gene specifying a major protein of the outer cell envelope membrane of *Escherichia coli*. *Journal of Bacteriology*, **128**, 834–41.

FROST, G. E. & ROSENBERG, H. (1975). Relationship between the *tonB* locus and iron transport in *Escherichia coli*. *Journal of Bacteriology*, **124**, 704–12.

GHUYSEN, J.-M. & SHOCKMAN, G. D. (1973). Biosynthesis of peptidoglycan. In *Membranes and Walls of Bacteria*, ed. L. Leive, pp. 37–130, New York: Dekker.

HALLER, I. & HENNING, U. (1974). Cell envelope and shape of *Escherichia coli* K12. Crosslinking with dimethyl imidoesters of the whole cell wall. *Proceedings of the National Academy of Sciences, USA*, **71**, 2018–21.

HALLER, I., HOEHN, B. & HENNING, U. (1975). Apparent high degree of asymmetry of protein arrangement in the outer *Escherichia coli* cell envelope membrane. *Biochemistry*, **14**, 478–84.

HANCOCK, R. E. W. & BRAUN, V. (1976). Nature of the energy requirement for the irreversible adsorption of bacteriophages T1 and ϕ80 to *Escherichia coli*. *Journal of Bacteriology*, **125**, 409–15.

HANCOCK, R. E. W., HANTKE, K. & BRAUN, V. (1976). Iron transport in *Escherichia coli* K12: the involvement of the colicin B receptor and of a citrate-inducible protein. *Journal of Bacteriology*, **127**, 1370–5.

HANCOCK, R. E. W., HANTKE, K. & BRAUN, V. (1977). Iron transport in Escherichia coli K 12. 2,3-dihydroxybenzoate-promoted iron uptake. *Archives of Microbiology*, **114**, 231–9.

HANTKE, K. (1976). Phage T6 – colicin K receptor and nucleoside transport in *Escherichia coli. FEBS Letters*, **70**, 109–12.

HANTKE, K. & BRAUN, V. (1973). Covalent binding of lipid to protein. Diglyceride and amide-linked fatty acid at the N-terminal end of murein-lipoprotein of the *Escherichia coli* outer membrane. *European Journal of Biochemistry*, **34**, 284–96.

HANTKE, K. & BRAUN, V. (1975). Membrane receptor dependent iron transport in *Escherichia coli. FEBS Letters*, **49**, 301–4.

HARDY, K. G. (1975). Colicinogeny and related phenomena. *Bacteriological Reviews*, **39**, 464–515.

HARWOOD, J. P., GAZDAR, C., PRASAD, C. & PETERKOFSKY, A. (1976). Involvement of the glucose enzyme II of the sugar phosphotransferase system in the regulation of adenylate cyclase by glucose in *Escherichia coli. Journal of Biological Chemistry*, **251**, 2462–8.

HASEGAWA, Y., YAMADA, H. & MIZUSHIMA, S. (1976). Interactions of outer membrane proteins σ-8 and σ-9 with peptidoglycan sacculus of *Escherichia coli* K12. *Journal of Biochemistry (Tokyo)*, **80**, 1401–9.

HAZELBAUER, G. L. (1975). Maltose chemoreceptor of *Escherichia coli. Journal of Bacteriology*, **122**, 206–14.

HENNING, U. & HALLER, I. (1975). Mutants of *Escherichia coli* K12 lacking all 'major' proteins of the outer cell envelope membrane. *FEBS Letters*, **55**, 161–4.

HENNING, U., HÖHN, B. & SONNTAG, I. (1973). Cell envelope and shape of *Escherichia coli* K12: the ghost membrane. *European Journal of Biochemistry*, **39**, 27–36.

HIROTA, Y., SUZUKI, H., NISHIMURA, Y. & YASUDA, S. (1977). On the process of cellular division in *Escherichia coli*: a mutant of *Escherichia coli* lacking a murein-lipoprotein. *Proceedings of the National Academy of Sciences, USA*, **74**, 1417–20.

HOFNUNG, M. (1974). Divergent operons and the genetic structure of the maltose B region in *Escherichia coli. Genetics*, **76**, 169–84.

HOFFMANN-BERLING, H. & MAZÉ, R. (1964). Release of male-specific bacteriophages from surviving host bacteria. *Virology*, **22**, 305–13.

HENNING, U., HÖHN, B. & SONNTAG, I. (1973). Cell envelope and shape of *Escherichia coli* K12: the ghost membrane. *European Journal of Biochemistry*, **39**, 27–36.

INOUYE, M. (1975). Biosynthesis and assembly of the outer membrane proteins of *Escherichia coli*. In *Membrane Biogenesis, Mitochondria, Chloroplasts, and Bacteria*, ed. A. Tzagoloff, pp. 351–91. New York and London: Plenum.

INOUYE, M., SHAW, J. & SHEN, C. (1972). The assembly of a structural lipoprotein in the envelope of *Escherichia coli. Journal of Biological Chemistry*, **247**, 8154–9.

INOUYE, S., WANG, S. S., SEKIZAWA, J., HALEGOUA, S. & INOUYE, M. (1977). Amino acid sequence for the peptide extension on the prolipoprotein of the *Escherichia coli* outer membrane. *Proceedings of the National Academy of Sciences, USA*, **74**, 1004–8.

JAMES, R. & GUDAS, L. J. (1976). Cell cycle-specific incorporation of lipoprotein into the outer membrane of *Escherichia coli. Journal of Bacteriology*, **125**, 374–5.

KAMIO, Y. & NIKAIDO, H. (1976), Outer membrane of *Salmonella typhimurium*: accessibility of phospholipid head groups to phospholipase C and cyanogen bromide activated dextran in the external medium. *Biochemistry*, **15**, 2561–70.

KAMIO, Y. & NIKAIDO, H. (1977). Outer membrane of *Salmonella typhimurium*. Identification of proteins exposed on cell surface. *Biochimica et Biophysica Acta*, **464**, 589–601.

KELLERMANN, O. & SZMELCMAN, S. (1974). Active transport of maltose in *Escherichia coli* K12. Involvement of a 'periplasmic' maltose binding protein. *European Journal of Biochemistry*, **47**, 139–49.

KONISKY, J. & LIU, C.-T. (1974). Solubilization and partial characterization of the colicin I receptor of *Escherichia coli*. *Journal of Biological Chemistry*, **249**, 835–40.

KOPLOW, J. & GOLDFINE, H. (1974). Alterations in the outer membrane of the cell envelope of heptose deficient mutants of *Escherichia coli*. *Journal of Bacteriology*, **117**, 527–43.

LUGTENBERG, B., MEIJERS, J., PETERS, R., VAN DER HOEK, P. & VAN ALPHEN, L. (1975). Electrophoretic resolution of the 'major outer membrane protein' of *Escherichia coli* K12 into four bands. *FEBS Letters*, **58**, 254–8.

LUGTENBERG, B., PETERS, R., BERNHEIMER, H. & BERENDSEN, W. (1976). Influence of cultural conditions and mutations on the composition of the outer membrane proteins of *Escherichia coli*. *Molecular and General Genetics*, **147**, 251–62.

MCGREGOR, C. & SCHNAITMAN, C. (1974). Nitrate reductase in *Escherichia coli*: properties of the enzyme and in vitro reconstitution from enzyme-deficient mutants. *Journal of Supramolecular Structure*, **2**, 715–27.

MANNING, P. A., PUSPURS, A. & REEVES, P. (1976). Outer membrane of *Escherichia coli* K12: isolation of mutants with altered protein 3A by using host range mutants of bacteriophage K3. *Journal of Bacteriology*, **127**, 1080–4.

MÜHLRADT, P. F. (1976). Topography of outer membrane assembly in *Salmonella*. *Journal of Supramolecular Structure*, **51**, 103–8.

MÜHLRADT, P., MENZEL, J., GOLECKI, J. R. & SPETH, V. (1974). Lateral mobility and surface density of lipopolysaccharide in the outer membrane of *Salmonella typhimurium*. *European Journal of Biochemistry*, **43**, 533–9.

NAKAE, T. (1976a). Outer membrane of *Salmonella*. Isolation of protein complex that produces transmembrane channels. *Journal of Biological Chemistry*, **251**, 2176–8.

NAKAE, T. (1976b). Identification of the outer membrane protein of *Escherichia coli* that produces transmembrane channels in reconstituted vesicle membranes. *Biochemical and Biophysical Research Communications*, **71**, 877–84.

NAKAE, R. & NIKAIDO, H. (1975). Outer membrane as a diffusion barrier in *Salmonella typhimurium*. Penetration of oligo- and polysaccharides into isolated outer membrane vesicles and cells with degraded peptidoglycan layer. *Journal of Biological Chemistry*, **250**, 7359–65.

NAKAMURA, K. & MIZUSHIMA, S. (1976). Effects of heating in dodecyl sulphate solution on the conformation and electrophoretic mobility of isolated major outer membrane proteins of *Escherichia coli* K12. *Journal of Biochemistry (Tokyo)*, **80**, 1411–22.

NANNINGA, N. (1970). Ultrastructure of the cell envelope of *Escherichia coli* B after freeze-etching. *Journal of Bacteriology*, **101**, 297–303.

NIKAIDO, H., TAKEUCHI, Y., OHNISHI, S. I. & NAKAE, T. (1977). Outer membrane of *Salmonella typhimurium*: electron spin resonance studies. *Biochimica et Biophysica Acta*, **465**, 152–64.

NURMINEN, M., LOUNATMAA, K., SARVAS, M., MÄKELÄ, P. H. & NAKAE, T. (1976). Bacteriophage-resistant mutants of *Salmonella typhimurium* deficient in two major outer membrane proteins. *Journal of Bacteriology*, **127**, 941–55.

OHKI, M. (1972). Correlation between metabolism of phosphatidylglycerol and membrane synthesis in *Escherichia coli*. *Journal of Molecular Biology*, **68**, 249–64.

OSBORN, M. J., GANDER, J. E., PARISI, E. & CARSON, J. (1972). Mechanism of assembly of the outer membrane of *Salmonella typhimurium*. Isolation and characterization of cytoplasmic and outer membrane. *Journal of Biological Chemistry*, **247**, 3962–72.

OVERATH, P., BRENNER, M., GULIK-KRZYWICKI, T., SHECHTER, E. & LETELLIER, L. (1975). Lipid phase transitions in cytoplasmic and outer membrane of *Escherichia coli*. *Biochimica et Biophysica Acta*, **389**, 358–69.

PALVA, E. T. & RANDALL, L. L. (1976). Nearest neighbour analysis of *Escherichia coli* outer membrane proteins, using cleavable cross-links. *Journal of Bacteriology*, **127**, 1558–60.

PAYNE, J. W. & GILVARG, C. (1968). Size restriction of peptide utilization in *Escherichia coli*. *Journal of Biological Chemistry*, **243**, 2691–9.

PETERKOFSKY, A. & GAZDAR, C. (1975). Interaction of Enzyme I of the phosphoenol pyruvate: sugar phosphotransferase system with adenylate cyclase of *Escherichia coli*. *Proceedings of the National Academy of Sciences, USA*, **72**, 2920–4.

PUGSLEY, A. P. & REEVES, P. (1976). Iron uptake in colicin B-resistant mutants of *Escherichia coli*. *Journal of Bacteriology*, **126**, 1052–62.

PUGSLEY, A. P. & REEVES, P. (1977). The role of colicin receptors in the uptake of ferrienterochelin by *Escherichia coli* K12. *Biochemical and Biophysical Research Communications*, **74**, 903–11.

RANDALL-HAZELBAUER, L. & SCHWARTZ, M. (1973). Isolation of the bacteriophage lambda receptor from *Escherichia coli*. *Journal of Bacteriology*, **166**, 1436–46.

ROGERS, H. J. (1970). Bacterial growth and the cell envelope. *Bacteriological Reviews*, **34**, 194–214.

ROSENBERG, H. & YOUNG, J. G. (1974). Iron transport in the enteric bacteria. In *Microbial Iron Metabolism*, ed. J. B. Neilands, pp. 67–82. New York and London: Academic Press.

ROSENBUSCH, J. P. (1974). Characterization of the major envelope protein from *Escherichia coli*. Regular arrangement on the peptidoglycan and unusual dodecyl sulphate binding. *Journal of Biological Chemistry*, **249**, 8019–29.

RYTER, A., SHUMAN, H. & SCHWARTZ, M. (1975). Integration of the receptor for bacteriophage lambda in the outer membrane of *Escherichia coli*: coupling with cell division. *Journal of Bacteriology*, **122**, 295–301.

SABET, S. F. & SCHNAITMAN, C. A. (1973). Purification and properties of the colicin E3 receptor of *Escherichia coli*. *Journal of Biological Chemistry*, **248**, 1797–806.

SCANDELLA, D. & ARBER, W. (1974). An *Escherichia coli* mutant which inhibits the injection of phage λ DNA. *Virology*, **58**, 504–13.

SCHLEIFER, K. H. & KANDLER, O. (1972). Peptidoglycan types of bacterial cell walls and their taxonomic implications. *Bacteriological Reivews*, **36**, 407–77.

SCHMITGES, C. J. & HENNING, U. (1976). The major protein of the *Escherichia coli* outer cell-envelope membrane. Heterogeneity of protein I. *European Journal of Biochemistry*, **63**, 47–52.

SCHNAITMAN, C. (1974). Outer membrane proteins of *Escherichia coli*. Differences in outer membrane proteins due to strain and cultural differences. *Journal of Bacteriology*, **118**, 454–64.

SCHNAITMAN, C., SMITH, D. & DE SALSAS, M. F. (1975). Temperate bacteriophage which causes the production of a new major outer membrane protein by *Escherichia coli*. *Journal of Virology*, **15**, 1121–30.

SCHWARTZ, M. (1976). The adsorption of coliphage λ to its host: effect of variation in the surface density of receptor and in phage-receptor affinity. *Journal of Molecular Biology*, **103**, 521–36.

SCHWARTZ, U., RYTER, A., RAMBACH, A., HELLIO, R. & HIROTA, Y. (1975). Process of cellular division in *Escherichia coli*: differentiation of growth zones in the sacculus. *Journal of Molecular Biology*, **98**, 749–59.

SCHWEIZER, M. & HENNING, U. (1977). Action of a major outer cell envelope membrane protein in conjugation of *Escherichia coli* K12. *Journal of Bacteriology*, **129**, 1651–2.

SCHWEIZER, M., SCHWARTZ, H., SONNTAG, I. & HENNING, U. (1976). Mutational change of membrane architecture. Mutants of *Escherichia coli* K12 missing major proteins of the outer cell envelope membrane. *Biochimica et Biophysica Acta*, **448**, 474–91.

SHEN, B. H. P. & BOOS, W. (1973). Regulation of the β-methylgalactoside transport system and the galactose-binding protein by the cell cycle of *Escherichia coli*. *Proceedings of the National Academy of Sciences, USA*, **70**, 1481–85.

SKURRAY, R. A., HANCOCK, R. E. W. & REEVES, P. (1974). Con⁻ mutants: class of mutants in *Escherichia coli* K12 lacking a major cell wall protein and defective in conjugation and adsorption of a phage. *Journal of Bacteriology*, **119**, 726–35.

SMIT, J., KAMIO, Y. & NIKAIDO, H. (1975). Outer membrane of *Salmonella typhimurium*: chemical analysis and freeze-fracture studies with lipopolysaccharide mutants. *Journal of Bacteriology*, **124**, 942–58.

SPENCER, M. E. & GUEST, J. R. (1974). Proteins of the inner membrane of *Escherichia coli*: changes in composition associated with anaerobic growth and fumarate reductase amber mutation. *Journal of Bacteriology*, **117**, 954–9.

SUZUKI, H., NISHIMURA, Y., IKETANI, H., CAMPISI, J., HIRASHIMA, A., INOUYE, M. & HIROTA, Y. (1976). Novel mutation that causes a structural change in a lipoprotein of the outer membrane of *Escherichia coli*. *Journal of Bacteriology*, **127**, 1494–501.

SZMELCMAN, S. & HOFNUNG, M. (1975). Maltose transport in *Escherichia coli* K12: involvement of the bacteriophage lambda receptor. *Journal of Bacteriology*, **124**, 112–8.

SZMELCMAN, S., SCHWARTZ, M., SILHAVY, T. J. & BOOS, W. (1976). Maltose transport in *Escherichia coli* K12. A comparison of transport kinetics in wild type and λ-resistant mutants with the dissociation constants of the maltose-binding protein as measured by fluorescence quenching. *European Journal of Biochemistry*, **65**, 13–9.

TANAKA, N., ISEKI, M., MIYOSHI, T., AOKI, H. & IMANAKA, H. (1976). Mechanism of action of bicyclomycin. *Journal of Antibiotics*, **29**, 155–68.

TOMITA, T., IWASHITA, S. & KANEGASAKI, S. (1976). Role of cell surface mobility and bacteriophage infection: translocation of Salmonella phages to membrane adhesions. *Biochemical and Biophysical Research Communications*, **73**, 807–13.

TORTI, S. V. & PARK, J. T. (1976). Lipoprotein of gram-negative bacteria is essential for growth and division. *Nature (London)*, **263**, 323–6.

VERKLEIJ, A. J., LUGTENBERG, E. J. J. & VERVERGAERT, P. H. J. Th. (1976). Freeze-etch morphology of outer membrane mutants of *Escherichia coli* K12. *Biochimica et Biophysica Acta*, **426**, 581–6.

WAYNE, R., FRICK, K. & NEILANDS, J. B. (1976). Siderophore protection against colicins, M, B, V and Ia in *Escherichia coli*. *Journal of Bacteriology*, **126**, 7–12.

WAYNE, R. & NEILANDS, J. B. (1975). Evidence for a common binding site for ferrichrome compounds and bacteriophage φ80 in the cell envelope of *Escherichia coli*. *Journal of Bacteriology*, **121**, 497–503.

WEIGAND, R. A., VINCI, K. D. & ROTHFIELD, L. I. (1976). Morphogenesis of the bacterial division septum: a new class of septation-defective mutants. *Proceedings of the National Academy of Sciences, USA*, **73**, 1882–6.

WICKNER, W. (1976). Asymmetric orientation of phage M13 coat protein in *Escherichia coli* cytoplasmic membranes and in synthetic lipid vesicles. *Proceedings of the National Academy of Sciences, USA*, **73**, 1159–63.

WOHLHIETER, J. A., GEMSKI, P. & BARON, L. S. (1975). Extensive segments of the *Escherichia coli* K12 chromosome in *Proteus mirabilis* diploids. *Molecular and General Genetics*, **139**, 93–101.

WU, H. C. & LIN, J. J.-C. (1976). *Escherichia coli* mutants altered in murein-lipoprotein. *Journal of Bacteriology*, **126**, 147–56.

YAMAMOTO, S. & LAMPEN, J. O. (1976). Membrane penicillinase of *Bacillus licheniformis* 749/C: sequence and possible repeated tetrapeptide structure of the phospholipopeptide region. *Proceedings of the National Academy of Sciences, USA*, **73**, 1457–61.

ZUKIN, R. S. & KOSHLAND, D. E. (1976), Mg^{2+}, Ca^{2+}-dependent adenosine triphosphatase as receptor for divalent cations in bacterial sensing. *Nature (London)*, **193**, 405–8.

STRUCTURE AND GROWTH OF THE WALLS OF GRAM-POSITIVE BACTERIA

HOWARD J. ROGERS, J. BARRIE WARD AND
IAN D. J. BURDETT

Division of Microbiology,
National Institute for Medical Research,
Mill Hill, London NW7 1AA, UK

INTRODUCTION

The study of events occurring during the growth of an individual bacterium and its eventual division to form two almost exactly similar cells has gathered considerable momentum over the last ten years. Early painstaking measurements, followed by highly sophisticated mathematical treatment of the results (Errington, Powell & Thompson, 1965; Harvey, Marr & Painter, 1967; and references cited therein) had established much about the degree of precision with which bacteria replicated both in their dimensions and in time. Representatives of a number of species had been examined but the growth conditions and techniques available for these early observations were limited. The introduction of the electronic particle counter, of better methods for synchronising cells and the development of high resolution electron microscopy together with the availability of suitable mutant strains, have all helped to open the field, even if they have each introduced their own problems and uncertainties.

Bacterial replication can be, and has been, approached from many different angles, all of which will eventually have to be integrated. Among them is a study of the growth and behaviour of the outer layers of the cell that constitute the so-called envelope. In producing two cells from an individual, these layers must approximately double in volume and surface area. Moreover, in order to form the septum either in Gram-positive species such as *Bacillus subtilis* (see Plate 1*a*), and *Streptococcus faecium* (see p. 155), or in Gram-negative forms such as *Escherichia coli* (Burdett & Murray, 1974), they must behave in a specific way so that the cytoplasm is divided into two equal biologially distinct units. It is on these aspects of the process of cell growth and division that this article will concentrate. Biosynthesis of specific wall polymers must be involved and since the later stages in these processes are catalysed by intrinsic membrane proteins, the organisation and biogenesis of the membrane is fundamental to the behaviour of the wall.

One legitimate aim of work on bacterial growth and division is to describe growth of the wall in molecular terms; that is in terms of the biosynthesis, and its regulation, of wall polymers of known structure and conformation, at defined topological sites on the membrane. To do this we need to know both the structures of the individual polymers and the nature of any linkages occurring between them. Currently there is little information available on conformational aspects to test the existing speculations. We also need, and have, some information about biosynthesis of the polymers, but know very little about regulation or about the surface topology of these processes except by indirect deductions. This article will not attempt to give detailed information about the structure of peptidoglycans and the other polymers present in the walls of Gram-positive species of bacteria, since these have been the subject of recent reviews (Rogers, 1974; Schleifer & Kandler, 1972; Archibald, 1973; Ghuysen & Shockman, 1973; Baddiley, 1972a, b). Likewise the biosynthesis of the peptidoglycans and some of the other polymers associated with them have been reviewed (Osborn, 1969; Ghuysen & Shockman, 1973; Baddiley, 1972a, b; Archibald, 1973; Braun & Hantke, 1974) in considerable detail. Most attention will be given to wall growth, and how far the structure and biosynthesis of wall polymers can be interpreted to explain the observations. Only the general ideas of structure and biosynthesis will be outlined.

MORPHOLOGICAL AND CHEMICAL DESCRIPTION OF THE ENVELOPE

The walls of Gram-positive bacteria, as seen either by examining sections of the micro-organism or shadowed replicas of freeze-etched material, appear as a thick, essentially structureless layer. In many species of exponentially growing bacteria the wall is about 30 nm thick (e.g. Plate 1) although in some organisms it may be much thicker, up to 80 nm (Glauert & Thornley, 1969). The thickness of the wall probably varies according to the growth rate and greatly increases in a number of species when protein synthesis is inhibited either by the addition of antibiotics such as chloramphenicol or by omitting essential amino acids (Shockman, 1965; Giesbrecht & Ruska, 1968; Hughes, Tanner & Stokes, 1970). When the bacteria are fixed and stained with osmium tetroxide, potassium permanganate or lead citrate, the walls of many species appear to be tribanded with darker inner and outer layers (Nermut, 1967; Glauert, Brieger & Allen, 1961; Higgins & Shockman, 1970b). Even when not stained at all but only fixed with glutaraldehyde, a difference in electron density between the inner and outer layers of the wall can be seen if high contrast conditions for photography are used (Weibull, 1973). The significance of this banded appearance has been discussed in terms of the distribution of the peptidoglycan and the polymers, particularly teichoic acids, associated with it (Nermut, 1967; Glauert & Thornley, 1969). Further evidence on this matter is, however, still desirable. Other more intricate structures within the wall have also been described (Hurst & Stubbs, 1969, Giesbrecht & Wecke, 1971) which are sometimes related, as in staphylococci, to division processes. It is, however, difficult in most instances to decide how far preparative procedures such as negative staining, have modified the appearance of the micro-organisms to create some of these structures.

Immediately underlying the wall and in sections of most preparations, appearing to be very closely applied to it, is the cytoplasmic membrane. Indeed it is usually difficult to distinguish the outer layer of the membrane from the inner layer of the wall. The cytoplasmic membrane itself shows no exceptional ultrastructural characteristics and the exact status of the mesosomes, which appear as infoldings of this membrane in sections of all Gram-positive bacteria examined after standard fixation and staining procedures or after negative staining, is questionable (Nanninga, 1971; Higgins, Tsien & Daneo-Moore, 1976). They are certainly not directly involved in wall formation (Reusch & Burger, 1973) but their relatively specific localisation (Burdett & Rogers, 1972, Higgins & Shockman, 1970a, b) suggests some difference about the membrane in the region of the organism concerned with septum formation. Thus they appear to mark the first evidence of the position at which septation and division of the cell will occur.

Chemically the walls of all the Gram-positive organisms examined consist of peptidoglycan together with a variety of other associated polymers, including the phosphorus containing teichoic acids, the commonest of which are the glycosylated-glycerol or ribitol phosphates, polymerised with either the sugars included in the chain or covalently attached to the polyol units. The proportion of peptidoglycan amounts usually to about 50% of the dry weight of wall preparations although in at least one example (the walls of *Micrococcus luteus*) it forms 80–90% of the weight. All the evidence to date supports the idea that the main structural and shape-maintaining polymer in the wall is peptidoglycan. However, recent work (see pp. 167–8) has drawn attention to the importance, in regulating the morphogenesis of the envelope, of the other acidic polymers attached to peptidoglycan, whether they are the teichoic acids or acidic polysaccharides such as the teichuronic acids.

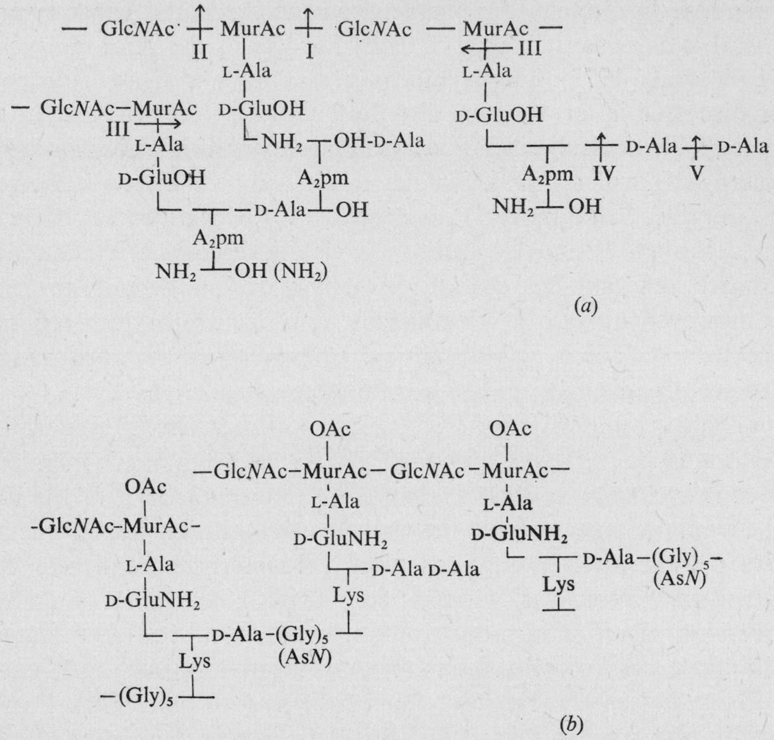

$$— \text{GlcNAc} \uparrow \text{MurAc} \uparrow \text{GlcNAc} — \text{MurAc} —$$

```
      — GlcNAc ┼ MurAc ┼ GlcNAc  — MurAc —
           II  |    I                ←┼III
              L-Ala                  L-Ala
  — GlcNAc—MurAc  D-GluOH            |
      III→                          D-GluOH
         L-Ala    └ NH₂ ┬OH-D-Ala      └ ┬    ┼ D-Ala ┼ D-Ala
      D-GluOH          A₂pm              A₂pm IV      V
        └ ┬ —D-Ala ┴ OH               NH₂ ┴ OH
        A₂pm
      NH₂ ┴ OH (NH₂)
```

(a)

```
                        OAc              OAc
                         |                |
            —GlcNAc—MurAc— GlcNAc ‑MurAc—
      OAc              L-Ala            L-Ala
       |                 |                |
 -GlcNAc-MurAc-       D-GluNH₂         D-GluNH₂
       |               └ ┬D-Ala D-Ala    └ ┬D-Ala—(Gly)₅—
      L-Ala              Lys              Lys    (AsN)
       |                                  ┴
    D-GluNH₂             ┴
       └ ┬D-Ala–(Gly)₅ ┴
        Lys              (AsN)
  –(Gly)₅ ┴
```

(b)

Fig. 1. Generalised structures of the peptidoglycans. (a) The Gram-negative bacteria and many bacilli. The D,D-carboxypeptidases in these organisms, hydrolysing bonds IV and V, frequently remove the terminal D-alanine residues. Other autolysins described in the text hydrolysed the bonds I, II and III. (b) From staphylococci and *S. faecalis*. In *S. aureus* the cross bridge is a peptide of five glycine residues and the *N*-acetylmuramic acid is *O*-acetylated. In *S. faecalis* cross-links are effected through a single asparagine residue.

Structure of the peptidoglycans

As the structures (Fig. 1) show, these polymers consist of glycan chains consisting of alternating units of glucosamine and muramic acid. In most organisms these are *N*-acetylated, although in the mycobacteria *N*-glycolylmuramic acid residues are present (Azuma *et al.*, 1970). In some strains of *B. cereus* the glucosamine appears to be largely unsubstituted (Araki, Nakatani, Nakayama & Ito, 1972) although in others this is unlikely to be so (Hughes, 1971). The glycan chains are linked together by short peptides containing L- and D-amino acids. The peptidoglycans of Gram-positive bacteria are distinguished from those of Gram-negative species by the involvement of a greater variety of amino acids and of different methods of linking the peptides together. Peptidoglycan from any one species contains no more than four or five amino acids but over the range of species examined, a much greater number have been found (Schleifer & Kandler, 1972). This variety of peptides has allowed their classification into five chemo-types (Ghuysen, 1968) although more complicated schemes are possible (Schleifer & Kandler, 1972). So far no clear correlations are to be seen between the primary structure of the peptidoglycans and the growth and morphogenesis of the micro-organisms. There are still considerable areas of uncertainty about peptidoglycan structure and arrangement including the size of the polymer as it exists in the wall, the distribution of uncross-linked peptides and the nature and extent of the intra- and intermolecular secondary valency linkages. It is of importance to gain knowledge in all three of these areas in order to understand the regulation of growth and morphogenesis of the envelope.

The size of the peptidoglycans

After the peptides have been cross-linked (see p. 144) functional $-NH_2$ and $-COOH$ groups remain free. In peptidoglycans containing α-2,6-diaminopimelic acid, for example, five carboxyl and one free amino group could theoretically remain after the formation of the cross-link between the sub-terminal D-alanyl residue of one pentapeptide and the amino group of diaminopimelic acid present in a second peptide chain. In the majority of organisms studied certain of the free carboxyl groups become amidated although the amino group remains free. Thus the potential for the formation of complex peptides exists through subsequent utilisation of these free amino groups. In bacilli this does not happen to any considerable extent whereas trimer, tetramer and higher oligomeric peptides are present in the peptidoglycans of *S. aureus*, *S. epidermidis* (Tipper & Berman, 1969) and *S. faecalis* (Deželée & Shockman, 1975). A few years ago the glycan chains were generally thought to be relatively short with the exception of those in the peptidoglycan from *Arthrobacter crystallopoietes* (see Rogers, 1970 for collection of data). Re-examination (Ward, 1973) of the problem, taking care to inactivate the potentially dangerous autolytic enzymes among which there are glycanases, has led to a dramatic upward revision of these figures. For example, whereas the average length of the chains in bacilli such as *B. subtilis* and *B. licheniformis* had been thought to be 13–14 disaccharides, re-examination showed them to be about 40–50. Significantly perhaps, they were even longer (70–80 disaccharides) in an autolytic enzyme deficient mutant. Fractionation of these heterogenous mixtures has shown them to contain glycan chains several hundred disaccharides in length (Fox, Ward & Sargent, 1977). The presence of such chains clearly affects the range of possibilities when considering the arrangement of the polymer in the walls of the living organism. *B. subtilis* is about 3 μm in length with a circumference of about 2 μm. Hence, the longest of the glycan chains biosynthesised (500 disaccharides with a length of 0.5 μm) would still not be long enough either to run from end to end or to wrap around the bacillus. Such a glycan, however, could not lie radially in the wall. Moreover, the average chain length of the glycans stayed constant with a three-fold variation in cell length brought about by altering the growth rate of the organisms (Fox *et al.*, 1977). Thus so far there is no evidence for the presence of peptidoglycan polymers large enough to cover the cell in the way envisaged by early models (Weidel & Pelzer, 1964). However, there is no difficulty in imagining ways in which the glycan strands may be joined together by peptides to form covalent layers over the cell surface (Rogers, 1974). Recent X-ray studies (Burge, Fowler & Reaveley, 1977) on oriented layers of peptidoglycan from the walls of a number of bacterial species support the idea that the glycan chains lie within the plane of the wall.

The distribution of cross-links

Although in the peptidoglycan from almost all species of bacteria (a notable exception is *M. luteus*) the muramic acid carboxyl groups in the glycan chains are all substituted by peptides, only a proportion of these peptides form cross-links with other contiguous peptides. The remainder, frequently losing some or all their terminal D-alanine residues (Hughes, 1970a, Weidel & Pelzer, 1964), presumably by the hydrolytic action of D-alanyl carboxypeptidases, (hydrolysing bonds IV and V, Fig. 1a) remain as uncross-linked tri- or tetra-peptides. The proportion and distribution of any regions containing such uncross-linked peptides could clearly modify the orientation and behaviour of the glycan chains compared with other chains constrained by cross-linkage. In many peptidoglycans about 30–40% of the peptides are engaged in cross-linkages but owing to ambiguities in the determinations of the proportions exact figures are not justified and to quote them would be misleading. Nothing is known about the distribution of the uncross-linked material and should there be any topological unevenness of distribution within the wall such knowledge would be very helpful in interpreting mechanisms for the processes of growth and division of bacterial cells.

The conformation of peptidoglycans

In order to understand more about the growth of the wall of a microorganism, an understanding of the conformation of peptidoglycans is becoming a matter of urgency. Some speculations have been published

(Tipper, 1970, Kelemen & Rogers, 1971; Braun, Gnirke, Henning & Rehn, 1973; Oldmixon, Glauser & Higgins, 1974; Formanek, Formanek & Wawra, 1974). Facts upon which to judge these suggestions are however accumulating only slowly. X-ray diffraction studies of properly oriented films of peptidoglycan is giving some information (Formanek *et al.*, 1974; Burge *et al.*, 1977) about the possibility or otherwise, of the chitin-like arrangement of the glycan strands, assumed in all the models; the most recent work (Burge *et al.*, 1977) would suggest that it is excluded. Infrared examination (Formanek *et al.*, 1976) showed the presence of considerable hydrogen bonding in peptidoglycans but gave no evidence for a β-conformation of the peptide as had been suggested on the basis of the formation of maximum hydrogen bonding to meet minimum free-energy requirements (Kelemen & Rogers, 1971). Density measurements made by suspending preparations in agents such as strong urea or guanidine–HCl solutions, also support the presence of some secondary valency forces, the weakening of these forces leading to an expansion of the structures and a reduction in density of the preparation (Ou & Marquis, 1972). These density measurements agree with those of Gerhardt & Judge (1964) but disagree with those of Formanek *et al* (1974) and have been used (Ou & Marquis, 1972) to argue that only a small part of the wall can be strictly ordered or 'crystalline'. However such disagreements and arguments may well be resolved by allowing for the degrees of hydration or expansion and contraction of the peptidoglycans that have been shown to be possible (Preusser, Heilman & Martin, 1969; Ou & Marquis, 1970). The observed alterations in the volume occupied by isolated walls and peptidoglycan that result from changes in the pH and ionic strength of the suspending fluid raise questions as to the meaning of determinations of absolute conformation and about the behaviour of peptidoglycan in growing bacteria. The flexibility of the peptide side chains for example, may allow them to adopt various conformations, some more compact than others, thus allowing for expansion and contraction during division. Such mobility may mean that in the living organisms, no one conformation is always present. Considerable change might result from the surface location of charges affected by both the pH and ionic strength of the culture medium. Similarly, proton flow through the underlying cytoplasmic membrane during uptake and excretion of metabolites might also be expected to exert some effects. The glycan chains are clearly more rigid but may be able to bend or move laterally to allow for the expansion of some or all of the peptide chains, during the events of the cell cycle.

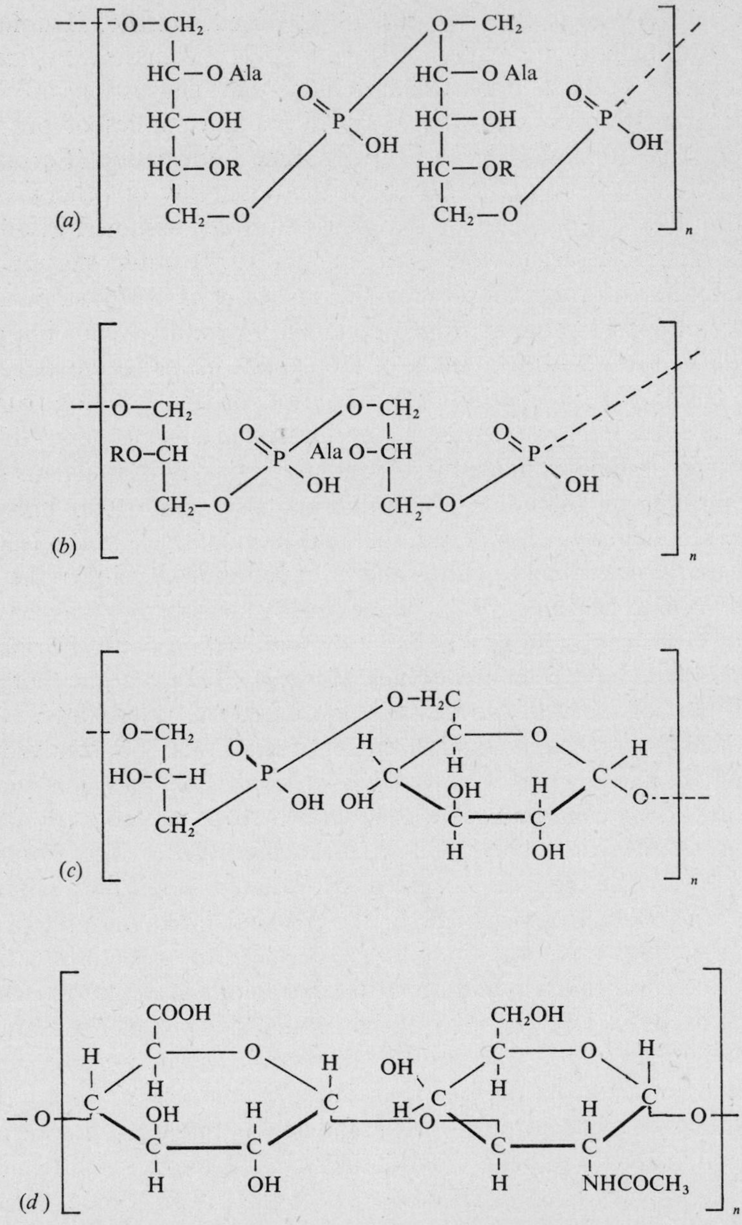

Fig. 2. The structures some polymers attached to peptidoglycan. (*a*) The ribitol teichoic acids. R = glycosyl e.g. α or β-linked *N*-acetylglucosaminyl in *S. aureus*, or β-glucosyl in *B. subtilis* strain W23. (*b*) The glycerol teichoic acids. R = glycosyl e.g. glucose in *B. subtilis* strain 168. (*c*) The glucosylglycerol phosphate teichoic acid found in the wall of *B. licheniformis* ATCC 9945. In other organisms galactosyl or disaccharide units may be present instead of the glycosyl residue. (*d*) A teichuronic acid isolated from *B. licheniformis* strain NCTC 6346.

Polymers attached to peptidoglycan

A complete list of the various teichoic acids, teichuronic acids and other polysaccharides found attached to peptidoglycan would be out of place in a review such as this; some examples and generalised structures are shown in Fig. 2. They appear to be attached by a single covalent bond, indirect evidence suggesting that this linkage is through a phosphodiester bond to the 6-OH group of N-acetylmuramic acid residues in the glycan chain. The details of the way this linkage is effected have provided a number of surprises and the story is not yet complete. Studies have been limited to the attachment of teichoic acids in bacilli and *S. aureus*. In *S. aureus* for example, which has a teichoic acid consisting of polyribitol phosphate substituted by either α- or β-linked N-acetylglucosamine residues, both biosynthetic and chemical evidence (Bracha & Glaser, 1976; Hancock & Baddiley, 1976; Heckels, Archibald & Baddiley, 1975) has established the presence of a glycerol phosphate trimer interposed between the teichoic acid and peptidoglycan. A similar situation exists in *B. subtilis* W23 where biosynthetic evidence suggests the presence of an additional molecule of N-acetylglucosamine in the linkage unit (Wyke & Ward, 1977*b*). In *B. licheniformis* the phosphate moiety on newly-synthesised muramic acid phosphate has been shown to be derived from UDP–N-acetylglucosamine and not CDP–glycerol, using cell-free preparations to catalyse the synthesis of linked peptidoglycan and polyglycerol phosphate teichoic acid (Wyke & Ward, 1977*a*). Chemical studies have recently established the presence of N-acetylglucosamine in the linkage unit of *S. aureus* (Coley, Archibald & Baddiley, 1977). Thus, N-acetylglucosamine phosphate is involved in the linkage of both polyribitol and polyglycerol phosphate teichoic acids. It is interesting to note that in the former case a glycerol phosphate trimer is involved as well as an amino sugar which may suggest the participation of a common mechanism for linking the polymers together.

When compared with the glycan chains both the teichoic and teichuronic acids are of a relatively small size. Average chain lengths of 20–30 residues have been reported for teichoic acids of *S. aureus*, *S. lactis* (Archibald, Baddiley & Heckels, 1973) and various bacilli and of about 40 residues for teichuronic acids from *B. licheniformis* (Hughes, 1970*b*) and *B. subtilis* W23 (Wright & Heckels, 1975). Three properties of all these polymers would seem to be relevant from the point of view of the present rather general discussion. First they are all hydrophilic, flexible, linear molecules; secondly they are negatively charged, although the teichoic acids may be partially substituted with ester-linked D-alanine, the free amino group of which will impart certain amphoteric properties. Finally as described above, they all appear to be linked to peptido-glycan by a single terminal covalent bond. This must give them con-siderable freedom to adopt a variety of conformations which will be restrained only by the three-dimensional structure of the peptidoglycan, where linkage occurs in the depth of the wall. Thus by analogy, one has attached to the peptidoglycan a series of negatively charged flexible bristles and when these negative charges are expressed the molecules will tend to be extended by mutual repulsion. If however, the charges are neutralised either by lower pH or in the presence of increased salt concentrations, the bristles will tend to collapse. Such reversible exten-sion and collapse of isolated teichoic acid molecules in salt solutions of varying concentration has been described (Doyle *et al.*, 1974). Since these polymers are concerned with holding Mg^{2+} in the walls of Gram-positive bacteria (Heckels, Lambert & Baddiley, 1977) and possibly with the supply of divalent cations to the membrane (Hughes, Hansock, & Baddiley, 1973) this behaviour may be relevant to their functions.

BIOSYNTHESIS OF WALL POLYMERS

To understand how the wall of a bacterium grows, we clearly need information on the biosynthesis of the individual polymers and their linkage together. Our knowledge about the biosynthesis of both peptido-glycan and certain of the accessory polymers, summarised in Fig. 3, is by now reasonably good. The major areas of uncertainty concern the mechanisms whereby the accessory polymers become linked to the peptidoglycan and the exact steps by which new peptidoglycan chains grow and are attached to the existing wall. The former of these problems, studied only for linkage of teichoic acids, is discussed briefly in the preceeding section and will not be considered further. However, the growth and attachment of newly synthesised peptidoglycan is so relevant to wall growth that it justifies a detailed examination of the evidence. Two basic methods for the accretion of new wall are possible. New disaccharide peptide units might be transferred from the isoprenoid intermediate to existing peptidoglycan by either glycosylation or trans-peptidation. These processes would occur by addition to acceptor groups in the wall. Such groups may be either the ends of the existing glycan chains or if the addition is by transpeptidation, the free NH_2 groups or D-alanyl-D-alanine termini in the peptide side chains of the existing peptidoglycan. Until recently the idea of accretion of wall by

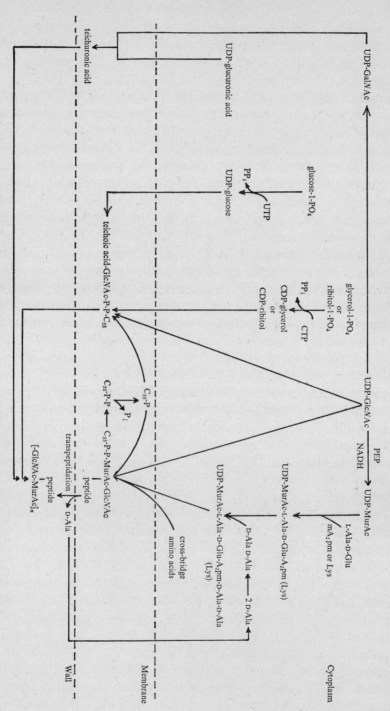

Fig. 3. A generalised pathway for the biosynthesis of both peptidoglycan and other accessory polymers.

the transglycosylation of newly synthesised disaccharide peptide units was generally assumed to be the most likely. However other results to be discussed below have now cast some doubt on the validity of this assumption. Before considering these, some attention must be paid to the nature of the various in-vitro preparations used to study peptido-glycan biosynthesis. Membrane preparations from several species of Gram-positive organisms, exceptions being *B. megaterium* (Reynolds, 1971*a*, *b*; Wickus & Strominger, 1972*a*, *b*), *B. stearothermophilus* (Linnett & Strominger, 1974) and *Sporosarcina ureae* (Linnett, Roberts & Strominger, 1974), catalyse polymerisation reactions but not the formation of cross-links, i.e. they make uncross-linked peptido-glycan. Recently preparations consisting of crude walls with membrane still attached (Mirelman & Sharon, 1972, Mirelman, Bracha & Sharon, 1972; Ward, 1974; Ward & Perkins, 1974; Hammes & Kandler, 1976; Weston, Ward & Perkins, 1977) have been shown both to polymerise material and to catalyse transpeptidation reactions to form cross-linked material. Both types of preparation have been used to study the mechanism of the extension of glycan chains. Using membranes from *B. licheniformis* (Ward & Perkins, 1973) and wall+membranes from *M. luteus* (Weston *et al.*, 1977) it has been shown that glycan chains are extended by addition of new units to their reducing ends. Moreover, the chains remain attached, by a linkage labile to mild acid hydrolysis, to a molecule presumed to be the isoprenoid lipid intermediate. Thus at least in cell-free preparations, peptidoglycan synthesis is analogous to the synthesis of the lipopolysaccharides found in the Gram-negative bacterial wall (Osborn, 1969; Nikaido, 1973) rather than to the synthesis of many other polysaccharides which proceed from the non-reducing terminus.

The importance of the transpeptidation reaction in the linkage of newly synthesised peptidoglycan to existing wall was established by the finding that attachment was inhibited by low concentrations of β-lactam antibiotics (Mirelman & Sharon, 1972; Mirelman *et al.*, 1972; Ward, 1974). The direction of the transpeptidation reaction was then investigated (Ward & Perkins, 1974) using wall+membrane preparations from *B. licheniformis*. To prevent newly synthesised material acting as the acceptor in transpeptidation, the free NH_2 group in the precursor UDP-MurAc-L-Ala-D-isoGlu-meso-A_2pm-D-Ala-D-Ala was blocked by acetylation with [^{14}C]acetic anhydride. Incorporation of radioactivity from the modified precursor into the wall and the isolation of [^{14}C]acetyl-labelled cross-linked peptides, established that the newly synthesised units acted as the donor peptide in the transpeptidation reaction. Other work with *S. faecalis* (Dezélée & Shockman, 1975) and with soluble transpeptidase-D, D-carboxypeptidase preparations from *Streptomyces* R61 (Frère, Ghuysen, Zeiger & Perkins, 1976) supports the idea that nascent peptide acts as the donor and the existing peptide as the acceptor, as in the bacillus system. Recently however, evidence for an inversion of the donor role of the precusor in transpeptidation has emerged. In *Gaffkya homari* (Hammes & Kandler, 1976; Hammes, 1976) the modified precursor UDP-MurAc-L-Ala-D-isoGlu-L-Lys-D-Ala is utilised as effectively as the normal pentapeptide precursor for the in-vitro synthesis of peptidoglycan. The absence of the terminal D-alanyl-D-alanine dipeptide clearly renders it incapable of acting as the donor peptide.

As described above, evidence that transpeptidation plays a major role in attaching newly synthesised material to the wall is provided by the effectiveness with which penicillins stop the process. Penicillin-resistant incorporation amounting to 25–30% of the uninhibited value, has however been found in wall+membrane preparations from *M. luteus* (Mirelman *et al.*, 1972; Weston *et al.*, 1977). This, presumably, is due to incorporation by transglycosylation rather than transpeptidation and appears to be specific to *M. luteus* among the organisms studied.

If the results for the growth of peptidoglycan in cell-free preparations can be applied to that in the growing organism, then the picture for peptidoglycan synthesis is fascinating. It suggests that uncross-linked peptidoglycan while still attached to the membrane may be 'zipped' into the existing wall by transpeptidation. One indication that wall synthesis by whole organisms is similar to that found in isolated membrane preparations is the finding that in the presence of penicillins uncross-linked peptidoglycan is secreted by the whole cells into the culture medium (Mirelman, Bracha & Sharon, 1974; Tynecka & Ward, 1975). Moreover, the average length of such glycan chains is also similar to that found in the wall (Tynecka & Ward, 1975). The mechanism of release of these glycan chains from their anchorage in the membrane is unknown and the possibility remains that it may be related to the method used to terminate growth of a particular glycan chain in the growing organism.

WALL 'TURNOVER'

The hydrolysis and renewal of a polymer such as peptidoglycan obviously involves both the biosynthetic process and the hydrolytic autolysins. However, the concept of 'turnover' of a two- or three-dimension-

ally organised exo-cellular polymer like peptidoglycan cannot be as
straightforward as that for the turnover of small molecules within the
cell, such as the peptidoglycan precursors. It is, however, an important
process involved in the interpretation of both the detailed studies of
wall growth and of the general nature of wall accretion. 'Turnover' of
peptidoglycan has been measured by first incorporating components
such as [^{3}H]-2,6-diaminopimelic acid, N-acetyl-[^{14}C]glucosamine or
[^{14}C]lysine under conditions such that the wall is more or less specifically
labelled and then following either the rate of decay of the radioactivity
in the peptidoglycan when the labelled compound in the medium is
replaced by unlabelled material, or looking for the release of soluble
labelled wall fragments into the culture supernatant. A second perhaps
more debatable method is to follow the rate of replacement of peptidoglycan and the accessory polymers in the walls of organisms when
the limiting nutrient is changed in chemostat cultures. Cultures of
bacilli for example, switched from K^+ or Mg^{2+} to PO_4^{3-} limitation lose
their wall teichoic acids which are replaced by teichuronic acids (Ellwood
& Tempest, 1969; Forsberg, Wyrick, Ward & Rogers, 1973; Mauck &
Glaser, 1972; Archibald & Coapes, 1976). By comparing the theoretical
dilution rate in the chemostat with the observed rate of change, the rate
of 'turnover' of the wall can be calculated. This method can be extended
by labelling the peptidoglycan differentially with say a ^{14}C-labelled
compound during Mg^{2+} or K^+ limitation and a ^{3}H-labelled compound
during PO_4^{3-} limitation. Examination of the peptidoglycan fragments
attached to either the teichuronic acid or teichoic acids, then allows
some deduction as to what is happening to the peptidoglycan during
'turnover' under these particular and perhaps exceptional circumstances (Mauck & Glaser, 1972).

Studies on streptococci and various bacilli have shown that whereas
the peptidoglycan and hence the walls of the bacilli are subject to 'turnover' (Chaloupka, 1967; Mauck, Chan & Glaser, 1971; Mauck &
Glaser, 1972; Boothby et al., 1973; Pooley, 1976a, b; Glaser & Lindsay,
1977) those of the streptococci are not (Boothby et al., 1973). Furthermore 'turnover' occurs equally well in B. subtilis with two autolysins
(hydrolysing bonds II and III in Fig. 1a) as in Lactobacillus acidophilus
with only a single autolysin of lysozyme-like specificity (hydrolysing
bond I, Fig. 1a) (Coyette & Ghuysen, 1970). The specificity of this
latter enzyme is the same as the autolysin of S. faecalis which does not
show 'turnover' (Boothby et al., 1973). Thus the number and nature
of the wall-hydrolysing enzymes does not appear to be of particular
importance.

Some of the most interesting studies of wall 'turnover' were those of Pooley (1976a, b). By following the release of soluble peptidoglycan fragments from steady-state labelled wall of *B. subtilis*, he determined that 'turnover' occurred at a rate of 8–10% per generation, a figure considerably less than most previous estimates. He then studied the kinetic behaviour of a pulse of *N*-acetyl-[^{14}C]glucosamine as the bacilli were grown at a steady exponential rate in the presence of unlabelled *N*-acetylglucosamine. The observations made resolved the paradox of the apparent long lag (Mauck *et al.*, 1971) which occurred before release of radioactivity could be demonstrated from whole organisms. Apparently, the new peptidoglycan, synthesised at the wall-membrane interface, moves outward during growth eventually to form a thin sheet on the surfaces of new bacilli. Only then is it removed in a soluble form from the walls. Approximately two generations are required for the radioactivity to traverse the wall during which time the original bacillus has become four daughter cells. Hence the kinetics suggest a spreading of the labelled peptidoglycan before it is lost from the surface.

Most interesting comparative results have been provided in a study of the binding of bacteriophage SP50 to *B. subtilis* W23 (Archibald & Coapes, 1976). The absorption of this phage depends upon the presence of teichoic acid on the surface of the wall. When the absorption of phage was studied during loss of teichoic acid, by growth of the organisms under phosphate limitation, phage binding sites only disappeared when the amount of wall teichoic acid had dropped to 20% or less of that originally present. Likewise in passing from PO_4^{3-} to K^+ limitation, when the walls were regaining teichoic acid, the reverse situation occurred. The phage binding sites returned very much more slowly than the wall teichoic acid increased. The results of these and Pooley's (1976a, b) experiments are entirely consistent with the idea that new wall, being fed out from the membrane, forms spreading layers that only reach the surface after 1–2 generations of growth. If this interpretation is correct then such a mechanism of wall growth will necessarily have profound effects upon the conclusions that can be drawn from methods designed to study the sites of cell extension, particularly those depending on reactions with superficial wall layers. Clearly the results obtained at the cell surface may not reflect the areas on the membrane at which synthesis of the peptidoglycan had occurred.

More recently reservations have been placed on the idea that loss of wall during turnover occurs only from the exterior surface (Glaser & Lindsay, 1977). In *B. subtilis* turnover was found to be inhibited under conditions where wall thickening was taking place (i.e. in the presence of chloramphenicol or in the absence of protein amino acids) but not by pretreatment of the walls with concanavalin A. This lectin binds to glucosyl-substituents of the teichoic acid presumably at the surface of the wall. Earlier Herbold and Glaser (1975*b*) had shown that the presence of teichoic acid was necessary for tight and effective binding of the *N*-acetylmuramyl-L-alanine amidase to isolated walls. Thus for turnover to occur at the surface the teichoic acid with lectin already bound would have to be capable of simultaneously binding the amidase. If this is possible then the observations of Glaser and Lindsay (1977) would not conflict with those of Pooley (1976*a, b*) and Archibald and Coapes (1976). This represents an area where further evidence is required before a definite conclusion can be drawn.

TOPOGRAPHY OF THE GROWTH OF CELL WALLS

We now turn from biochemical considerations to the morphological aspects of wall growth, particularly to the problems involved in the pattern of surface enlargement and the replication of cell shape. The most detailed descriptions of both these aspects of envelope growth have been obtained from studies of the genus *Streptococcus* (Shockman, Daneo-Moore & Higgins, 1974). More recently, a technique has been devised from which a cycle of pole formation in this organism may be reconstructed from measurements taken from electron micrographs, a technique that has also been applied to bacilli, and holds promise for further advances in our knowledge of the growth of Gram-positive rod-shaped species in which the pattern of surface growth has been particularly difficult to establish. This section will describe the application of quantitative methods to problems of wall growth in streptococci and bacilli.

Sites of wall growth in streptococcus

Cole and Hahn (1962) used a fluorescent antibody technique to show that new wall appeared to be incorporated solely at sites of division in *S. haemolyticus*. Similar results were obtained by applying this technique to *S. pneumoniae* (Wagner, 1964) and also by radioactively labelling the teichoic acid in the walls of this organism and following the segregation of label by autoradiography during subsequent growth (Tomasz, Westphal, Briles & Fletcher, 1975).

The impression of a single, central growth-zone in streptococci has been confirmed and considerably extended through studies of naturally

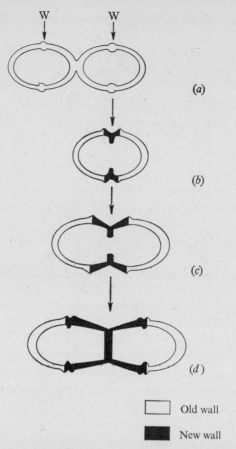

Fig. 4. Diagram of a cycle of wall growth in *Streptococcus faecalis* (based on Higgins and Shockman, 1970*b*). (*a*) The potential site for enlargement of the cell surface is located under a raised wall band (W) at the coccal equator; the cycle of growth (*b–d*) is followed in one coccus. (*b–d*) Following splitting of the wall band, interpreted as a peeling apart of the cross-wall at its base, new peripheral wall is bilaterally issued from the septal region. At cross-wall closure (*d*), new wall enlargement sites are initiated at the centre of daughter cocci.

occurring surface markers detected by electron microscopy. The markers (wall bands) are visible in negatively stained walls (Shockman & Martin, 1968) and in thin sections of *S. faecalis* (Higgins & Shockman, 1970*a*, *b*) and appear to define the junction between 'old' surface and newly synthesised wall material. Using wall bands to delimit the surface in this way, Higgins and Shockman (1970*b*) proposed a model for the growth of the wall in *S. faecalis* (Fig. 4). New peripheral wall was assumed to be generated by the peeling apart of the cross-wall at its base, thereby bilaterally displacing the wall bands during enlargement of the polar surfaces. Newly synthesised wall was presumed to be incorporated exclusively at the leading edges of the cross-wall.

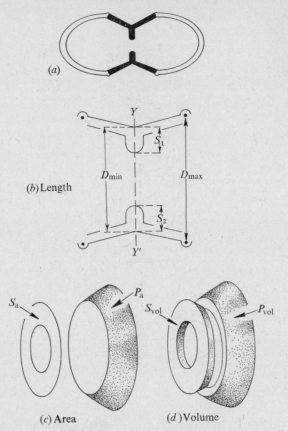

Fig. 5. Diagram to illustrate the principle of the rotation technique (see text) applied to the 'growth zone'. (a) Cell profile in longitudinal section; newly synthesised wall, delimited by wall bands, is shown in black. (b) Section through 'growth zone'; S_1, S_2, length of cross-wall; D_{min}, width between septum bases; D_{max}, cell diameter from centres (●) of wall bands. By mathematical rotation of the profile in (b), areas and volumes are obtained. For re-construction purposes, measurements are made in each half of the growth zone, i.e. in the halves bisected by the line YY' passing through the bases of the cross-wall. (c) S_a, area of a plane passing through the centre of the cross-wall (at YY'); P_a, surface area of the nascent pole from base of the cross-wall to the wall bands. (d) S_{vol}, the volume of the septal annulus (on one side of the cell); P_{vol}, volume of wall contained in the nascent pole. Additional measurements of wall thickness and curvature are also made (see Fig. 6).

Quantitative studies of pole formation in streptococci

The essentially qualitative model for the growth of *S. faecalis* has been refined by means of an analogue rotation technique (Higgins, 1976; Higgins & Shockman, 1976). A series of lines and trapezoids are fitted to electron micrographs of central, longitudinal sections of the bacteria; assuming the cells to be circular in cross-section, areas and volumes are obtained by mathematical rotation of the image. Whole-cell reconstructions are thus obtained from rotation of axial sections. The principal

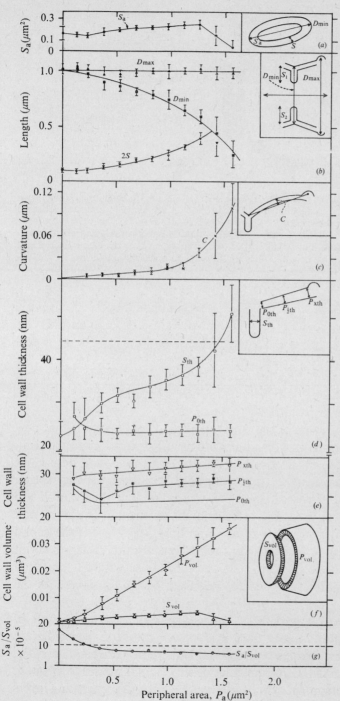

Fig. 6. Reconstruction of the geometry of a nascent pole of *S. faecalis* during a round of envelope synthesis. All parameters are plotted as functions of P_a (the surface area of a nascent pole); the points represent the mean ± one standard deviation. The diagrams show where measurements are taken. (*a*) The area of the septal annulus (S_a) remains at a plateau level until two thirds of the cycle and cross-wall closure occurs at P_a of 1.3 μm^2; (*b*) D_{min} width of the outer diameter of cross-wall; D_{max}, diameter at wall bands; S_1, S_2, length of cross-wall. D_{min} is gradually reduced during a cycle and the intersection of S_1+S_2 (i.e. $2S$) with D_{min} indicates the point of cross-wall closure; (*c*) estimates of curvature of nascent peripheral wall; (*d, e*) wall thickness taken at the cross-wall (S_{th}) and at 3 locations (P_{0th}, $P_{1/2th}$, P_{xth}) along the gradient of thickness at the separating layers of peripheral wall. Note that P_{0th} remains approximately constant whereas S_{th} increases continuously; (*f*) S_{vol} (the volume of the transverse half of the cross-wall) remains relatively constant and P_{vol} (the peripheral wall volume) increases linearly with respect to P_a. Reductions in D_{min} are thus met by increases in $2S$ and in S_{th}, maintaining S_{vol} at a constant level. From Higgins and Shockman (1976).

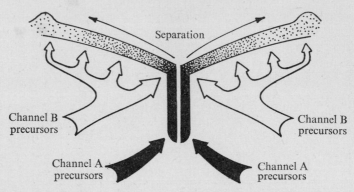

Fig. 7. Cross-wall separation model for *S. faecalis* (from Higgins & Shockman, 1976). Two channels of newly-synthesised wall material are proposed to regulate wall growth. Channel A would be involved in the synthesis of the cross-wall, and would be applied continuously until septal closure. Channel B precursors would bring the separating layers of the cross-wall to a constant thickness and would be used to thicken and expand the nascent peripheral wall. The figure is diagrammatic and does not attempt to precisely locate sites of wall incorporation.

features measured are shown in Fig. 5. To simulate a cycle of envelope growth, the surface areas of the nascent poles are used as an index of developmental age. From the interrelationships between the measurements, a rather detailed model for pole construction has emerged. We shall not attempt to discuss all aspects of the model but rather try to emphasise only those basic features which appear diagnostic of the streptococcal growth pattern.

A fundamental feature of the streptoccal mode of growth is that new polar surface is formed as a result of a cross-wall separation mechanism. Thus, following the start of pole formation, there is a gradual but continuous decrease in D_{min} (the width of the cell between the septum bases, see Fig. 5). This is equivalent to saying that thin annuli are removed successively from the exterior diameter around the cross-wall as pole formation proceeds. At the same time, the area of the septal annulus (S_a, Fig. 6) is maintained at a nearly constant value for about two thirds of the cycle. The implication of this observation is that wall precursors are continuously fed to the edges of the cross-wall as the poles enlarge.

From geometrical considerations, it is apparent that conversion of the cross-wall (a flat annular disc) into two oblate hemispheroids (the nascent poles) must involve a considerable increase in area. Therefore, Higgins & Shockman (1976) amended their original model to propose that wall precursors might also be intercalated into the separating layers of polar wall (Fig. 7).

In *S. faecalis*, the septal annulus closes relatively late in the cycle and at this stage there is an apparent decrease in S_a as the cells approach the point of separation (Fig. 6). Cross-wall closure was determined by finding the intersection of the line for the sum of the septal lengths (i.e. $2S = S_1 + S_2$, Fig. 6) with the curve for D_{min}.

Growth of the wall in B. subtilis

A bacillus may be regarded as a cylinder capped by hemispherical poles. The work done in attempts to examine the topology of growth of bacilli has all shown that the poles differ in some way from the wall in the cylinder. Studies of wall thickening and the subsequent formation of wall of normal thickness (Frehel, Beaufils & Ryter, 1971; Boothby *et al.*, 1973), bacteriophage fixation during changes in nutritional limitation (Archibald & Coapes, 1976) and studies of partial autolysis of wall (Fan & Beckman, 1973; Fan, Pelvit & Cunningham, 1972) all suggest that the cylindrical wall grows by diffuse intercalation of new material, that no part of it is conserved and that it is more susceptible to autolytic breakdown whereas the poles (presumably completed poles) appear to be conserved and are relatively resistant to autolysis.

It has been possible to apply the analogue rotation technique described above to the formation of cell poles in *B. subtilis* (Burdett & Higgins, 1977) because as in *S. faecalis*, surface markers have also been found. These markers appear different from those of streptococci in the sense that they do not apparently originate from a raised wall band. Instead, the markers appear rather as edges of a tear in an outer layer at the base of the cross-wall (Plate 1, Figs. 2–4). Their presence may indicate some layered infra-structure of the wall in this region. In high resolution scanning micrographs of *B. subtilis*, the bands are visible, running around the cell (Amako & Umeda, 1977). Markers of similar appearance have also been reported in sections of *B. cereus* (Chung, 1973) and perhaps also in *A. crystallopoietes* (Krulwich & Pate, 1971). The wall bands have been used to delimit the septal zone from the adjacent cylindrical portions of the cell (Burdett & Higgins, 1977). In the following summary, it must be remembered that the events described concern only the wall assembled at the division site, amounting to 10–20% of the total volume of wall. A further site(s) must presumably exist elsewhere for the extension of the cylindrical wall (see below).

In *B. subtilis* as in *S. faecalis*, the cycle of pole construction also appears to proceed by a mechanism involving separation of the cross-wall. The width of the division furrow (D_{min}, Fig. 5) decreases during pole formation and septal closure occurs relatively earlier than in *S*.

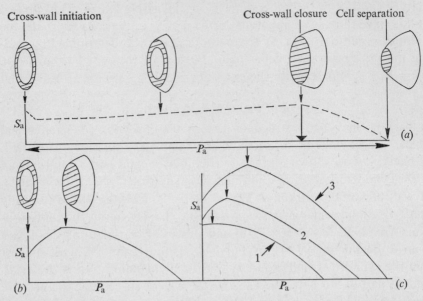

Fig. 8. Relationships between the area of the septal annulus (S_a, see Fig. 5) and the surface area (P_a) of the nascent pole in *S. faecalis* (*a*) and *B. subtilis* (*b*, *c*). (*a*) In *S. faecalis*, a cycle of envelope growth is described by the formation of a cell pole. For two thirds of the cycle S_a is maintained at a plateau level until cross-wall closure (arrow), which occurs at P_a of about 1.3 μm^2 (see Fig. 6). (*b*) Pole construction in *B. subtilis* (not to same scale as (*a*), illustrating the curvilinear relationship between S_a and P_a. Septal closure (arrow) occurs relatively much earlier in the cycle of pole formation than in *S. faecalis*. (*c*) Hypothetical plots (curves 1–3) of S_a and P_a in a bacillus-like organism. The shape of the relationship would be influenced by the stage of cross-wall closure (arrows), which progressively occur later in the cycle (curves 2, 3).

faecalis. From the regression line used to determine the curve for D_{min}, it is possible to calculate the amount of expansion in area involved in the conversion of the cross-wall into polar wall. Initially the wall expands four-fold but this value is progressively reduced during the separation process.

The relation between S_a and P_a in *B. subtilis* is quite different from that in *S. faecalis*. In *B. subtilis* this relationship is typically curvilinear (Fig. 8). Prior to the start of separation, there occurs a considerable increase in S_a, presumably by centripetal addition of wall to the leading edge of the cross-wall. By this interpretation, wall precursors are applied in a continuous but decreasing amount once the cross-wall has closed.

A major dimension of interest in the comparison of streptococci and bacilli is added by preliminary results from the application of the rotation technique to *rod* mutants of *B. subtilis* (I. D. J. Burdett, P. F. Thurman & H. J. Rogers, unpublished results). These are conditional morphological mutants that grow either as rods or as coccal-shaped cells (Rogers, McConnell & Burdett, 1971). Mutants of the *rod* B class have been examined (Karamata, McConnell & Rogers, 1972; Rogers, Thurman & Buxton, 1976; Rogers & Thurman, 1977).

The principal change which takes place on a shift-up from 20 °C to 42 °C is an increase in cell volume. Most of this increase occurs rather specifically as a result of an increase in width at the centre of the cells. After 45–60 min at 42 °C, the average length of the cells remains relatively constant whereas the width is nearly doubled in comparison to the diameter of the rods at 20 °C.

Because wall bands of the type described above are present (Plate 2*a–c*), we have been able to analyse the sequence of pole construction. In these organisms (unlike streptococci) the septum is closed before separation and new polar surface is probably produced by separation and subsequent expansion of the cross-wall. Thus, in the plot of S_a and P_a, the relationship is of a curvilinear type. The polar surface area is about four-fold greater in the cocci than in the rods grown at 20 °C. Moreover, in the rods the volume of wall assembled at the septum amounts to about 15% of the total wall volume, whereas it is increased to about 44% in the cocci.

One hypothesis to explain these results is that there exists a labile or repressible extension site which is gradually inactivated on incubation of the cells at the restrictive temperature. Thus, as the cells cease to extend in length, and increase in diameter, a greater volume of wall is assembled at the septal site. As the cells adopt a coccus-like morphology, the septal growth zone becomes the primary means of producing new cell surface.

Evaluation of growth models

Formation of a cell pole both in *S. faecalis* and *B. subtilis* appears to depend upon cleavage of the cross-wall and subsequent expansion in area of the separating layers. The separation of cells is likely to be closely regulated by the action of autolytic enzymes. For example, in autolysin-deficient mutants of *B. licheniformis* (Forsberg & Rogers, 1974) and of *B. subtilis* (Fein & Rogers, 1976), the cross-walls are completed but fail to split. Both expansion and separation are also inhibited in cultures of *S. faecalis* in which autolysin content has been reduced by inhibiting protein synthesis either by threonine starvation (Higgins, Pooley & Shockman, 1971) or by the addition of specific antibiotics (Higgins, Daneo-Moore, Boothby & Shockman, 1974).

The mechanism whereby the cross-wall expands in area upon separation is open to speculation. Conceivably, this might occur by addition of wall at specific sites or by intercalation over the surface, involving a combination of biosynthetic and autolytic processes (Higgins & Shockman, 1976; Burdett & Higgins, 1977). Alternatively, modification of the surface could also occur by remodelling and movement of wall after incorporation. These various possibilities illustrate the difficulty of identifying sites of wall growth. Most of the expansion probably occurs as the cross-wall is separating (Higgins & Shockman, 1976; Burdett & Higgins, 1977). That secondary expansions may also occur after cell separation is suggested by measurements of axial diameters of nascent and mature poles of *S. faecalis* (Higgins & Shockman, 1976). The length of the separated, mature poles appears to increase slightly without change of diameter. In an organism in which there is no detectable turnover of peptidoglycan, secondary expansions of the surface may be a means of modifying the shape of the cell. Where wall turnover is extensive, as in *B. subtilis*, such modifications may be considerable and are presumably responsible for the adoption of the increasingly hemispherical shape of the mature pole.

Both these aspects of growth (i.e. separation and expansion) can be used to examine possible models of surface growth. Consider first the formation of new surface by separation at the site of a cross-wall. The relationship between S_a and P_a (Fig. 8) is perhaps the most sensitive indicator of the mode of surface enlargement. It is likely to be considerably influenced by the stage of septal closure and by the amount of expansion occurring in the polar wall. In *S. faecalis*, where cross-wall closure occurs relatively late in the cycle, S_a is maintained at a plateau level until the septum closes (Fig. 8a). Thus, for every increase in S_a (and reduction in D_{min}), a proportionately greater amount of polar surface area (P_a) is produced by cell separation and expansion; this relation is also true for declining values of S_a (at $P_a > 1.3 \mu m^2$; Higgins & Shockman, 1976). If cross-wall closure occurred earlier, so that cross-wall assembly was favoured, the graph might be expected to show a steeper gradient of increase in S_a.

When the septum closes prior to cell separation (as in *B. subtilis*), S_a is related to P_a in a curvilinear manner (Fig. 8b). This type of relationship may be regarded as a modification of the terminal segment (at $P_a > 1.3 \mu m^2$) of the curve for S_a and P_a in *S. faecalis* (Fig. 8a). Thus, in the *rod* B mutants of *B. subtilis* (see p. 160) nearly as much polar surface area may be generated in this manner as by a streptococcal mechanism. Both mechanisms have the common property that the amount of expansion undergone by the polar wall as the septum is cleaved, is greater than the quantity of wall removed by cell separation.

Clearly, the curvilinear relationship between S_a and P_a is specifically related to pole construction, once the cells have entered the sequence of cross-wall development. Variations in the timing of septal closure and the amount of polar surface produced by a cell of given diameter would affect the shape of the curve (Fig. 8c).

If pole construction depends upon a separation mechanism, what of the means for extending the cylindrical portion of the cell in rod-shaped organisms? As yet there is no evidence to suggest that the cylinder in growing organisms is in any way morphologically differentiated, but the apparent uniformity of the surface may be illusory and the problem requires further analysis. Both in *E. coli* (Marr, Harvey & Trentini, 1966) and the *rod* B mutants of *B. subtilis* grown at the permissive temperature (I. D. J. Burdett, P. F. Thurman & H. J. Rogers, unpublished results) regression analysis of cell diameter upon length shows the line to have zero slope. Therefore, to a first approximation, we may assume that the width does not increase as the cells grow in length. Using this observation, and assuming it were possible to delimit the cylindrical growth zone, a sequence linking extension of the cylinder to the development of the cell poles might be as shown in Fig. 9. In this hypothetical scheme, there is no apparent reduction in D_{min}^* or in S_a^* (see Fig. 9 for explanation of symbols) until septum formation. One possible explanation is that the mode of cylindrical wall extension is fundamentally different from pole construction. We have noted (p. 159) that previous attempts to localise growth zones in bacilli have resulted in the impression that wall is intercalated at many sites or perhaps the surface is enlarged by the sequential addition 'spreading' of layers (Pooley, 1976a, b). This might result in maintaining the apparently constant values for D_{min}^* and S_a^*. Extension of the cylinder might occur by an 'intercalatory' mechanism rather than by a process involving the separation of wall at a growth site (as in streptococci). The common feature would be the resulting expansion of the assembled surfaces. A separation-based mechanism may simply be an adaptation to handling wall placed under different geometric and physical constraints (for example, the amount of stress).

Our hypothetical scheme (Fig. 9) invokes the idea, previously discussed by Sargent (1975a), that the process of septation may be regarded as the culmination of a cycle of growth terminating in a specific developmental sequence. This proposal is clearly favoured by the process of surface growth apparent in streptocci. Thus, the hypothetical scheme bears close resemblance to the sequence outlined for *S. faecalis*, with the exception that only the stage for pole formation would involve a

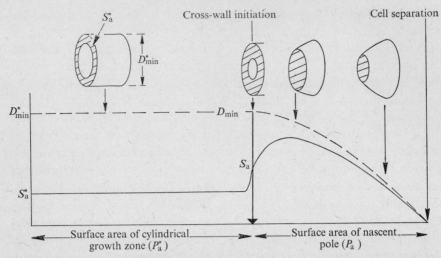

Fig. 9. Hypothetical growth sequence in *B. subtilis*, illustrating a possible relationship between synthesis of cylindrical wall and the formation of the cross-wall. During the cell cycle the cylindrical 'growth zone' is assumed to assemble a given surface area (P_a^*) of new wall. The location of the growth site is unspecified, but it is assumed that cell extension occurs by an 'intercalatory' mechanism (see text) and not by separation process. Therefore, S_a^* (the area of wall formed by an annulus in a plane passing through the cylinder) and D_{min}^* (the outer diameter of the cell) are assumed to remain constant. Cross-wall formation (arrow) would occur by centripetal synthesis of wall (an increase of S_a) at the centre of the cylinder. Nascent poles would be produced by separation of the cross-wall at its base, resulting in a decrease in D_{min}. The curvilinear relationship between S_a and P_a (the nascent pole surface area) would describe the terminal events of the cycle, involving closure and subsequent conversion of the cross-wall into two layers of polar wall.

separation mechanism. In the morphological mutants when they are undergoing their change in shape, the expected growth pattern, could it be reconstructed in its entirety, might appear as shown in Fig. 10. As the amount of surface generated by cross-wall separation increases, the curvilinear portion of the graph would be shifted towards the origin. There would be a concurrent reduction in the amount of surface generated prior to cross-wall initiation (Fig. 10). It is also possible to elaborate some of these features into more complex patterns of growth. For example, Sargent (1975*a*) has suggested that the surface of bacilli may be organised into rather specific growth zones at the junction of old and new cell surface determined by patterns of nuclear segregation. These models have been discussed in detail (Sargent, 1975*a*, *b*).

We have only discussed geometric features in terms of length, area and volume and not in terms of the relative rates at which the wall might grow at different locations. Ultimately, it should prove possible to fit the data to a time axis related to other major events of the cell cycle (such as rounds of chromosome replication). In this manner it might also be possible to interpret the graphical relationships in terms

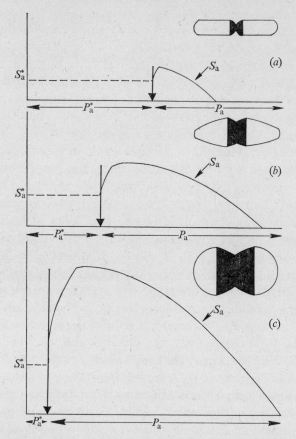

Fig. 10. Hypothetical sequence of wall growth in *rod* B mutants of *B. subtilis* (see text) using the concept and notation shown in Fig. 9. (*a*) Postulated growth pattern of rods at the permissive temperature. (*b*, *c*) After a shift-up to the restrictive conditions of growth, there occurs a gradual increase in width at the centre of the cell (shown by an increase in S_a^*). As the organisms become more coccus-like, there might be a gradual reduction in the surface area (P_a^*) of wall assembled by the cylindrical growth zone and a greater reliance upon septal growth as a means of producing new surface (increase in P_a). These features are illustrated by the schematic cell profiles, where the surface area of the nascent poles is shown in black. New polar surface would be generated by separation and expansion in area of the completed cross-wall. The arrows show the point of cross-wall closure.

of the regulation in the flow of precursors to the membrane and the activity of the enzymes therein that feed the growing wall as it changes in shape.

HELICAL GROWTH

The growth of living organisms and the arrangement of cell components in helices is ubiquitous in the biological kingdom. Apart from the well-known double helix of DNA, examples of helical growth can be found in genera as widely separated as mycoplasmas (Cole, Tully, Popkin & Bové, 1973) and in the stem form of the common convolvulus. In plant and fungal cells, supporting wall polymers, like cellulose and chitin, have long been known to be arranged helically, the direction of succeeding layers of fibrils being opposed. It is only recently, however, that attention has been drawn to helical forms of common rod-shaped bacteria (Mendelson, 1976; Tilby, 1977). Once observed, it is possible to see other examples. A penicillinase negative strain of *B. licheniformis* grown in penicillin (Highton & Hobbs, 1971), a novobiocin resistant strain of this organism grown in novobiocin (Robson & Baddiley, 1977); or *B. subtilis* growing under Mg^{2+} limitation in a chemostat after a reduction in dilution rate (H. J. Rogers, unpublished observation), all grow as helices. The formation of two sorts of helix by bacteria seems to be possible: a regular loose one and a tight rather irregular one. In the example of the former with *B. subtilis* (Mendelson, 1976) the angle of the helix is 70 °C. One hypothesis to account for this particular type of growth is that a helical orientation of surface expansion of a rod-shaped organism leads to rotation of the poles in opposite directions. If the poles are not free to rotate, a torque develops, so Mendelson (1976) argues, that would distort the chain of cells to a helix. For this to happen the author suggests that two conditions must be met. (1) The bacteria should be unable to separate from each other and (2) the resulting chain of cells should arise from two initials with the poles fixed. The necessity for these two conditions can be tested by constructing double mutants of *B. subtilis* containing the *lyt* and *rod* genes (P. J. Piggot, C. Taylor, P. F. Thurman, H. J. Rogers, unpublished work). The former has an autolytic deficient, non-cell-separating, non-flagellated phenotype, whereas the latter, as already described, is a conditional morphogenic mutation such that the strain can grow either as cocci or as rods. Thus the double mutant can be grown as long strings of unseparated cocci which when growth conditions are suitably modified change to rods emerging from the unseparated cocci. Long regular helices of cells are formed. Since neither single mutation alone leads to helical growth (Fein & Rogers, 1976; Rogers & Thurman, 1977), this may be further argument for a helical component in the cellular growth of the wild-type as well as for Mendelson's (1976) ideas of the origin of one form of the helix. Tilby (1977) favours the possibility of opposed helices of wall polymers with unequal stress in them to explain the tighter helical growth which appears to reverse in direction within the same chain of cells.

It is clearly too early to be dogmatic about the molecular origins of helical growth in bacteria but it must be pointed out that actual helical arrangements of wall components is not a necessary condition for helical growth and, if present, does not necessarily lead to a helical cell form. All that is required is a helical or rotational component in cell growth. It is also possible that the two quite different forms of helical growth described in bacilli have different origins.

ATTEMPTS AT INTEGRATION: CONCLUDING REMARKS

Any integration of existing biochemical and morphological knowledge must as yet be highly speculative. Nevertheless, areas worth closer study can be seen, among which the autolysins would appear to be particularly good candidates. We have described in previous sections the formation of the peripheral walls of streptococci where the autolysin would be expected to act by peeling apart the septal cross-wall. However, morphological evidence from autolysing cells shows most activity at the leading (inner) edge of the septum (Higgins, Pooley & Shockman, 1970), whereas the logical site of action for such a mechanism would be at the base (outer edge) of the septum or in its central region. On the other hand absence or a deficiency of autolytic activity is undoubtedly correlated with the failure of the individual organisms to separate (Rogers, 1970; Forsberg & Rogers, 1974; Fein & Rogers, 1976; Tomasz, 1968; Pooley, Shockman, Higgins & Porres-Juan, 1972), although cocci grow and remain spherical whilst rods remain rods. Further analysis of the surface growth patterns of autolysin-deficient mutants may provide evidence to clarify the function of these enzymes during growth. Similarly, few autolysins have been isolated and studied but the glimpses both of their substrate specificities and of their regulation by proteins, lipids and lipoteichoic acids (Herbold & Glaser, 1975a, b; Cleveland, Daneo-Moore, Wicken & Shockman, 1976; Höltje & Tomasz, 1975) are tantalising. It would be of interest to have a greater number of the enzymes in a purified state, and to explore in detail the underlying mechanisms of their substrate and inhibitor specificities.

The morphogenic role of the negatively charged wall polymers such as the techoic and teichuronic acids particularly in relation to the supply of Mg^{2+} to the membrane is another area of promise. Bacilli lacking these polymers grow as spheres instead of rods (Forsberg et al., 1973; Forsberg & Rogers, 1974; Boylan, Mendelson, Brookes & Young, 1972; Rogers, Thurman, Taylor & Reeve, 1974). Other conditional mutants showing the same morphological change have normal wall teichoic acid but have a temperature-sensitive property dependent on the supply of Mg^{2+} and suitable anions (Rogers *et al.*, 1976; Rogers & Thurman, 1977). A major suggested function of the negatively charged wall polymers is to regulate the supply of Mg^{2+} to the membrane-bound biosynthetic enzymes (Heckels *et al.*, 1977; Heptinstall, Archibald & Baddiley, 1970). Thus the proper functioning of the length-extension site in rod-shaped organisms may depend upon a localised supply of Mg^{2+} normally mediated by the negatively charged wall polymers.

Finally, the identification and specification of the growth sites in the surface in molecular terms, although perhaps a problem of a different order of difficulty is now open to new considerations. The precise positioning of mesosomes (Burdett & Rogers, 1972; Higgins & Shockman, 1970*a*, *b*) even though these may be artifactual, suggests that membranous material in a condensable form exists at the positions where the cell will eventually septate. Maybe this pre-membrane contains enzymes or enzyme activators which will start the process of wall extension and septation. The increasing knowledge of the biosynthesis of peptidoglycan and the mechanisms for the addition of new material to the existing wall should allow a more meaningful search for the positions on the cell surface at which the long glycan chains were initiated. A new attitude towards the meaning of growth sites and growth zones may need to be evolved.

We should like to thank the American Society for Microbiology, Dr M. L. Higgins and Dr G. D. Shockman of the Microbiology Department, Temple University, Philadelphia for permission to reproduce Figs. 6 and 7.

REFERENCES

AMAKO, K. & UMEDA, A. (1977). Bacterial surfaces as revealed by the high resolution scanning electron microscope. *Journal of General Microbiology*, **98**, 297–9.

ARAKI, Y., NAKATANI, T., NAKAYAMA, K. & ITO, E. (1972). Occurrence of *N*-nonsubstituted glucosamine residues in peptidoglycan of lysozyme resistant cell walls from *Bacillus cereus*. *Journal of Biological Chemistry*, **247**, 6312–22.

ARCHIBALD, A. R. (1973). The structure, biosynthesis and function of teichoic acid. *Advances in Microbial Physiology*, **11**, 53–95.

ARCHIBALD, A. R., BADDILEY, J. & HECKELS, J. E. (1973). Molecular arrangement of teichoic acid in the cell wall of *Staphylococcus lactis*. *Nature New Biology*, **241**, 29–31.

ARCHIBALD, A. R. & COAPES, H. E. (1976). Bacteriophage SP50 as a marker for cell wall growth in *Bacillus subtilis*. *Journal of Bacteriology*, **125**, 1195–206.

AZUMA, I., THOMAS, D. W., ADAM, A., GHUYSEN, J. M., BONALY, R., PETIT, J. & LEDERER, E. (1970). Occurrence of *N*-glycolylmuramic acid in bacterial cell walls: a preliminary survey. *Biochimica et Biophysica Acta*, **208**, 444–51.

BADDILEY, J. (1972*a*). Bacterial cell wall biosynthesis. In *Polymerisation in Biological Systems, Ciba Foundation Symposium*, **7**, pp. 87–107.

BADDILEY, J. (1972b). Teichoic acids in cell walls and membranes. In *Essays in Biochemistry*, **8**, ed. P. N. Campbell and F. Dickens, pp. 35–77. New York: Academic Press.

BOOTHBY, D., DANEO-MOORE, L., HIGGINS, M. L., COYETTE, J. & SHOCKMAN, G. D. (1973). Turnover of bacterial cell wall peptidoglycan. *Journal of Biological Chemistry*, **248**, 2161–9.

BOYLAN, R. J., MENDELSON, N. H., BROOKS, D. & YOUNG, F. E. (1972). Regulation of the bacterial cell wall: autolysis of a mutant of *Bacillus subtilis* defective in biosynthesis of teichoic acid. *Journal of Bacteriology*, **110**, 281–90.

BRACHA, R. & GLASER, L. (1976). *In vitro* system for the synthesis of teichoic acid linked to peptidoglycan. *Journal of Bacteriology*, **125**, 872–9.

BRAUN, V., GNIRKE, H., HENNING, U. & REHN, K. (1973). Model for the structure of the shape-maintaining layer of the *Escherichia coli* cell envelope. *Journal of Bacteriology*, **114**, 1264–70.

BRAUN, V. & HANTKE, K. (1974). Biochemistry of bacterial cell envelopes. *Annual Reviews of Biochemistry*, **43**, 89–121.

BURDETT, I. D. J. & HIGGINS, M. L. (1977). Comparison of the assembly of the poles in *Bacillus subtilis* and *Streptococcus faecalis* by computer reconstruction of septal growth zones seen in central longitudinal thin sections of cells. *Journal of Bacteriology*, in press.

BURDETT, I. D. J. & MURRAY, R. G. E. (1974). Septum formation in *Escherichia coli*: characterisation of septal structure and effects of antibiotics on cell division. *Journal of Bacteriology*, **119**, 303–24.

BURDETT, I. D. J. & ROGERS, H. J. (1972). The structure and development of mesosomes studied in *Bacillus licheniformis* strain 6346. *Journal of Ultrastructure Research*, **38**, 113–38.

BURGE, R. E., FOWLER, A. G. & REAVELEY, D. A. (1977). The structure of the peptidoglycan of bacterial cell walls. *Journal of Molecular Biology*, in press.

CHALOUPKA, J. (1967). Synthesis and degradation of surface structures by growing and non-growing *Bacillus megaterium*. *Folia Microbiologica*, Praha, **12**, 264–73.

CHUNG, K. L. (1973). Influence of cell wall thickness on cell division: electron microscope study with *Bacillus cereus*. *Canadian Journal of Microbiology*, **19**, 217–21.

CLEVELAND, R. F., DANEO-MOORE, L., WICKEN, A. J. & SHOCKMAN, G. D. (1976). Effect of lipoteichoic acid and lipids on lysis of intact cells of *Streptococcus faecalis*. *Journal of Bacteriology*, **127**, 1582–4.

COLE, R. M. & HAHN, J. J. (1962). Cell wall replication in *Streptococcus pyogenes*. *Science*, **135**, 722–3.

COLE, R. M., TULLY, J. G., POPKIN, T. S. & BOVE, J. M. (1973). Morphology, ultrastructure and bacteriophage infection of the helical mycoplasma-like organism. (*Spiroplasma citri* gen. nov. sp. nov.) cultured from 'stubborn' disease of citrus. *Journal of Bacteriology*, **115**, 367–83.

COLEY, J., ARCHIBALD, A. R. & BADDILEY, J. (1977). The presence of N-acetylglucosamine-1-phosphate in the linkage unit that connects teichoic acid to peptidoglycan in *Staphylococcus aureus*. *FEBS Letters*, in **80**, 405–7.

COYETTE, J. & GHUYSEN, J-M. (1970). Wall autolysin of *Lactobacillus acidophilus* strain 63 AM Gasser. *Biochemistry*, **9**, 2952–5.

DEZÉLÉE, P. & SHOCKMAN, G. D. (1975). Studies of the formation of peptide cross-links in the cell wall peptidoglycan of *Streptococcus faecalis*. *Journal of Biological Chemistry*, **250**, 6806–16.

DOYLE, R. J., McDANNEL, M. L., STREIPS, U. N., BIRDSELL, D. C. & YOUNG, F. E. (1974). Polyelectrolyte nature of bacterial teichoic acids. *Journal of Bacteriology*, **118**, 606–15.

ELLWOOD, D. C. & TEMPEST, D. W. (1969). Control of teichoic acid and teichuronic acid biosynthesis in chemostat cultures of *Bacillus subtilis* var niger. *Biochemical Journal*, 111, 1–5.

ERRINGTON, F. P., POWELL, E. O. & THOMPSON, N. (1965). Growth characterisation of some Gram-negative bacteria. *Journal of General Microbiology*, 39, 109–23.

FAN, D. & BECKMAN, B. E. (1973). Structural difference between walls from hemispherical caps and partial septa of *Bacillus subtilis*. *Journal of Bacteriology*, 114, 790–7.

FAN, D., PELVIT, M. C. & CUNNINGHAM, W. P. (1972). Structural difference between walls from ends and sides of the rod-shaped bacterium *Bacillus subtilis*. *Journal of Bacteriology*, 109, 1266–72.

FEIN, J. E. & ROGERS, H. J. (1976). Autolytic enzyme deficient mutants of *Bacillus subtilis* 168. *Journal of Bacteriology*, 127, 1427–42.

FORMANEK, H., FORMANEK, S. & WAWRA, H. (1974). A three-dimensional atomic model of the murein layer of bacteria. *European Journal of Biochemistry*, 46, 279–94.

FORMANEK, H., SCHLEIFER, K. H., SEIDL, H. P., LINDEMANN, R. & ZUNDEL, G.: (1976). Three-dimensional structure of peptidoglycan of bacterial cell walls infra-red investigations. *FEBS Letters*, 70, 150–4.

FORSBERG, C. W. & ROGERS, H. J. (1974). Characterisation of *Bacillus licheniformis* 6346 mutants which have altered lytic enzyme activities. *Journal of Bacteriology*, 118, 358–68.

FORSBERG, C. W., WYRICK, P. B., WARD, J. B. & ROGERS, H. J. (1973). Effect of phosphate limitation on the morphology and wall composition of *Bacillus licheniformis* and its phosphoglucomutase-deficient mutants. *Journal of Bacteriology*, 113, 969–84.

FOX, S. M., WARD, J. B. & SARGENT, M. G. (1977). Glycan chain lengths of *Bacillus subtilis*. *Proceedings of the Society for General Microbiology*, 4, 90.

FREHEL, C., BEAUFILS, A. M. & RYTER, A. (1971). Étude en microscope electronique de la croissance de la paroi chez *Bacillus subtilis* et *Bacillus megaterium*. *Annals of the Pasteur Institute*, 121, 139–48.

FRÈRE, J-M., GHUYSEN, J-M., ZEIGER, A. R. & PERKINS, H. R. (1976). The direction of peptide trimer synthesis from the donor–acceptor substrate N^α-(acetyl)-N-$^\epsilon$(glycyl)-L-lysyl-D-alanyl-D-alanine by the exocellular DD-carboxypeptidase-transpeptidase of *Streptomyces* R61. *FEBS Letters*, 63, 112–16.

GERHARDT, P. & JUDGE, J. A. (1964). Porosity of cell walls of *Saccharomyces cerevisiae* and *Bacillus megaterium*. *Journal of Bacteriology*, 87, 945–51.

GHUYSEN, J-M. (1968). Use of bacteriolytic enzymes in determination of wall structure and their role in cell metabolism. *Bacteriological Reviews*, 32, 425–64.

GHUYSEN, J-M. & SHOCKMAN, G. D. (1973). Biosynthesis of peptidoglycan. In *Bacterial Cell Walls and Membranes*, ed. L. Leive, pp. 132–208, New York: Marcel Dekker.

GIESBRECHT, P. & RUSKA, H. (1968). Uber Veränderung der Feinstrukturen von Bakterien unter Einwirkung von Chloramphenicol. *Klinische Wochenschrift*, 46, 575–82.

GIESBRECHT, P. & WECKE, J. (1971). Zur Morphogenese der Zellwand von Staphylokokken. 1. Querwandbildung und Zelltrennung. *Cytobiologie*, 4, 349–68.

GLASER, L. & LINDSAY, B. (1977). Relation between cell wall turnover and cell growth in *Bacillus subtilis*. *Journal of Bacteriology*, 130, 610–19.

GLAUERT, A. M., BRIEGER, E. M. & ALLEN, J. M. (1961). The fine structure of vegetative cells of *Bacillus subtilis*. *Experimental Cell Research*, 22, 73–85.

GLAUERT, A. M. & THORNLEY, M. J. (1969). The topography of the bacterial cell wall. *Annual Reviews of Microbiology*, 23, 159–98.

HAMMES, W. P. (1976). Biosynthesis of peptidoglycan in *Gaffkya homari*. The mode of action of penicillin G and mecillinam. *European Journal of Biochemistry*, **70** 107–13.

HAMMES, W. P. & KANDLER, O. (1976). Biosynthesis of peptidoglycan in *Gaffkya homari*. The incorporation of peptidoglycan into the cell wall and the direction of transpeptidation. *European Journal of Biochemistry*, **70**, 97–106.

HANCOCK, I. C. & BADDILEY, J. (1976). *In vitro* synthesis of the unit that links, teichoic acid to peptidoglycan. *Journal of Bacteriology*, **125**, 880–6.

HARVEY, R. J., MARR, A. G. & PAINTER, P. R. (1967). Kinetics of growth of individual cells of *Escherichia coli* and *Azotobacter agilis*. *Journal of Bacteriology*, **93**, 605–17.

HEKCKELS, J. E., ARCHIBALD, A. R. & BADDILEY, J. (1975). Studies on the linkage between teichoic acid and peptidoglycan in a bacteriophage-resistant mutant of *Staphylococcus aureus* H. *Biochemical Journal* **149**, 637–47.

HECKELS, J. E., LAMBERT, P. A. & BADDILEY, J. (1977). Binding of magnesium and cell walls of *Bacillus subtilis* W23 containing teichoic acid or teichuronic acid. *Biochemical Journal*, **162**, 359–65.

HEPTINSTALL, S., ARCHIBALD, A. R. & BADDILEY, J. (1970). Teichoic acid and membrane function in bacteria. *Nature, London*, **225**, 519–21.

HERBOLD, D. R. & GLASER, L. (1975a). *Bacillus subtilis* N-acetylmuramic acid L-alanine amidase. *Journal of Biological Chemistry*, **250**, 1676–82.

HERBOLD, D. R. & GLASER, L. (1975b). Interaction of N-acetylmuramic acid L-alanine amidase with cell wall polymer. *Journal of Biological Chemistry*, **250**, 7231–8.

HIGGINS, M. L. (1976). Three-dimensional reconstruction of whole cells of *Streptococcus faecalis* from thin sections of cells. *Journal of Bacteriology*, **127**, 1337–45.

HIGGINS, M. L., DANEO-MOORE, L. BOOTHBY, D. & SHOCKMAN, G. D. (1974). Effect of inhibition of deoxyribonucleic acid and protein synthesis on the direction of cell wall growth in *Streptococcus faecalis* (ATCC 9790). *Journal of Bacteriology*, **118**, 681–92.

HIGGINS, M. L., POOLEY, H. M. & SHOCKMAN, G. D. (1970). Site of initiation of cellular autolysis in *Streptococcus faecalis* as seen by electron microscopy. *Journal of Bacteriology*, **103**, 504–12.

HIGGINS, M. L., POOLEY, H. M. & SHOCKMAN, G. D. (1971). Reinitiation of cell wall growth after threonine starvation of *Streptococcus faecalis*. *Journal of Bacteriology*, **105**, 1175–83.

HIGGINS, M. L. & SHOCKMAN, G. D. (1970a). Model for cell wall growth of *Streptococcus faecalis*. *Journal of Bacteriology*, **101**, 643–8.

HIGGINS, M. L. & SHOCKMAN, G. D. (1970b). Early changes in the ultrastructure of *Streptococcus faecalis* after amino acid starvation. *Journal of Bacteriology*, **103**, 244–54.

HIGGINS, M. L. & SHOCKMAN, G. D. (1976). Study of a cycle of cell wall assembly in *Streptococcus faecalis* by three-dimensional reconstructions of thin sections of cells. *Journal of Bacteriology*, **127**, 1346–58.

HIGGINS, M. L., TSIEN, H. O. & DANEO-MOORE, L. (1976). Organisation of mesosomes in fixed and unfixed cells. *Journal of Bacteriology*, **127**, 1519–23.

HIGHTON, P. J. & HOBBS, D. G. (1971). Penicillin and cell wall synthesis: a study of *Bacillus licheniformis* by electron microscopy. *Journal of Bacteriology*, **106**, 646–58.

HOLTJE, J-V. & TOMASZ, A. (1975). Lipoteichoic acid: a specific inhibitor of autolysin activity in *Pneumococcus*. *Proceedings of the National Academy of Sciences, USA*, **72**, 1690–4.

HUGHES, A. H., HANCOCK, I. C. & BADDILEY, J. (1973). The function of teichoic acids in cation control in bacterial membranes. *Biochemical Journal*, **132**, 83–93.

HUGHES, R. C. (1970a). Autolysis of isolated walls of *Bacillus licheniformis* NCTC 6346, and *Bacillus subtilis* Marburg strain 168. Separation of the products and characterisation of the mucopeptide fragments. *Biochemical Journal*, **119**, 849–60.

HUGHES, R. C. (1970b). The cell wall of *Bacillus licheniformis* NCTC 6346: linkage between the teichuronic acid and mucopeptide components. *Biochemical Journal*, **117**, 431–9.

HUGHES, R. C. (1971). Autolysis of *Bacillus cereus* walls and the isolation of structural components. *Biochemical Journal*, **121**, 791–802.

HUGHES, R. C., TANNER, P. J. & STOKES, E. (1970). Cell wall thickening in *Bacillus subtilis*. Comparison of thickened and normal walls. *Biochemical Journal*, **120**, 159–70.

HURST, A. & STUBBS, J. M. (1969). Electron microscopic study of membranes and walls of bacteria and changes during growth initiation. *Journal of Bacteriology*, **97**, 1466–79.

KARAMATA, D., McCONNELL, M. & ROGERS, H. J. (1972). Mapping of *rod* mutants of *Bacillus subtilis*. *Journal of Bacteriology*, **111**, 73–9.

KELEMEN, M. V. & ROGERS, H. J. (1971). Three-dimensional molecular models of cell wall mucopeptides (peptidoglycans). *Proceedings of the National Academy of Sciences, USA*, **68**, 992–6.

KRULWICH, T. A. & PATE, J. L. (1971). Ultrastructural explanation for snapping post-fission movements in *Arthrobacter crystallopoietes*. *Journal of Bacteriology*, **105**, 408–12.

LINNETT, P. E., ROBERTS, R. J. & STROMINGER, J. L. (1974). Biosynthesis and cross-linking of the γ-glutamylglycine containing peptidoglycan of vegetative cells of *Sporosarcina ureae*. *Journal of Biological Chemistry*, **249**, 2497–506.

LINNETT, P. E. & STROMINGER, J. L. (1974). Amidation and cross-linking of the enzymatically synthesised peptidoglycan of *Bacillus stearothermophilus*. *Journal of Biological Chemistry*, **249**, 2489–96.

MARR, A. G., HARVEY, R. J. & TRENTINI, W. C. (1966). Growth and division of *Escherichia coli*. *Journal of Bacteriology*, **91**, 2388–9.

MAUCK, J., CHAN, L. & GLASER, L. (1971). Turnover of the cell wall of Gram-positive bacteria. *Journal of Biological Chemistry*, **246**, 1820–7.

MAUCK, J. & GLASER, L. (1972). On the mode of *in vivo* assembly of the cell wall of *Bacillus subtilis*. *Journal of Biological Chemistry*, **247**, 1180–7.

MENDELSON, N. H. (1976). Helical growth of *Bacillus subtilis*: a new model of cell growth. *Proceedings of the National Academy of Sciences, USA*, **73**, 1740–4.

MIRELMAN, D., BRACHA, R. & SHARON, N. (1972). Role of the penicillin-sensitive transpeptidation reactions in attachment of newly-synthesised peptidoglycan to cell walls of *Micrococcus luteus*. *Proceedings of the National Academy of Sciences, USA*, **69**, 3355–9.

MIRELMAN, D., BRACHA, R. & SHARON, N. (1974). Penicillin-induced secretion of a soluble uncross-linked peptidoglycan by *Micrococcus luteus* cells. *Biochemistry*, **13**, 5045–53.

MIRELMAN, D. & SHARON, N. (1972). Biosynthesis of peptidoglycan by a cell wall preparation of *Staphylococcus aureus* and its inhibition by penicillin. *Biochemical and Biophysical Research Communications*, **46**, 1909–17.

NANNINGA, N. (1971). The mesosomes of *Bacillus subtilis* as affected by chemical and physical fixation. *Journal of Cell Biology*, **48**, 219–24.

NERMUT, M. V. (1967). The ultrastructure of the cell wall of *Bacillus megaterium*. *Journal of General Microbiology*, **49**, 503–12.

NIKAIDO, H. (1973). Biosynthesis and assembly of lipopolysaccharide and the outer membrane layer of Gram-negative cell wall. In *Bacterial Membranes and Walls*, ed. L. Leive, pp. 132–208. New York: Marcel Dekker.

OLDMIXON, E. H., GLAUSER, S. & HIGGINS, M. L. (1974). Two proposed general configurations for bacterial cell wall peptidoglycans shown by space filling molecular models. *Biopolymers*, 13, 2037–60.

OSBORN, M. J. (1969). Structure and biosynthesis of the bacterial cell wall. *Annual Review of Biochemistry*, 38, 501–38.

OU, L-T., & MARQUIS, R. E. (1970). Electromechanical interactions in cell walls of Gram-positive cocci. *Journal of Bacteriology*, 101, 92–101.

OU, L-T. & MARQUIS, R. E. (1972). Coccal wall compaction and the swelling action of denaturants. *Canadian Journal of Microbiology*, 18, 623–9.

POOLEY, H. M. (1976a). Turnover and spreading of old wall during surface growth of *Bacillus subtilis*. *Journal of Bacteriology*, 125, 1127–38.

POOLEY, H. M. (1976b). Layered distribution, according to age, within the cell wall of *Bacillus subtilis*. *Journal of Bacteriology*, 125, 139–47.

POOLEY, H. M., SHOCKMAN, G. D., HIGGINS, M. L. & PORRES-JUAN, J. (1972). Some properties of two autolytic-defective mutants of *Streptococcus faecalis* ATCC 9790. *Journal of Bacteriology*, 109, 423–31.

PREUSSER, H. J., HEILMAN, H. D. & MARTIN, H. H. (1969). Die Wirkung von Schermetallionen auf der Murein-Sacculus von *Spirillum serpens*. *Cytobiologie*, 1, 187–93.

REUSCH, V. M. & BURGER, M. M. (1973). The bacterial mesosome. *Biochimica et Biophysica Acta*, 300, 79–104.

REYNOLDS, P. E. (1971a). Peptidoglycan synthesis in bacilli. I. Effect of temperature on the *in vitro* system from *Bacillus megaterium* and *Bacillus stearothermophilus*. *Biochimica et Biophysica Acta*, 237, 239–54.

REYNOLDS, P. E. (1971b). Peptidoglycan synthesis in bacilli. II. Characteristics of protoplast membrane preparations. *Biochimica et Biophysica Acta*, 237, 255–72.

ROBSON, R. L. & BADDILEY, J. (1977). Morphological changes associated with novobiocin resistance in *Bacillus licheniformis*, *Journal of Bacteriology*, 129, 1045–50.

ROGERS, H. J. (1970). Bacterial growth and the cell envelope. *Bacteriological Reviews*, 34, 194–214.

ROGERS, H. J. (1974). Peptidoglycans (mucopeptides): structure, function and variations. *Annals of the New York Academy of Sciences*, 235, 29–51.

ROGERS, H. J., MCCONNELL, M. & BURDETT, I. D. J. (1971). The isolation and characterisation of mutants of *Bacillus subtilis* and *Bacillus licheniformis* with disturbed morphology and cell division. *Journal of General Microbiology*, 61, 155–71.

ROGERS, H. J. & THURMAN, P. F. (1977). The temperature-sensitive nature of the *rod* B mutation in *Bacillus subtilis*. *Journal of Bacteriology*, in press.

ROGERS, H. J., THURMAN, P. F. & BUXTON, R. S. (1976). Magnesium and anion requirements of *rod* B mutants of *Bacillus subtilis*. *Journal of Bacteriology*, 125, 556–64.

ROGERS, H. J., THURMAN, P. F., TAYLOR, C. & REEVE, J. N. (1974). Mucopeptide synthesis by *rod* mutants of *Bacillus subtilis*. *Journal of General Microbiology*, 85, 335–50.

SARGENT, M. G. (1975a). Control of cell length in *Bacillus subtilis*. *Journal of Bacteriology*, 123, 7–19.

SARGENT, M. G. (1975b). Anucleate cell production and surface extension in a temperature-sensitive chromosome initiation mutant of *Bacillus subtilis*. *Journal of Bacteriology*, 123, 1218–34.

SCHLEIFER, K. H. & KANDLER, O. (1972). Peptidoglycan types of bacterial cell walls and their taxononic significance. *Bacteriological Reviews*, **36**, 407–62.

SHOCKMAN, G. D. (1965). Symposium on the fine structure and replication of bacteria and their parts. IV. Unbalanced cell wall synthesis: autolysis and cell wall thickening. *Bacteriological Reviews*, **29**, 345–58.

SHOCKMAN, G. D., DANEO-MOORE, L. & HIGGINS, M. L. (1974). Problems of cell wall and membrane growth enlargement and division. *Annals of the New York Academy of Sciences*, **235**, 161–97.

SHOCKMAN, G. D. & MARTIN, J. T. (1968). Autolytic enzyme system of *Streptococcus faecalis*. IV. Electron microscopic observations of autolysin and lysozyme action. *Journal of Bacteriology*, **96**, 1803–10.

TILBY, M. J. (1977). Helical shape and wall synthesis. *Nature, London*, **266**, 450–2.

TIPPER, D. J. (1970). Structure and function of peptidoglycans. *International Journal of Systematic Bacteriology*, **20**, 361–77.

TIPPER, D. J. & BERMAN, M. F. (1969). Structures of the cell wall peptidoglycan of *Staphylococcus epidermidis* Texas 26 and *Staphylococcus aureus* Copenhagen. I. Chain length and average sequence of cross-bridge peptides. *Biochemistry*, **8**, 2183–92.

TOMASZ, A. (1968). Biological consequences of the replacement of choline by ethanolamine in the cell wall of pneumococcus: chain formation loss of transformability and loss of autolysis. *Proceedings of the National Academy of Sciences, USA*, **59**, 86–93.

TOMASZ, A., WESTPHAL, M., BRILES, E. B. & FLETCHER, P. (1975). On the physiological functions of teichoic acids. *Journal of Supramolecular Structure*, **3**, 1–16.

TYNECKA, Z. & WARD, J. B. (1975). Peptidoglycan synthesis in *Bacillus licheniformis*. The inhibition of cross-linking by benzylpenicillin and cephaloridine *in vivo* accompanied by the formation of soluble peptidoglycan. *Biochemical Journal*, **146**, 253–67.

WAGNER, M. (1964). Studien mit fluoreszierenden Antikorpern an wachsenden Bakterien. I. Die Neubilding der Zellwend bei *Diplococcus pneumoniae*. *Zentralblatt für Bakteriologie, Parasitenkunde Infektionskrankheiten und Hygiene (Abteilung I)*, **195**, 87–93.

WARD, J. B. (1973). The chain length of the glycans in bacterial cell walls. *Biochemical Journal*, **133**, 395–8.

WARD, J. B. (1974). The synthesis of peptidoglycan in an autolysin-deficient mutant of *Bacillus licheniformis* NCTC 6346 and the effect of β-lactam antibiotics, bacitracin and vancomycin. *Biochemical Journal*, **141**, 227–41.

WARD, J. B. & PERKINS, H. R. (1973). The direction of glycan synthesis in a bacterial peptidoglycan. *Biochemical Journal*, **135**, 721–8.

WARD, J. B. & PERKINS, H. R. (1974). Peptidoglycan synthesis by preparations from *B. licheniformis*: cross-linking of newly-synthesised chains to preformed cell wall. *Biochemical Journal*, **139**, 781–4.

WEIBULL, C. (1973). Electron microscope studies on aldehyde-fixed unstained microbial cells. *Journal of Ultrastructure Research*, **43**, 150–9.

WEIDEL, W. & PELZER, H. (1964). Bag-shaped macromolecules – a new outlook in bacterial cell walls. *Advances in Enzymology*, **26**, 193–232.

WESTON, A., WARD, J. B. & PERKINS, H. R. (1977). Biosynthesis of peptidoglycan in wall plus membrane preparations from *Micrococcus luteus*: direction of chain extension, length of chains and effect of penicillin on cross-linking. *Journal of General Microbiology*, **99**, 171–81.

WICKUS, G. G. & STROMINGER, J. L. (1972a). Penicillin-sensitive transpeptidation during peptidoglycan biosynthesis in cell-free preparations from *Bacillus megaterium*. *Journal of Biological Chemistry*, **247**, 5297–306.

WICKUS, G. G. & STROMINGER, J. L. (1972*b*). Penicillin-sensitive transpeptidation during peptidoglycan biosynthesis in cell-free preparations from *Bacillus megaterium*. II. Effect of penicillins and cephalosporins on bacterial growth and *in vitro* transpeptidation. *Journal of Biochemistry*, **247**, 5307–11.

WRIGHT, J. & HECKELS, J. E. (1975). The teichuronic acid of cell walls of *Bacillus subtilis* W23 grown in a chemostat under phosphate limitation. *Biochemical Journal*, **147**, 187–9.

WYKE, A. W. & WARD, J. B. (1977*a*). The biosynthesis of muramic acid phosphate in *Bacillus licheniformis*. *FEBS Letters*, **73**, 159–63.

WYKE, A. W. & WARD, J. B. (1977*b*). The biosynthesis of wall polymers in *Bacillus subtilis* W23. *Journal of Bacteriology*, **130**, 1055–63.

EXPLANATION OF PLATES

PLATE 1

(*a*) Electron micrograph of central longitudinal section of *B. subtilis* showing incipient cross-wall; n, nuclear bodies.

(*b*)–(*d*). Electron micrographs of a simulated sequence of cross-wall development and pole formation in *B. subtilis*. A split (arrow, *b*) develops at the outer surface of the partially constructed cross-wall prior to the formation of polar wall. Wall bands (W) appear to delimit the junction of newly synthesised polar wall and the cylindrical portion of the cell (*c*). As the poles enlarge, the wall bands are displaced bilaterally from the base of the cross-wall (*c, d*).

PLATE 2

Electron micrographs of central, longitudinal sections of *rod* B mutants of *B. subtilis*.

(*a*) Section through rod-shaped cell at 20 °C (permissive growth conditions); cross-wall closure occurs prior to cell separation.

(*b, c*). After 75 min (*b*) and 150 min (*c*) at the restrictive temperature (42 °C), the organisms enlarge in diameter at the cell equator. The limit of nascent polar wall appears to be marked by wall bands (arrows); there is also a gradient of wall thickness, increasing from the base of the cross-wall to the band markers.

PLATE 1

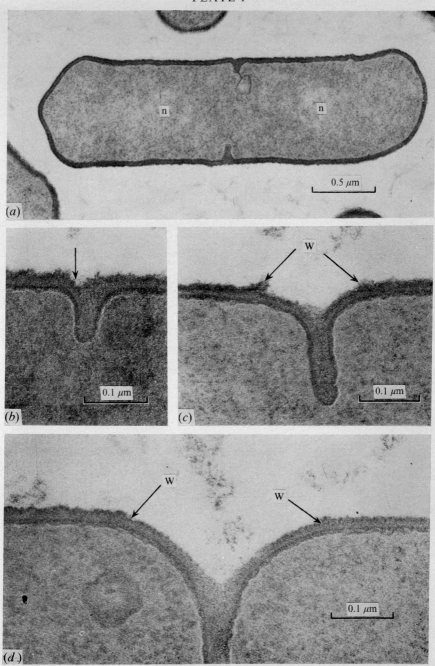

(a)

(b)

(c)

(d)

PLATE 2

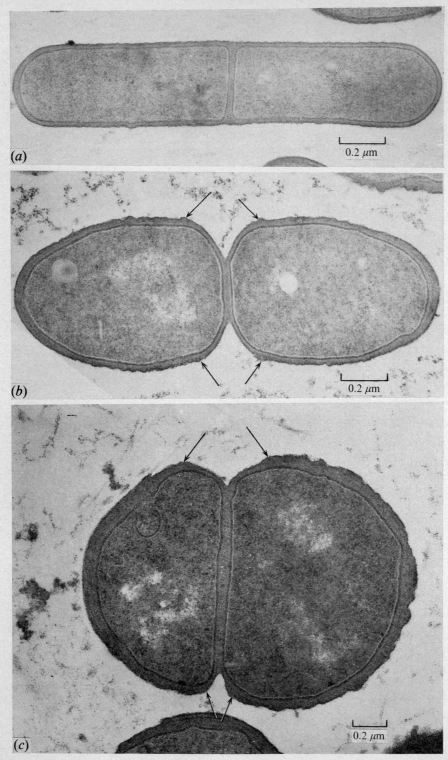

(a)

0.2 μm

(b)

0.2 μm

(c)

0.2 μm

STRUCTURE AND IMMUNOSTIMULANT PROPERTIES OF MYCOBACTERIAL CELL WALLS

JEAN-FRANÇOIS PETIT* AND EDGAR LEDERER*†

* *Institut de Biochimie,*
Université Paris-Sud,
91405 Orsay, France

† *Institut de Chimie des Substances Naturelles,*
CNRS 91190 Gif/Yvette, France

INTRODUCTION

Injection of live mycobacteria can not only protect against tuberculosis but can also modify the immune status of the animal host in other ways (Mackaness, Auclair & Lagrange, 1973; Mackaness, Lagrange & Ishibashi, 1974), as expressed in its response to an antigen (Lewis & Loomis, 1924; Dienes & Schonheit, 1928) or to a challenge by pathogenic bacteria (Dubos & Schaedler, 1957; Sulitzeanu, Bekierkunst, Groto & Loebel, 1962; Blanden, Lefford & Mackaness, 1969; Howard *et al.*, 1959) or tumour cells (Old, Clarke & Benaceraff, 1959; Biozzi, Stiffel, Halpern & Mouton, 1959; Zbar, Rapp & Ribi, 1972; Laucius, Bodurtha, Mastrangelo & Creech, 1974). Live BCG is used with varying success in man for the immunotherapy of leukaemia (Mathé *et al.*, 1974) and melanomas (Nathanson, 1972; Pinski, Hirshaut & Oettgen, 1972).

Immunostimulation can be achieved with killed mycobacteria injected by a particular route in an appropriate vehicle: addition of mycobacteria (or nocardiae) to Freund's Incomplete Adjuvant, which consists of an emulsion of mineral oil and a water phase containing the antigen, increases the concentration of circulating antibodies and induces a cellular immunity to this antigen (Freund, 1956). Killed BCG cells can increase resistance to infections (Dubos & Schaedler, 1957; Parant, Parant, Chedid & Le Minor, 1975) and have an anti-tumour activity (Bekierkunst, Wang, Toubiana & Lederer, 1974; Yarkoni, Wang & Bekierkunst, 1974; Bekierkunst, 1975).

The identification of the components responsible for the vaccinating and immunostimulating properties of living or killed mycobacteria has been one of the main goals of chemical research in this field. Most of the active components have been located in the wall which therefore represents the main immunomodifier of the mycobacterium.

We shall give a general outline of the chemistry and biological properties of mycobacterial walls and of their components (for recent reviews see: Lederer, 1971, 1976; Lederer *et al.*, 1975; Barksdale & Kim, 1977) and focus on recent results and tentative interpretations of the relations between chemical structure and biological activity.

CHEMICAL STRUCTURE OF MYCOBACTERIAL CELL WALLS

Mycobacterial wall preparations are obtained by the usual procedures: namely, disruption of the bacteria, differential centrifugations, and repeated washings with buffers and water (see for instance Ribi *et al.*, 1966; Wietzerbin-Falszpan *et al.*, 1973). They contain several components: (1) the so-called 'covalent skeleton' (Fig. 1), made up of two covalently linked polymers, peptidoglycan and an arabinogalactan mycolate (Misaki, Yukawa, Tsuchiya & Yamasaki, 1966; Kanetsuna, 1968; Kanetsuna & San Blas, 1970; Wietzerbin-Falszpan *et al.*, 1973; Azuma, Ribi, Meyer & Zbar, 1974); (2) free lipids, i.e. lipids which can be removed by neutral solvents (Asselineau, 1966; Goren, 1972; Lederer *et al.*, 1975); (3) peptides which can be removed by proteolytic enzymes (Misaki *et al.*, 1966; Wietzerbin-Falszpan *et al.*, 1973; Azuma *et al.*, 1974) and a partly amidated poly-L-glutamic acid (present in both pathogenic and vaccinating strains) (Wietzerbin-Falszpan *et al.*, 1973); and (4) a glucan in some strains (Misaki & Yukawa, 1966; Amar Nacasch & Vilkas, 1970).

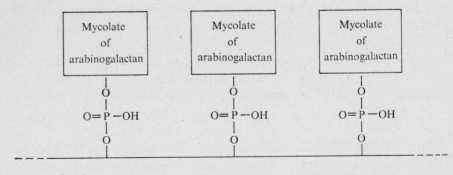

Fig. 1. Schematic representation of the 'cell wall skeleton'.

Treatment of whole walls with proteolytic enzymes followed by lipid solvents gives 'purified walls', in which the main constituent is the covalent skeleton, accompanied by a few amino acids characteristic of proteins and in some strains poly-L-glutamic acid and/or glucan. Mild acid hydrolysis of 'purified walls' followed by treatment with lipid solvents gives the 'basal layer' of the cell walls. Its main constituent is peptidoglycan (Misaki *et al.*, 1966; Wietzerbin-Falszpan *et al.*, 1973), but it also contains poly-L-glutamic acid, when present, and a few amino acids other than those in the peptidoglycan.

THE COVALENT SKELETON

The peptidoglycan

The general structure of all bacterial peptidoglycans consists of glycan strands substituted by partly cross-linked tri- or tetrapeptides. Three features distinguish those from mycobacteria: (1) The presence of *N-glycolylmuramic* acid instead of the usual *N*-acetyl derivative; this was established by mass spectrometry of the disaccharide of the repeating subunit (Fig. 2) isolated by sequential enzymatic degradation of the basal layer of various species (Adam *et al.*, 1969; Azuma *et al.*, 1970). (2) The presence of *two amide* groups, on both glu and meso-A$_2$pm, in

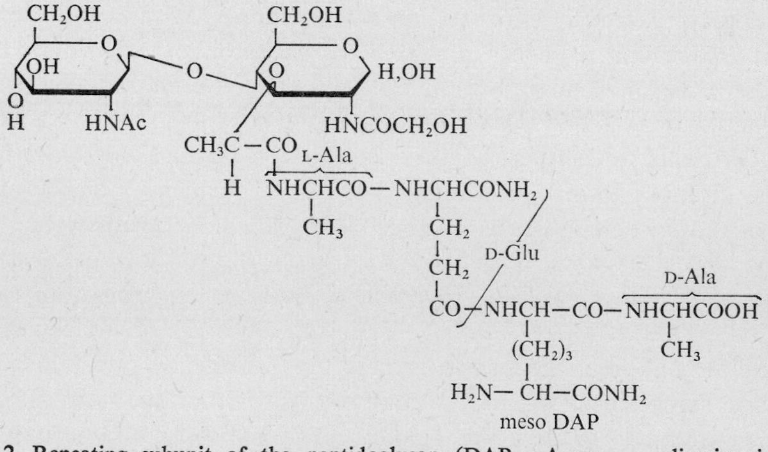

Fig. 2. Repeating subunit of the peptidoglycan (DAP = A$_2$pm = α,ε-diaminopimelic acid).

the tri- and tetrapeptides of the repeating subunit (Fig. 2). This was also established by mass spectrometry of the peptides isolated after sequential enzymatic hydrolyses (Wietzerbin-Falszpan *et al.*, 1970). (3) The presence of *two kinds of interpeptide linkages*: D-Ala-meso-A$_2$pm and meso-A$_2$pm-meso-A$_2$pm. The existence of D-Ala-meso-A$_2$pm inter-peptide linkages was established using a specific D,D carboxypeptidase isolated from *Streptomyces albus* G (Leyh-Bouille *et al.*, 1970). This enzyme leaves an unhydrolysed core of cross-linked peptides almost completely devoid of D-alanine. It was therefore necessary to postulate a new type of interpeptide linkage not involving D-alanine. Its nature was established by isolation, from partial acid hydrolysates of the core, of the dipeptide meso-A_2pm-meso-A_2pm and of the tripeptide meso-A_2pm-meso-A_2pm-meso-A_2pm identified by mass spectrometry (Wietzerbin *et al.*, 1974; Petit, Wietzerbin, Das & Lederer, 1975). These peptides have been found in the peptidoglycan of all *Mycobacteria* examined, but are absent from other A_2pm-containing peptidoglycans. Meso-A_2pm-meso-A_2pm interpeptide linkages seem therefore restricted to mycobacteria. The presence of two kinds of interpeptide linkages in mycobacterial peptidoglycan implies that either there are two kinds of transpeptidases or a single transpeptidase can synthesise both linkages. The presence of an unusual transpeptidase might account for the resistance of mycobacteria to β-lactam antibiotics (penicillins and cephalosporins).

The arabinogalactan mycolate

The arabinogalactan is a highly branched polymer containing D-arabinose and D-galactose in a ratio of approximately 5:2 (Misaki & Yukawa, 1966). Its terminal branches are linear arabinose oligosaccharides which represent the main immunogenic determinant of the molecule (Misaki, Seto & Azuma, 1974). Arabinose (Misaka & Yukawa, 1966; Misaki *et al.*, 1974) and at least part of the galactose (Vilkas *et al.*, 1973*a*, *b*) are in the furanose form. Most linkages are $1 \rightarrow 5$, branching occurring at C-3 of arabinose and probably C-6 of the galactose residues. The size of the molecules of the arabinogalactan from the cell wall of the BCG strain of *M. bovis* has been established, by ultracentrifugation, to be around 30000 daltons (Misaki & Yukawa, 1966). These molecules are covalently linked to the peptidoglycan, presumably by phosphodiester linkages between the C-6–OH of one out of ten muramic acid residues and an arabinose of the arabinogalactan: by appropriate hydrolyses of mycobacterial cell walls, muramic acid-6-phosphate (Liu & Gotschlich, 1967; Kanetsuna, 1968) and arabinose phosphate (Amar & Vilkas, 1973) have been isolated.

About one in ten of the arabinose residues of the arabinogalactan is esterified by a molecule of mycolic acid. The linkage is on the 5′-OH of a terminal arabinose residue, as is proved by the isolation of arabinose-5-*O*-mycolate or diarabinose mycolate after mild acid hydrolysis of whole bacilli (Azuma & Yamamura, 1962, 1963; Acharya, Senn & Lederer, 1967), of cell walls, or of wax D (see below)

(Azuma, Yamamura & Fukushi, 1968; Amar-Nacasch & Vilkas, 1970). Mycolic acids are α-branched β-hydroxy-acids with 60–90 carbon atoms (Asselineau, 1966; Etemadi, 1967; for a review, see Lederer, 1976). The α-substituent is a normal $C_{22}H_{45}$ chain in *M. phlei, M. smegmatis* and *M. avium*, and $C_{24}H_{49}$ in *M. bovis* and *M. tuberculosis*. The main chain contains either double bonds (as in *M. smegmatis*) or cyclopropane rings, methyl or methoxy substituents, or a keto group, or even a second carboxyl function (in *M. avium* and *M. phlei*): some examples are given in Fig. 3. In a given strain, one usually finds two to five types of mycolic acids, the main chain of each type being a mixture of homologues.

Mycolic acid	Strain	Formula [a]
α-Smegmamycolic acid 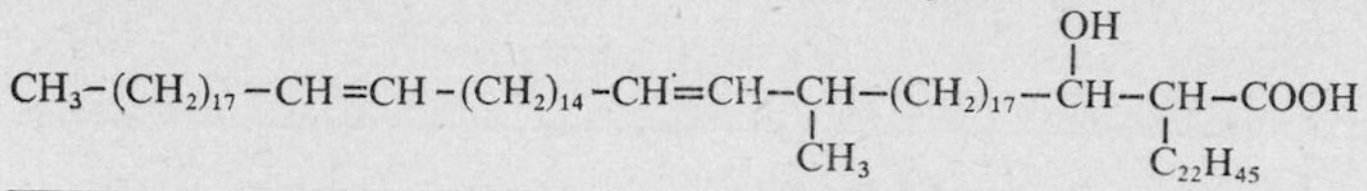	*M. smegmatis*	$C_{79}H_{154}O_3$

$$CH_3-(CH_2)_{17}-CH=CH-(CH_2)_{14}-CH=CH-CH-(CH_2)_{17}-\overset{OH}{\underset{}{CH}}-CH-COOH$$
$$\underset{CH_3}{|} \qquad \underset{C_{22}H_{45}}{|}$$

| α-Kansamycolic acid 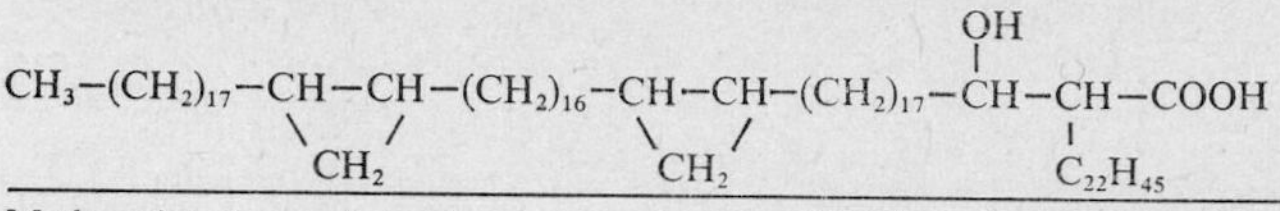 | *M. kansasii* | $C_{80}H_{156}O_3$ |

$$CH_3-(CH_2)_{17}-CH-CH-(CH_2)_{16}-CH-CH-(CH_2)_{17}-\overset{OH}{CH}-CH-COOH$$
$$\overset{\diagdown\diagup}{CH_2} \qquad \overset{\diagdown\diagup}{CH_2} \qquad \underset{C_{22}H_{45}}{|}$$

| Methoxylated mycolic acid 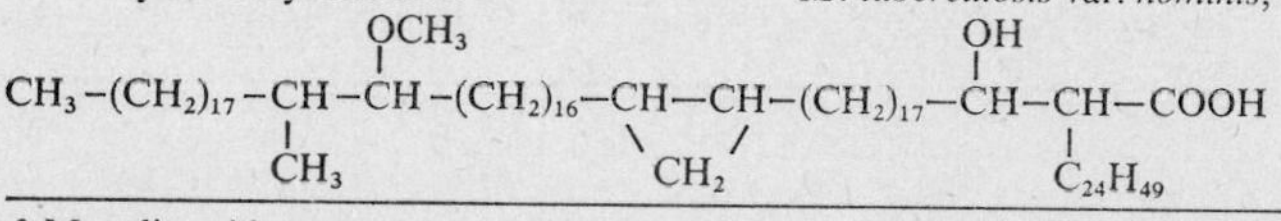 | *M. tuberculosis* var. *hominis*, strain Test | $C_{85}H_{168}O_4$ |

$$CH_3-(CH_2)_{17}-\overset{OCH_3}{CH}-CH-(CH_2)_{16}-CH-CH-(CH_2)_{17}-\overset{OH}{CH}-CH-COOH$$
$$\underset{CH_3}{|} \qquad \overset{\diagdown\diagup}{CH_2} \qquad \underset{C_{24}H_{49}}{|}$$

| β-Mycolic acid 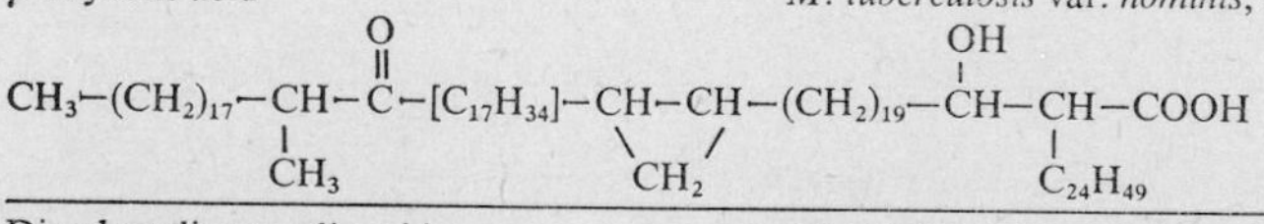 | *M. tuberculosis* var. *hominis*, strain Test | $C_{87}H_{170}O_4$ |

$$CH_3-(CH_2)_{17}-CH-\overset{O}{\overset{||}{C}}-[C_{17}H_{34}]-CH-CH-(CH_2)_{19}-\overset{OH}{CH}-CH-COOH$$
$$\underset{CH_3}{|} \qquad \overset{\diagdown\diagup}{CH_2} \qquad \underset{C_{24}H_{49}}{|}$$

| Dicarboxylic mycolic acid 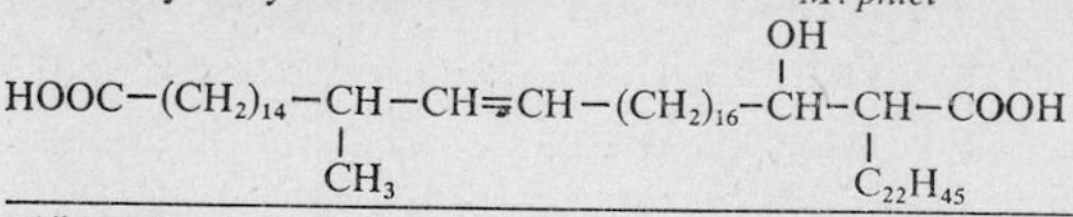 | *M. phlei* | $C_{60}H_{116}O_5$ |

$$HOOC-(CH_2)_{14}-CH-CH=CH-(CH_2)_{16}-\overset{OH}{CH}-CH-COOH$$
$$\underset{CH_3}{|} \qquad \underset{C_{22}H_{45}}{|}$$

[a] All mycolic acids are mixtures of homologues. The molecular formulae and the structures given in this table are those of the principal members of the homologous series.

Fig. 3. Structures of some mycolic acids of mycobacteria, from Etemadi (1967).

A covalent skeleton with the same general structure also exists in the related genera *Nocardia* and *Corynebacterium*. Cell walls of *N. asteroides* and *C. diphtheriae* contain an arabinogalactan immunologically related to the mycobacterial arabinogalactan (Misaki *et al.*, 1974). Fatty acids having the same α-branched β-hydroxystructure as mycolic acids are present in both genera: the nocardomycolic acids of *Nocardia* spp.

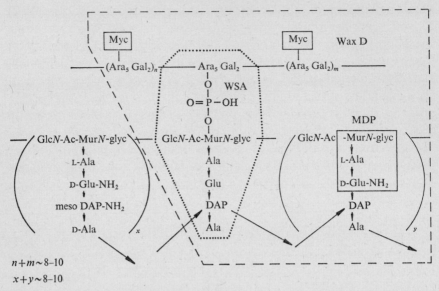

$n+m \sim 8\text{–}10$
$x+y \sim 8\text{–}10$

Fig. 4. Chemical structure of the 'cell wall skeleton' and of its derivatives which are adjuvants (DAP = A$_2$pm = α,ϵ-diaminopimelic acid).

range from C-40 to C-60 with a C-14 or C-16 side chain; the corynemycolic acids of *Corynebacterium* spp. range from C-28 to C-36 with a C-14 side chain (Lederer & Pudlès, 1951; Asselineau, 1966; Lederer *et al.*, 1975). The chemical nature of the mycolic acids permits an unambiguous generic identification (Lechevalier, Horan & Lechevalier, 1971; Lechevalier, Lechevalier & Horan, 1973; Lechevalier, 1977).

THE LIPIDS
Free lipids

Free lipids (i.e. lipids which can be extracted by neutral solvents) account for 25–30% of the weight of mycobacterial cell walls. Most studies on mycobacterial lipids have been conducted using whole cells as starting material. However, there is a consensus of opinion that lipids not hitherto found in cytoplasmic membranes are likely to belong to the cell wall. This has been confirmed for most of those classified as cell wall lipids.

Wax D

Wax D of *M. tuberculosis* strains contains an arabinogalactan esterified by mycolic acids and linked to a peptidoglycan containing N-acetylglucosamine, N-glycolylmuramic acid, L- and D-alanine, meso-α,ϵ-diaminopimelic acid and D-glutamic acid (Asselineau, 1966; Amar & Vilkas, 1973). Migliore & Jollès (1968, 1969) have described a tetra-saccharide heptapeptide isolated from wax D of these strains. Wax D is now considered to be an autolysis product of the cell wall (Markovits, Vilkas & Lederer, 1971) which is in agreement with the finding of Azuma, Kimura & Yamamura (1968) that the arabinogalactans of cell walls and wax D are immunologically identical.

Figure 4 shows how such wax D fractions could be formed. The action of a phosphodiesterase would be expected to yield a hydrophobic arabinogalactan mycolate (wax D of 'non-human' strains), whereas lysozyme and a D,D-carboxypeptidase would yield peptidoglycan containing wax D fractions similar to those obtained from human strains and *M. kansasii* (for more details on wax D see Lederer, 1967; Goren, 1972).

Cord factors

These are 6,6'-dimycolates of α,α'-D-trehalose and their general structure is given in Fig. 5. The first was isolated by petrol–ether extraction of pathogenic mycobacteria that form 'cords' when grown on the surface of a liquid medium (Bloch, 1950). The structure of the cord factor from *M. tuberculosis* strain Brévannes was established by Noll, Bloch, Asselineau & Lederer (1956). Since then, dimycolates of trehalose have been extracted from a wide range of mycobacterial strains such as *M. tuberculosis* strain Peurois, BCG (Adam, Senn, Vilkas, & Lederer 1967), *M. phlei* (Promé, Lacave, Ahido-Coffy & Savagnac, 1976) and *M. smegmatis* (Mompon, Fédérici, Toubiana & Lederer, 1977). Such material has also been isolated under the name of P_3 from lipid fractions

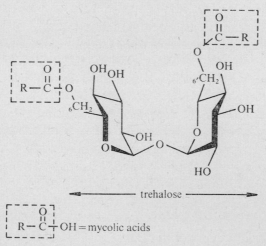

Fig. 5. General formula of cord factor (trehalose 6,6'-dimycolate).

of BCG that are able to restore the biological properties of NaOH-treated cell walls (Azuma *et al.*, 1974) and from *M. tuberculosis* strain Aoyama B (Ribi *et al.*, 1976). The mycolic acids present in cord factor vary from strain to strain; in any given strain a mixture of various of the acids is found. Lower homologues of cord factors have been isolated from *Corynebacterium* spp. (Ioneda, Lenz & Pudlès, 1963), *Arthrobacter* spp. (Saito *et al.*, 1976) and *Nocardia* spp. (Ioneda, Lederer & Rozanis, 1970). They contain C-32 to C-56 mycolic acids. 6,6′-Diesters of trehalose can be synthesised by various methods (Polonsky, Soler & Varenne, 1977; Toubiana & Toubiana, 1973, 1975; Tocanne, 1975) and some of them have biological properties similar to those of natural cord factors (Yarkoni *et al.*, 1973).

Other free lipids such as mycosides and sulpholipids have been reviewed (Goren, 1972; Lederer *et al.*, 1975) and will not be considered here.

'NON-PEPTIDOGLYCAN' AMINO ACIDS

Peptide chains

The non-peptidoglycan amino acids of mycobacterial cell walls represent 10–15% of their weight, most of which can be removed by treatment with proteolytic enzymes (Misaki *et al.*, 1966; Wietzerbin-Falszpan *et al.*, 1973; Azuma *et al.*, 1974). The peptide chains hydrolysed by these enzymes contain the usual amino acids of proteins. Their linkage to the other cell wall constituents is unknown. However, the basal layer always contains small amounts of amino acids which do not belong to the peptidoglycan: they are presumed to represent the point of attachment to the peptidoglycan of the peptide chains hydrolysed by proteolytic enzymes. The peptide chains thus removed have a role in the biological activities of the cell wall. Misaki *et al.* (1966) have shown that treatment with proteolytic enzymes suppresses the ability of BCG cell walls to elicit a delayed-type hypersensitivity response in guinea pigs sensitised to tuberculin. Azuma *et al.* (1969) have isolated, by partial acid hydrolysis of crude cell walls, a 'tuberculin active peptide' devoid of A_2pm and composed of 'non-peptidoglycan amino acids'. It has also been shown that treatment with proteolytic enzymes diminishes the anti-tumour activity of BCG cell walls (Zbar *et al.*, 1974).

Poly-α-L-glutamic acid

Purified cell walls of pathogenic mycobacteria (Migliore, Acharya & Jollès, 1966) and of *M. bovis* BCG (Wietzerbin-Falszpan *et al.*, 1973) contain larger quantities of glutamic acid than could be accommodated by the peptidoglycan structure. This 'excess' glutamic acid, which is the L-isomer, remains in the basal layer; it can be extracted by partial acid hydrolysis (Wietzerbin-Falszpan et al., 1973) in the form of a partly amidated poly-α-L-glutamic acid (Wietzerbin, Lederer & Petit, 1975). This polymer is absent from saprophytic strains but is present in pathogenic and most strains of mycobacteria used for vaccination. Its amount varies from 8% of the weight of the basal layer in M. tuberculosis strain Brévannes to 2% in BCG strain Pasteur (Wietzerbin et al., 1973; Phiet et al., 1976). Its presence seems thus related to virulence but its exact role is not yet clear.

IMMUNOSTIMULANT PROPERTIES OF CELL WALLS AND THEIR COMPONENTS

As stated in the introduction, most of the early studies on the chemistry of mycobacteria were oriented towards the preparation of a non-living vaccine against tuberculosis and towards the identification of the components of the mycobacterial cell able to modify the immune response to antigens unrelated to mycobacteria. Experiments in vivo are now complemented by studies both of the effects of active products on isolated cells in vitro and of the mechanism of action of purified mycobacterial constituents on well-defined subpopulations of the cellular immune system.

We shall discuss the following biological activities: (1) adjuvant activity; (2) induction of specific and non-specific resistance to infections; (3) anti-tumour activity.

Adjuvant activity

An adjuvant is a substance which increases and/or changes the characteristics of the immune response to an antigenically unrelated antigen. Freund's Complete Adjuvant (FCA), most frequently used experimentally, is prepared by adding killed mycobacteria to 'Freund's Incomplete Adjuvant' (FIA: a water-in-oil emulsion containing an antigen in the water phase). Comparison of the immune response to an antigen injected either in FIA or in FCA shows that the addition of killed mycobacteria increases the humoral response (circulating antibodies) and induces a cellular immunity to the antigen which can be detected by a test of delayed hypersensitivity, such as the skin test. To identify the active structure in the mycobacterial cell, the ability of defined fractions to replace killed mycobacteria in Freund's adjuvant has usually been evaluated using guinea pigs as experimental animals. It was first shown by White, Bernstock, Johns & Lederer (1958) and White, Jollès, Samour & Lederer (1964) that wax D containing a peptidoglycan moiety (i.e. wax D of human strains) was active, whereas wax D devoid of peptidoglycan was not so. The covalent skeleton of mycobacterial cell walls, which has the same overall composition as active wax D, is also adjuvant (Azuma, Kishimoto, Yamamura & Petit, 1971; Adam, Ciorbaru, Petit & Lederer, 1972). The next step was the preparation of water-soluble adjuvants by lysozyme treatment of purified cell walls of *M. smegmatis* (Adam *et al.*, 1972). The first water-soluble adjuvant (WSA) obtained was a peptidoglycan-arabinogalactan. of MW 20000 daltons which thus had the structure of a demycolated wax D. WSA is devoid of the toxic properties of whole mycobacteria; it does not induce tuberculin hypersensitivity, polyarthritis, nor increase susceptibility to endotoxins (Chedid *et al.*, 1972). Products similar to WSA were obtained by extraction of delipidated cells of *M. tuberculosis* (Migliore-Samour & Jollès, 1972), by hydrogenolysis under acidic conditions of delipidated BCG cells (Hiu, 1972) or by lysozyme treatment of delipidated cells of either *M. smegmatis* or *N. opaca* (Adam *et al.*, 1973).

The last step in the identification of the adjuvant structure was the demonstration that peptidoglycans and their soluble fragments, whether isolated from mycobacteria or from other sources, were able to replace killed mycobacteria in Freund's adjuvant. Nauciel, Fleck, Martin & Mock (1973) showed that insoluble peptidoglycans of Gram-negative bacteria can induce a delayed hypersensitivity to ABA-tyrosine. Adam *et al.* (1974a) showed that soluble peptidoglycan fragments from mycobacteria and *E. coli* were adjuvant, as was a tetra-saccharide heptapeptide obtained by Migliore-Samour & Jollès (1973) by acetolysis of wax D. Kotani *et al.* (1975a) reported the adjuvant activity of peptidoglycan fragments obtained by the action of various bacteriolytic enzymes on the cell walls from 20 Gram-positive species. The way was open to define the minimum structure active as adjuvant.

The monomeric subunit of the peptidoglycan of mycobacteria (see Fig. 2) (Adam *et al.*, 1974b) and that of *E. coli*, which differs from the former by the presence of a *N*-acetylated muramic acid instead of *N*-glycolylmuramic acid and by the absence of amides in the peptide, are active (Adam *et al.*, 1974b; Nauciel, Fleck, Mock & Martin, 1973) as is the monomeric subunit of the lysine-containing peptidoglycans prepared from *Micrococcus roseus* (Ellouz, Adam, Ciorbaru & Lederer, 1974) and *Staphylococcus aureus* (Kotani *et al.*, 1975c). Removal of the terminal D-alanine of the disaccharide tetrapeptide of mycobacterial origin does not reduce its activity (Adam *et al.*, 1974*a*). *N*-acetylglucosamine can also be removed as the *N*-acetylmuramyl-tripeptides which can accumulate in cells of *S. epidermidis* or *B. megaterium* treated with D-cycloserine are also adjuvant (Ellouz *et al.*, 1974; Adam *et al.*, 1975).

Finally, the minimal adjuvant structure was obtained by chemical synthesis (Merser, Sinaÿ & Adam, 1975) as *N*-acetylmuramyl-L-alanyl-D-isoglutamine (MDP) (Fig. 6) (Ellouz *et al.*, 1974; Adam *et al.*, 1975), *N*-acetylmuramyl-D-alanine being inactive (Ellouz *et al.*, 1974). These results have been confirmed by Kotani *et al.* (1975*c*).

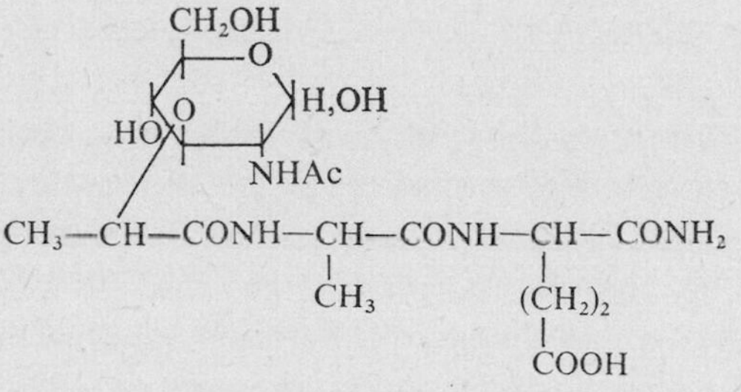

CH₂OH
O
HO O H₂OH
NHAc
CH₃—CH—CONH—CH—CONH—CH—CONH₂
CH₃   (CH₂)₂
COOH

Fig. 6. Chemical structure of the minimal adjuvant structure (MDP: *N*-acetylmuramyl-L-alanyl-D-isoglutamine).

The identification of MDP as the minimal adjuvant structure opened the way to a systematic study of the relations between structure and adjuvant activity of peptidoglycan derivatives. Two aspects were investigated: the ability of a series of *N*-acetylmuramyl-oligopeptides to replace mycobacteria in Freund's adjuvant, and the activity of these compounds in other tests and in other species. The activity of *N*-acetyl-muramyl-dipeptides in Freund's adjuvant using the guinea-pig test is critically dependent upon the presence of an amide group on the α-carboxyl of the isoglutamyl residue so long as the γ-carboxyl group of this amino acid is free. The non-amidated muramyl-dipeptide is almost completely inactive (Kotani *et al.*, 1975*b*; Adam *et al.*, 1976; Audibert, Chedid, Lefrancier & Choay, 1976; Chedid *et al.*, 1976; Audibert *et al.*, 1977) but methylation of the amide group leaves the activity intact (Chedid *et al.*, 1976). The dimethyl ester is active, but is less so than MDP (Adam *et al.*, 1976; Chedid *et al.*, 1976). Substitution of the γ-carboxyl group of glutamic acid by L-lysine, by meso-diaminopimelic acid or by a longer sequence as in the *N*-acetylmuramyl-tri- and -pentapeptides obtained by hydrolysis of cell wall precursors, also give active products (Ellouz *et al.*, 1974). Replacement of L-alanine by L-serine which occurs in the peptidoglycan of some bacteria (Schleifer

& Kandler, 1972) leaves the activity intact. Replacement by glycine, which also occurs in some natural peptidoglycans, gives a less active compound (Adam *et al.*, 1976). Replacement of D- by L-glutamic acid (Kotani *et al.*, 1975b; Chedid *et al.*, 1976), by D-aspartic acid (Kotani *et al.*, 1975b; Audibert *et al.*, 1977) or by D-norleucine (Audibert *et al.* 1977) gives inactive compounds. Replacement of L-alanine by the D-isomer gives a compound which diminishes the activity of FCA and of MDP in FIA and thus is an anti-adjuvant (Adam *et al.*, 1976).

It thus seems, that a structure close to that of the natural peptidoglycans is necessary for activity. Chedid *et al.* (1976) showed that MDP and some analogues were able to increase the humoral response to bovine serum albumin (BSA) in the mouse, when injected in saline. They are active when injected by routes other than that for the antigen and are so even when administered orally. Products active in the guinea pig test are also active in the mouse test. The non-amidated analogue of MDP and its α-methyl ester, which are inactive in the guinea pig test, are active in the mouse. The analogues of MDP where L-alanine is replaced by the D-isomer, which inhibits the adjuvant activity of MDP in Freund's adjuvant, also decreases the response of mice to BSA; this compound is thus an immunodepressor. It is then clear that the *N*-acylmuramyl-peptide structure present in all peptidoglycans is a powerful modulator of the immune response.

Some experiments have been done to identify the target cells of soluble adjuvants, in an in-vitro system involving the response of mouse spleen cells to sheep red blood cells (SRBC). Modolell, Luckenbach, Parant & Munder (1974) showed that WSA was probably acting via macrophages and Juy & Chedid (1975) showed that WSA and MDP both act on macrophages. Addition of these products to mouse peritoneal macrophages in vitro enables them to inhibit the growth of syngeneic tumour cells. Löwy, Bona & Chedid (1977) showed that the ability of MDP to increase the response of mice to SRBC in vivo was mediated by T-lymphocytes.

Besides their adjuvant properties, soluble peptidoglycan fragments from nocardiae have proved mitogenic for B-lymphocytes from mouse and rabbits (Ciorbaru *et al.*, 1976). The minimal structure required for mitogenic activity is more complex than that required for adjuvant activity. Peptidoglycan fragments with intact glycan chains obtained by the action of *Streptomyces albus* endopeptidase (Ghuysen *et al.*, 1969) are mitogenic and adjuvant whereas the fragments obtained by lysozyme action are only adjuvant (Ciorbaru *et al.*, 1976).

Induction of homologous or heterologous resistance to infections

Intravenous injection of crude cell walls of BCG on the surface of oil droplets suspended in saline ('cell wall vaccine') protects mice and monkeys against a respiratory challenge by *M. tuberculosis* (Ribi *et al.*, 1966; Anacker *et al.*, 1969*a*; Ribi *et al.*, 1971). Delipidation and alkaline treatment of the walls destroys their protective capacity which can be restored by 'P$_3$', a mixture of trehalose dimycolates (cord factor) (Anacker *et al.*, 1969*b*, 1973). Cord factor alone, if injected intravenously in an emulsion containing 1% mineral oil, can protect mice against the same challenge (Bekierkunst *et al.*, 1969). The activity of cell walls and cord factor in this test has been related to the formation of granulomas in the lung. Cell walls and cord factor are also able to protect mice against infectious agents unrelated to mycobacteria – that is, they induce a non-specific resistance. Crude BCG cell walls in the presence of mineral oil or peanut oil injected intravenously can protect mice against a challenge by *Klebsiella pneumoniae* and *Listeria monocytogenes* injected by the same route; addition of an equal weight of cord factor increases this protective activity. Cord factor alone is active used either in an emulsion containing mineral or peanut oil or in suspension in saline prepared as described by Kato (1967) (Parant *et al.*, 1977). Yarkoni & Bekierkunst (1976) had shown that cord factor injected intraperitoneally in an emulsion containing mineral oil was able to protect mice against a challenge with *Salmonella typhi* or *S. typhimurium* injected 3–7 days later by the same route.

Chedid *et al.* (1977) have reported that MDP and four analogues (namely the non-amidated analogue, the α-methyl ester and the α,γ-dimethyl ester) protect mice against a challenge with *K. pneumoniae* injected 24 hours later. MDP and the analogues can be injected by various routes or even administered orally. WSA had been shown to prolong the length of the survival time but not to induce survival (Elin, Wolff & Chedid, 1976). Thus there are two principal types of chemically well-defined compounds which can stimulate non-specific resistance: cord factor and MDP and some of its analogues.

Anti-tumour activity

Tests of the anti-tumour activity of a preparation can be classified into three categories. (1) Prevention: to test this, the material is injected prior to the inoculation of the tumour cells. (2) Suppression: for this test the tumour cells are mixed with the material. (3) Regression: injection of the material is made in an already established tumour.

In the first type of test, purified cell walls of *M. kansasii* injected intraperitoneally protect mice against a syngeneic leukaemia and the Ehrlich carcinoma administered by the same route; addition of mineral oil increases the activity. The 'basal layer' (see above) is inactive injected as a saline suspension but is active in the presence of oil (Chedid *et al.*, 1973). Crude cell walls of BCG injected intraperitoneally in an emulsion containing mineral or peanut oil protect mice against the L1210 leukemia administered by the same route. This activity is increased by coating the cell wall with an equal weight of cord factor. Here again, as in the induction of non-specific resistance to infections, cord factor alone is active in the presence of mineral or peanut oil; MDP and its analogues are inactive (Leclerc *et al.*, 1976).

Cord factor injected intravenously in an oil containing emulsion can suppress urethan-induced lung adenomas in mice (Bekierkunst *et al.*, 1971) and injected intraperitoneally can prevent the growth of Ehrlich ascites carcinoma injected 3 days later by the same route (Yarkoni *et al.*, 1974). Analogues of cord factor (trehalose 6,6'-dipalmitate, sucrose 6,6'-dimycolate and trehalose monopalmitate) are active in this test (Yarkoni *et al.*, 1973).

The second type of test is usually done with tumours that are able to grow intradermally and the most widely used is the line 10 hepatoma of line 2 guinea pigs developed by Rapp (see for instance Zbar, Bernstein & Rapp, 1971). The 'cell wall vaccine' of Ribi (see above), prepared with crude cell walls from BCG (Zbar *et al.*, 1972), from *M. tuberculosis*, from *M. kansasii* (Zbar *et al.*, 1974) or from *M. phlei* (Gray *et al.*, 1975) is active. Purified cell walls ('cell wall skeleton') from BCG (Meyer, Ribi, Azuma & Zbar, 1974; Yamamura *et al.*, 1976*d*) and from *M. phlei* (Gray *et al.*, 1975), attached to oil droplets, are also active. Purified cell walls from various mycobacteria, nocardiae and corynebacteria are active in similar tests with mouse tumours (Yamamura *et al.*, 1976*a*).

In the third type of test which has also mainly been performed with the guinea pig system of Rapp, crude BCG cell walls are active in the form of 'cell wall vaccine', purified cell walls are poorly active but their activity is restored by addition of an equal weight of P_3 (cord factor) (Meyer *et al.*, 1974; Ribi *et al.*, 1976). Similar results have been obtained by Bekierkunst *et al.* (1974) using cell walls from *M. tuberculosis* H_{37}Ra. In this type of test, cord factor alone is inactive. It has recently been shown (Ribi *et al.*, 1976) that endotoxins from Gram-negative bacteria can replace purified cell walls in this system. Antigenicity of the endotoxin preparations seems to play an important role.

The requirements for prevention, suppression or regression are thus different. In all cases, the presence of oil is either necessary or at least greatly increases the activity. Cord factor or the purified cell walls are active in prevention. Purified cell walls induce suppression. To obtain regression of established tumours, however, both cord factor and purified cell walls are necessary, but the latter can be replaced by endotoxin. Strongly immunogenic preparations are more effective in the regression test.

Two other mycobacterial preparations (MER and IPM) have anti-tumour activity in the absence of oil. They contain the cell wall, cytoplasmic constituents and various lipids, among them cord factor. MER is the residue of extraction of BCG by acetone and hot methanol. It has been extensively studied by Weiss (1972) and is active in tumour prevention (Weiss, Stupp, Mang & Izak, 1975). IPM (inter phase material) prepared from *M. smegmatis* cells is active in prevention of syngeneic leukaemias in mice (Lamensans *et al.*, 1975). Clinical applications of MER (Yron *et al.*, 1973; Weiss *et al.*, 1975; Perloff, Holland, Lumb & Bekesi, 1977) and purified cell walls of BCG attached to oil droplets (Yamamura *et al.*, 1975*b*, 1976) have been undertaken with encouraging results in anti-cancer immunotherapy

REFERENCES

ACHARYA, N. P. V., SENN, M. & LEDERER, E. (1967). Sur la présence et la structure de mycolates d'arabinose dans les lipides liés de deux souches de Mycobactéries. *Comptes rendus de l'Académie des Sciences*, Série C, **264**, 2173–6.

ADAM, A., AMAR, C., CIORBARU, R., LEDERER, E., PETIT, J. F. & VILKAS, E. (1974*a*). Activité adjuvante des peptidoglycanes de Mycobactéries. *Comptes rendus de l'Académie des Sciences*, Série D, **278**, 799–801.

ADAM, A., CIORBARU, R., ELLOUZ, F., PETIT, J. F. & LEDERER, E. (1974*b*). Adjuvant activity of monomeric bacterial cell wall peptidoglycan. *Biochemical and Biophysical Research Communications*, **56**, 561–7.

ADAM, A., CIORBARU, R., PETIT, J. F. & LEDERER, E. (1972). Isolation and properties of a macromolecular water-soluble immunoadjuvant fraction from the cell wall of *Mycobacterium smegmatis*. *Proceedings of the National Academy of Sciences, USA*, **69**, 851–4.

ADAM, A., CIORBARU, R., PETIT, J. F., LEDERER, E., CHEDID, L., LAMENSANS, A., PARANT, F., PARANT, M., ROSSELET, J. P. & BERGER, F. M. (1973). Preparation and biological properties of water soluble adjuvant fractions from delipidated cells of *Mycobacterium smegmatis* and *Nocardia opaca*. *Infection and Immunity*, **7**, 855–61.

ADAM, A., DEVYS, M., SOUVANNAVONG, V., LEFRANCIER, P., CHOAY, J. & LEDERER, E. (1976). Correlation of structure and adjuvant activity of *N*-acetyl-muramyl-L-alanyl-D-isoglutamine (MDP), its derivatives and analogues. Anti-adjuvant and competition properties of stereoisomers. *Biochemical and Biophysical Research Communications*, **72**, 339–46.

ADAM, A., ELLOUZ, F., CIORBARU, R., PETIT, J. F. & LEDERER, E. (1975). Peptidoglycan adjuvants: minimal structure required for activity. *Zeitschrift für
Immunitätsforschung*, **149**, 341–8.

ADAM, A., PETIT, J. F., WIETZERBIN-FALSZPAN, J., SINAY, P., THOMAS, D. W. &
LEDERER, E. (1969). L'acide *N*-glycolyl-muramique, constituant des parois de
Mycobacterium smegmatis: identification par spectrométrie de masse. *FEBS
Letters*, **4**, 87–92

ADAM, A., SENN, M., VILKAS, E. & LEDERER, E. (1967). Spectrométrie de masse de
glycolipides. 2. Diesters de tréhalose naturels et synthétiques. *European Journal
of Biochemistry*, **2**, 460–8.

AMAR, C. & VILKAS, E. (1973). Isolement d'un phosphate d'arabinose à partir des
parois de *Mycobacterium tuberculosis* H37 Ra. *Comptes rendus de l'Académie
des Sciences*, Série D, **227**, 1949–51.

AMAR-NACASCH, C. & VILKAS, E. (1969). Etude des parois d'une souche humaine
virulente de *Mycobacterium tuberculosis* (1). Préparation et analyse chimique.
Bulletin de la Société de Chimie Biologique, Paris, **51**, 613–20.

AMAR-NACASCH, C. & VILKAS, E. (1970). Etude des parois de *Mycobacterium
tuberculosis* (2). Mise en évidence d'un mycolate d'arabinose et d'un glucane
dans les parois de *M-tuberculosis* H37 Ra. *Bulletin de la Société de Chimie
Biologique, Paris*, **52**, 145–51.

ANACKER, R. L., BARKLAY, W. R., BREHMER, W., GOODE, G., LIST, R. H., RIBI,
E. & TARMINA, D. F. (1969*a*). Effectiveness of cell walls of *Mycobacterium
bovis* strain BCG administered by various routes and different adjuvants in
protecting mice against airborne infection with *Mycobacterium tuberculosis*
strain H37 RV. *American Review of Respiratory Diseases*, **99**, 242–8.

ANACKER, R. L., BICKEL, W. D., BREHMER, W., NIWA, M., RIBI, E. & TARMINA,
D. F. (1969*b*). Immunization of mice by combinations of inactive fractions of
Mycobacterium tuberculosis strain BCG (33642). *Proceedings of the Society for
Experimental Biology and Medicine*, **130**, 723–9.

ANACKER, R. L., MATSUMOTO, J., RIBI, E., SMITH, R. F. & YAMAMOTO, K. (1973).
Enhancement of resistance of mice to tuberculosis by purified components of
mycobacterial lipid fractions, *Journal of Infectious Diseases*, **127**, 357–64.

ASSELINEAU, J. (1966). *The Bacterial Lipids*. Paris: Hermann. San Francisco:
Holden-Day.

AUDIBERT, F., CHEDID, L., LEFRANCIER, P. & CHOAY, J. (1976): Distinctive adjuvanticity of synthetic analogs of mycobacterial water-soluble components.
Cellular Immunology, **21**, 243–9.

AUDIBERT, F., CHEDID, L., LEFRANCIER, P., CHOAY, J. & LEDERER, E. (1977).
Relationship between chemical structure and adjuvant activity of some synthetic analogues of *N*-acetyl-muramyl-L-alanyl-D-isoglutamine (MDP). *Annales d'Immunologie* (*Institut Pasteur*), **128c**, 653–61.

AZUMA, I., KIMURA, H. & YAMAMURA, Y. (1968). Chemical and immunological
properties of polysaccharides of wax D extracted from *Mycobacterium tuberculosis* strain Aoyama B. *Journal of Bacteriology*, **96**, 567–8.

AZUMA, I., KISHIMOTO, S., YAMAMURA, Y. & PETIT, J. F. (1971). Adjuvanticity of
mycobacterial cell walls. *Japanese Journal of Microbiology*, **15**, 193–7.

AZUMA, I., RIBI, E., MEYER, T. & Zbar, B. (1974). Biologically active components
from mycobacterial cell walls. I. Isolation and composition of cell wall skeleton
and component P_3. *Journal of the National Cancer Institute*, **52**, 95–101.

AZUMA, I., THOMAS, D. W., ADAM, A., GHUYSEN, J. M., BONALY, R., PETIT, J. F.
& LEDERER, E. (1970). Occurrence of *N*-glycolylmuramic acid in bacterial cell
walls. A preliminary survey. *Biochimica et Biophysica Acta*, **208**, 444–51.

AZUMA, I. & YAMAMURA, Y. (1962). Studies on the firmly bound lipids of human tubercle bacillus. I. Isolation of arabinose mycolate. *Journal of Biochemistry, Tokyo*, **52**, 200–6.

AZUMA, I. & YAMAMURA, Y. (1963). Studies on the firmly bound lipids of human tubercle bacillus. II. Isolation of arabinose mycolate and identification of its chemical structure. *Journal of Biochemistry, Tokyo*, **53**, 275–81.

AZUMA, I., YAMAMURA, Y. & FUKUSHI, K. (1968). Fractionation of mycobacterial cell wall. Isolation of arabinose mycolate and arabinogalactan from cell wall fraction of *Mycobacterium tuberculosis* strain Aoyama B. *Journal of Bacteriology*, **96**, 1885–7.

AZUMA, I., YAMAMURA, Y., TAKARA, T., ONONE, K. & FUKUSHI, K. (1969). Isolation of tuberculin active peptides from cell wall fraction of human tubercle bacillus strain Aoyama B. *Japanese Journal of Microbiology*, **13**, 220–2.

BARKSDALE, L. & KIM, K. S. (1977). Mycobacterium. *Bacteriological Reviews*, **41**, 217–372.

BEKIERKUNST, A. (1975). Immunotherapy and vaccination against cancer with non-living BCG and cord factor (trehalose 6-6′-dimycolate). *International Journal of Cancer*, **16**, 442–7.

BEKIERKUNST, A., LEVIJ, I. S., YARKONI, E., VILKAS, E., ADAM, A. & LEDERER, E. (1969). Granuloma formation induced in mice by chemically defined mycobacterial fractions. *Journal of Bacteriology*, **100**, 95–102.

BEKIERKUNST, A., LEVIJ, I. S., YARKONI, E., VILKAS, E. & LEDERER, E. (1971). Suppression of urethan-induced lung adenomas in mice treated with trehalose 6-6′-dimycolate (cord factor) and living bacillus Calmette Guérin. *Science*, **174**, 1240.

BEKIERKUNST, A., WANG, L., TOUBIANA, R. & LEDERER, E. (1974). Immunotherapy of cancer with non-living BCG and fractions derived from mycobacteria: role of cord factor (trehalose-6,6′-dimycolate) in tumor regression. *Infection and Immunity*, **10**, 1044–50.

BIOZZI, G., STIFFEL, C., HALPERN, B. N. & MOUTON, D. (1959). Effet de l'inoculation du bacille de Calmette-Guerin sur le dévelopement de la tumeur ascitique d'Ehrlich chez la souris. *Comptes rendus des Séances de la Société de Biologie, Paris*, **153**, 987–9.

BLANDEN, R. V., LEFFORD, M. J. & MACKANNES, G. B. (1969). The host response to Calmette-Guérin bacillus infection in mice. *Journal of Experimental Medicine*, **129**, 1079–101.

BLOCH, H. (1950). Studies on the virulence of tubercle bacilli. Isolation and biological properties of a constituent of virulent organisms. *Journal of Experimental Medicine*, **91**, 197–218.

CHEDID, L., AUDIBERT, A., LEFRANCIER, P., CHOAY, J. & LEDERER, E. (1976). Modulation of the immune response by a synthetic adjuvant and analogs. *Proceedings of the National Academy of Sciences, USA*, **73**, 2472–5.

CHEDID, L., LAMENSANS, A., PARANT, F., PARANT, M., ADAM, A., PETIT, J. F. & LEDERER, E. (1973). Protective effect of delipidated mycobacterial cells and purified cell walls against Ehrlich carcinoma and a syngeneic lymphoid leukemia in mice. *Cancer Research*, **33**, 2187–95.

CHEDID, L., PARANT, M., PARANT, F., GUSTAFSON, R. H. & BERGER, F. M. (1972). Biological studies of a non-toxic, water-soluble immuno-adjuvant from mycobacterial cell walls. *Proceedings of the National Academy of Sciences, USA*, **69**, 855–8.

CHEDID, L., PARANT, M., PARANT, F., LEFRANCIER, P., CHOAY, J. & LEDERER, E. (1977). Enhancement of non-specific immunity to *Klebsiella pneumoniae* infection by a synthetic immunoadjuvant (*N*-acetyl-muramyl-L-alanyl-

D-isoglutamine) and several analogs. *Proceedings of the National Academy of Sciences, USA*, **74**, 2089–93.

CIORBARU, R., PETIT, J. F., LEDERER, E., ZISSMAN, E., BONA, C. & CHEDID, L. (1976). Presence and subcellular localization of two distinct mitogenic fractions in the cells of *Nocardia rubra* and *Nocardia opaca*: preparation of soluble mitogenic peptidoglycan fractions. *Infection and Immunity*, **13**, 1084–90.

DIENES, L & SCHONHEIT, E. W. (1928). Local hypersensitiveness. I. Sensitization of tuberculous guinea pigs with egg-white and Timothy-pollen. *Journal of Immunology*, **14**, 9–42.

DUBOS, R. & SCHAEDLER, R. W. (1957). Effects of cellular constituents of mycobacteria on the resistance of mice to heterologous infections. *Journal of Experimental Medicine*, **106**, 703–26.

ELIN, R. J., WOLFF, S. M. & CHEDID, L. (1976). Non-specific resistance to infection in mice induced by water-soluble adjuvant derived from *Mycobacterium smegmatis*. *Journal of Infectious Diseases*, **133**, 500–5.

ELLOUZ, F., ADAM, A., CIORBARU, R. & LEDERER, E. (1974). Minimal structural requirements for adjuvant activity of bacterial peptidoglycan derivatives. *Biochemical and Biophysical Research Communications*, **59**, 1317–25.

ETEMADI, A. H. (1967). Structure, biogenesis and phylogenetic interest of mycolic acids. *Exposés Annuels de Biochimie Médicale*, **28**, 77–109.

FREUND, J. (1956). The mode of action of immunologic adjuvant. *Advances in Tuberculology*, **7**, 130–48.

GHUYSEN, J. M., DIERICKX, L., COYETTE, J., LEYH-BOUILLE, M., GUINAND, M. & CAMPBELL, J. N. (1969). An improved technique for the preparation of *Streptomyces* peptidases and N-acetylmuramyl-L-alanine amidase active on bacterial wall peptidoglycans. *Biochemistry*, **8**, 213–22.

GOREN, M. (1972). Mycobacterial lipids: selected topics. *Bacteriological Reviews*, **36**, 33–64.

GRAY, G. R., RIBI, E., GRANGER, D. L., PARKER, R., AZUMA, I. & YAMAMOTO, K. (1975). Immunotherapy of cancer: tumor suppression and regression by cell walls of *Mycobacterium phlei* attached to oil droplets. *Journal of the National Cancer Institute*, **55**, 727–9.

HIU, I. J. (1972). Water-soluble and lipid-free fraction from BCG with adjuvant and antitumor activity. *Nature New Biology*, **238**, 241–2.

HOWARD, J. G., BIOZZI, G., HALPERN, B. N., STIFFEL, C. & MOUTON, D. (1959). The effect of *Mycobacterium tuberculosis* (BCG) infection on the resistance of mice to bacteria endotoxin and *Salmonella enteritidis* infection. *British Journal of Experimental Pathology*, **40**, 281–90.

IONEDA, T., LEDERER, E. & ROZANIS, J. (1970). Sur la structure des diesters de tréhalose ('cord factors') produits par *Nocardia asteroides* et *Nocardia rhodococcus*. *Chemistry and Physics of Lipids*, **4**, 375–92.

JUY, D. & CHEDID, L. (1975). Comparison between macrophage activation and enhancement of non-specific resistance to tumors by mycobacterial immunoadjuvants. *Proceedings of the National Academy of Sciences, USA*, **72**, 4105–9.

KANETSUNA, F. (1968). Chemical analysis of mycobacterial cell walls. *Biochimica et Biophysica Acta*, **158**, 130–8.

KANETSUNA, F. & SAN BLAS, G. (1970). Chemical analysis of a mycolic acid–arabino-galactan-mucopeptide complex of mycobacterial cell wall. *Biochimica et Biophysica Acta*, **208**, 434–43.

KATO, M. (1967). Procedure for the preparation of aqueous suspension of cord factor. *American Review of Respiratory Diseases*, **98**, 260–9.

KOTANI, S., NARITA, T., STEWART-TULL, D. E. S., SHIMONO, T., WATANABE, Y., KATO, K. & IWATA, S. (1975*a*). Immunoadjuvant activities of cell walls and their water-soluble fractions prepared from various Gram-positive bacteria. *Biken Journal*, **18**, 77–92.

KOTANI, S., WATANABE, Y., KINOSHITA, F., SHIMONO, T., MORISAKI, I., SHIBA, T., KUSUMOTO, S., TARUMI, Y. & IKENAKA, K. (1975*b*): Immunoadjuvant activities of synthetic *N*-acetylmuramyl-peptides or amino acids. *Biken Journal*, **18**, 105–11.

KOTANI, S., WATANABE, Y., SHIMONO, T., KINOSHITA, F., NARITA, T., KATO, K., STEWART-TULL, D. E. S., MORISAKI, I., YOKIGAWA, K. & KAWATA, S. (1975*c*). Immunoadjuvant activities of peptidoglycan subunits from the cell walls of., *Staphylococcus aureus* and *Lactobacillus plantarum. Biken Journal*, **18**, 93–103.

LAMENSANS, A., CHEDID, L., LEDERER, E., ROSSELET, J. P., LUDWIG, B., BERGER, F. M., GUSTAFSON, R. H. & SPENCER, H. J. (1975). Enhancement of immunity against murine syngeneic tumors by a fraction extracted from non-pathogenic mycobacteria. *Proceedings of the National Academy of Sciences, USA*, **72**, 3656–60.

LAUCIUS, J. F., BODURTHA, A. J., MASTRANGELO, M. J. & CREECH, R. H. (1974). Bacillus Calmette Guérin in the treatment of neoplastic disease. *Journal of the Reticuloendothelial Society*, **16**, 347–58.

LECHEVALIER, M. P. (1977). Lipids in bacterial taxonomy, A taxonomist's view. *Critical Reviews of Microbiology*, **15**, 109–210.

LECHEVALIER, M. P., HORAN, A. C. & LECHEVALIER, H. (1971). Lipid composition in the classification of nocardiae and mycobacteria, *Journal of Bacteriology*, **105**, 313–18.

LECHEVALIER, M. P., LECHEVALIER, H. & HORAN, A. C. (1973). Chemical characteristics and classification of nocardiae. *Canadian Journal of Microbiology*, **19**, 965–72.

LECLERC, C., LAMENSANS, A., CHEDID, L., DRAPIER, J. C., PETIT, J. F., WIETZERBIN, J. & LEDERER, E. (1976). Non-specific immunoprevention of L1210 leukemia by cord factor (6,6′-dimycolate of trehalose) administered in a metabolizable oil. *Cancer Immunology and Immunotherapy*, **1**, 227–32.

LEDERER, E. (1967). Glycolipids of mycobacteria and related microorganisms. *Chemistry and Physics of Lipids*, **1**, 294–308.

LEDERER, E. (1971). The mycobacterial cell wall. *Pure and Applied Chemistry*, **25**, 135–65.

LEDERER, E. (1976). Cord factor and related trehalose esters. *Chemistry and Physics of Lipids*, **16**, 91–106.

LEDERER, E., ADAM, A., CIORBARU, R., PETIT, J. F. & WIETZERBIN, J. (1975). Cell walls of mycobacteria and related organisms; chemistry and immunostimulant properties. *Molecular and Cellular Biochemistry*, **7**, 87–104.

LEDERER, E. & PUDLÈS, J. (1951). Sur l'isolement et la constitution chimique d'un hydroxy-acide ramifié du bacille diphtérique. *Bulletin de la Société Chimique Biologique*, **33**, 1003–16.

LEWIS, P. A. & LOOMIS, D. (1924). Allergic irritability. The formation of anti-sheep hemolytic amboceptor in the normal and tuberculosis guinea-pig. *Journal of Experimental Medicine*, **40**, 503–15.

LEYH-BOUILLE, M., GHUYSEN, J. M., BONALY, R., NIETO, M., PERKINS, H. R., & KANDLER, O. (1970). Substrate requirement of the *Streptomyces albus* G D-D-carboxypeptidase. *Biochemistry*, **9**, 2961–70.

LIU, T. Y. & GOTSCHLICH, E. C. (1967). Muramic acid phosphate as a component of the mucopeptide of Gram-positive bacteria. *Journal of Biological Chemistry*, **242**, 471–6.

LÖWY, I., BONA, C. & CHEDID, L. (1977). Target cells for the activity of a synthetic adjuvant, muramyl dipeptide. *Cellular Immunology*, **29**, 195–9.

MACKANESS, G. B., AUCLAIR, D. J. & LAGRANGE, P. H. (1973). Immunopotentiation with BCG. I. Immune response to different strains and preparations. *Journal of the National Cancer Institute*, **51**, 1655–67.

MACKANESS, G. B., LAGRANGE, P. H. & ISHIBASHI, T. (1974). The modifying effect of BCG on the immunological induction of T cells. *Journal of Experimental Medicine*, **139**, 1540–52.

MARKOVITS, J., VILKAS, E. & LEDERER, E. (1971). Sur la structure chimique des cires D, peptidoglycolipides macromoléculaires des souches humaines de *Mycobacterium tuberculosis. European Journal of Biochemistry*, **18**, 287–91.

MATHÉ, G., WEINER, R., POUILLART, P., SCHWARZENBERG, L., JASMIN, C., SCHNEIDER, M., HAYAT, M., AMIEL, J. L., DE VASSAL, F. & ROSENFELD, C. (1974). BCG in cancer immunotherapy: experimental and clinical trials of its use in treatment of leukemia minimal and/or residual disease. *National Cancer Institute, Monographs*, **39**, 165–76.

MERSER, C., SINAY, P. & ADAM, A. (1975). Total synthesis and adjuvant activity of bacterial peptidoglycan derivatives. *Biochemical and Biophysical Research Communications*, **66**, 1316–22.

MEYER, T. J., RIBI, E., AZUMA, I. & ZBAR, B. (1974). Biologically active components from mycobacterial cell walls. II. Suppression and regression of strain 2 guinea pig hepatoma. *Journal of the National Cancer Institute*, **52**, 103–11.

MIGLIORE, D., ACHARYA, N. & JOLLÈS, P. (1966). Caractérisation de quantités importantes d'acide glutamique dans les parois de Mycobactéries de souches humaines virulentes. *Comptes rendus de l'Académie des Sciences*, Series D, **263**, 846–8.

MIGLIORE, D. & JOLLÈS, P. (1968). Contribution to the study of the structure of adjuvant active waxes D from mycobacteria; isolation of a peptidoglycan. *FEBS Letters*, **2**, 7–9.

MIGLIORE, D. & JOLLÈS, P. (1969). Sur la structure de la partie azotée des cires D de *Mycobacterium tuberculosis* var *hominis. Comptes rendus de l'Académie des Sciences*, Série D **269**, 2268–70.

MIGLIORE-SAMOUR, D. & JOLLÈS, P. (1972). A hydrosoluble adjuvant active mycobacterial 'polysaccharide-peptidoglycan'. Preparation by a simple extraction technique of the bacterial cells (strain Peurois). *FEBS Letters* **25**, 301–4.

MIGLIORE-SAMOUR, D. & JOLLÈS, P. (1973). Hydrosoluble adjuvant active mycobacterial fractions of low molecular weight. *FEBS Letters*, **35**, 317–21.

MISAKI, A., SETO, N. & AZUMA, I. (1974). Structure and immunological properties of D-arabino-D-galactans isolated from cell walls of *Mycobacterium* species. *Journal of Biochemistry, Tokyo*, **76**, 15–27.

MISAKI, A. & YUKAWA, S. (1966). Studies on cell walls of mycobacteria. II. Constitution of polysaccharides from BCG cell walls. *Journal of Biochemistry, Tokyo*, **59**, 511–20.

MISAKI, A., YUKAWA, S., TSUCHIYA, K. & YAMASAKI, T. (1966). Studies on cell walls of mycobacteria. I. Chemical and biological properties of the cell walls and the mucopeptide of BCG. *Journal of Biochemistry, Tokyo*, **59**, 388–96.

MODOLELL, M. A., LUCKENBACH, G. A., PARANT, M. & MUNDER, P. (1974). The adjuvant activity of a mycobacterial water soluble adjuvant (WSA), *in vitro*. I. The requirement of macrophages. *Journal of Immunology*, **113**, 395–403.

MOMPON, B., FÉDÉRICI, C., TOUBIANA, R. & LEDERER, E. (1977). Isolation and structural determination of cord factor (trehalose-6,6'-dimycolate) from *Mycobacterium smegmatis. Chemistry and Physics of Lipids*, in press.

NATHANSON, L. (1972). Regression of intradermal malignant melanoma after intralesional injection of *Mycobacterium bovis* strain BCG. *Cancer Chemocherapy Reports*, **56**, 659–65.

NAUCIEL, C., FLECK, J., MARTIN, J. P. & MOCK, M. (1973): Activité adjuvante de peptidoglycanes de bactéries à gram négatif dans l'hypersensibilité de type retardé. *Comptes rendus de l'Academie des Sciences*, Serie D, **276**, 3499–500.

NAUCIEL, C., FLECK, J., MARTIN, J. P., MOCH, M. & NGUYEN-HUY, H. (1974). Adjuvant activity of bacterial peptidoglycans on the production of delayed hypersensitivity and on antibody response. *European Journal of Immunology*, **4**, 352–7.

NAUCIEL, C., FLECK, J., MOCK, M. & MARTIN, J. P. (1973). Activité adjuvante de fractions monomériques de peptidoglycanes bactériens dans l'hypersensibilité de type retardé. *Comptes rendus de l'Académie des Sciences*, Série D, **277**, 2841–4.

NOLL, H., BLOCH, H., ASSELINEAU, J. & LEDERER, E. (1956). Chemical structure of the cord factor of *Mycobacterium tuberculosis*. *Biochimica et Biophysica Acta*, **20**, 299–309.

OLD, L. J., CLARKE, D. A. & BENACERRAF, B. (1959). Effect of bacillus Calmette-Guérin infection on transplanted tumours in the mouse. *Nature, London*, **184**, 291–3.

PARANT, M., PARANT, F., CHEDID, L., DRAPIER, J. C., PETIT, J. F., WIETZERBIN, J. & LEDERER, E. (1977). Enhancement of non-specific immunity to bacterial infection by cord factor (6,6′-trehalose dimycolate). *Journal of Infectious Diseases*, **135**, 771–7.

PARANT, M., PARANT, F., CHEDID, L. & LE MINOR, L. (1975). Immunostimulants bactériens et protection de la souris infectée par *Klebsiella pneumoniae* résistante aux antibiotiques par mutation ou par transfert de plasmides. *Annales d'Immunologie*, **126C**, 319–26.

PERLOFF, M., HOLLAND, J. F., LUMB, G. J. & BEKESI, J. G. (1977). Effects of methanol extraction residue of bacillus Calmette-Guérin in humans. *Cancer Research*, **37**, 1191–6.

PETIT, J. F., WIETZERBIN, J., DAS, B. C. & LEDERER, E. (1975). Chemical structure of the cell wall of *Mycobacterium tuberculosis* var. *bovis*, strain BCG. *Zeitschrift für Immunitätsforschung*, **149**, 118–25.

PHIET, P. J., WIETZERBIN, J., ZISSMAN, E., PETIT, J. F. & LEDERER, E. (1976). Analysis of the cell wall of five strains of *Mycobacterium tuberculosis* BCG and of an attenuated human strain W_{115}. *Infection and Immunity*, **13**, 677–81.

PINSKI, C., HIRSHAUT, Y. & OETTGEN, H. (1972). Treatment of malignant melanoma by intratumoral injection of BCG. *Proceedings of the American Association for Cancer Research*, **13**, 21.

POLONSKY, J., SOLER, E. & VARENNE, J. (1977). Sur la synthèse du cord factor et de ses analogues. *Biochimica et Biophysica Acta*, in press.

PROMÉ, J. C., LACAVE, C., AHIDO-COFFY, A. & SAVAGNAC, A. (1976). Séparation et étude structurale des espèces moléculaires de monomycolates et de dimycolates de α-D-tréhalose présent chez *Mycobacterium phlei*. *European Journal of Biochemistry*, **63**, 543–52.

RIBI, E., LARSON, C., WICHT, W., LIST, R. & GOODE, G. (1966). Effective nonliving vaccine against experimental tuberculosis in mice. *Journal of Bacteriology*, **91**, 975–83.

RIBI, E., ANACKER, R. L., BARCLAY, W. R., BREHMER, W., HARRIS, S. C., LEIF, W. R. & SIMMONS, J. (1971). Efficacy of mycobacterial cell walls as a vaccine against airborne tuberculosis in the rhesus monkey. *Journal of Infectious Diseases*, **123**, 527–38.

RIBI, E., MILNER, C., GRANGER, D. L., KELLY, M. T., YAMAMOTO, K., BREHMER, W., PARKER, R., SMITH, R. F. & STRAIN, S. M. (1976). Immunotherapy with non-viable microbial components. *Annals of the New York Academy of Sciences*, **277**, 228–38.

SCHLEIFER, H. & KANDLER, O. (1972). Peptidoglycan types of bacterial cell walls and their taxonomic implications. *Bacteriological Reviews*, **36**, 407–77.

SULITZEANU, D., BEKIERKUNST, A., GROTO, L. & LOEBEL, J. (1962). Studies on the mechanism of non-specific resistance to *Brucella* induced in mice by vaccination with BCG. *Immunology*, **5**, 116–28.

TOUBIANA, R. & TOUBIANA, M. J. (1973). Synthèse d'analogues du cord factor. Partie II: Préparation de 6,6'-dipalmitate de tréhalose par transesterification. *Biochimie*, **55**, 575–8.

TOUBIANA, R. & TOUBIANA, M. J. (1975). Etude du cord factor et de ses analogues. Partie III: Synthèse du cord factor (6,6'-di-*O*-mycoloyl-α,α-tréhalose) et du 6,6' di-*O*-palmitoyl-α,α-tréhalose. *Carbohydrate Research*, **44**, 308–12.

TOCANNE, J. F. (1975). Sur une nouvelle voie de synthèse du cord factor glycolipide toxique de *Mycobacterium tuberculosis* (esters du tréhalose et d'acides gras α-ramifiés, β-hydroxylés). *Carbohydrate Research*, **44**, 301–7.

VILKAS, E., AMAR, C., MARKOVITS, J., VLIEGENTHART, J. F. G. & KAMERLING, J. P. (1973*a*). Occurrence of a galactofuranose disaccharide in immuno-adjuvant fractions of *Mycobacterium tuberculosis* (cell walls and wax D). *Biochimica et Biophysica Acta*, **297**, 423–35.

VILKAS, E., MARKOVITS, J., AMAR-NACASCH, C. & LEDERER, E. (1973*b*): Sur la présence d'unités de D-galactofuranose dans l'arabinogalactane des parois et des cires D de souches humaines de *Mycobacterium tuberculosis*. *Comptes rendus de l'Académie des Sciences*, Série C, **273**, 845–8.

WEISS, D. W. (1972). Non-specific stimulation and modulation of the immune response and of states of resistance by the methanol-extraction residue fraction of tubercule bacilli. *National Cancer Institute, Monographs*, **35**, 157–71.

WEISS, D. W., STUPP, Y., MANY, N. & IZAK, G. (1975). Treatment of acute myelocytic leukemia (AML) patients with the MER tubercule bacillus fraction. A preliminary report. *Transplantation Proceedings* **7** (Supplement 1), 545–52.

WHITE, R. G., BERNSTOCK, L., JOHNS, R. G. & LEDERER, E. (1958). The influence of components of *Mycobacterium tuberculosis* and other mycobacteria upon antibody production to ovalbumin. *Immunology*, **1**, 54–66.

WHITE, R. G., JOLLÈS, P., SAMOUR, D. & LEDERER, E. (1964). Correlation of adjuvant activity and chemical structure of wax D fractions of mycobacteria. *Immunology*, **7**, 158–71.

WIETZERBIN, J., DAS, B., PETIT, J. F., LEDERER, E., LEYH-BOUILLE, M. & GHUYSEN, J. M. (1974). Occurrence of D-alanyl-(D)-meso-diaminopimelic acid and meso-diaminopimelyl-meso-diaminopimelic acid interpeptide linkages in the peptidoglycan of mycobacteria. *Biochemistry*, **13**, 3471–6.

WIETZERBIN, J., LEDERER, E. & PETIT, J. F. (1975). Structural study of the poly-L-glutamic acid of the cell wall of *Mycobacterium tuberculosis* var. *hominis*, strain Brévannes. *Biochemical and Biophysical Research Communications*, **62**, 246–52.

WIETZERBIN-FALSZPAN, J., DAS, B. C., AZUMA, I., ADAM, A., [PETIT, J. F. & LEDERER, E. (1970). Isolation and mass spectrometric identification of the peptide subunits of mycobacterial cell walls. *Biochemical and Biophysical Research Communications*, **40**, 57–63.

WIETZERBIN-FALSZPAN, J., DAS, B., GROS, C., PETIT, J. F. & LEDERER, E. (1973). The amino acids of the cell walls of *Mycobacterium tuberculosis* var. *bovis*, strain BCG. Presence of a poly (L-glutamic acid). *European Journal of Biochemistry*, **32**, 525–32.

YAMAMURA, Y., AZUMA, I., TANIYAMA, T., SUGIMURA, K., HIRAO, F., TOKUZEN, R., OKABE, M., WARO N., YASUMOTO, K. & OHTA, M. (1976a). Immunotherapy of cancer with cell wall skeleton of *Mycobacterium bovis*–bacillus Calmette-Guérin: experimental and clinical results. *Annals of the New York Academy of Sciences*, **277**, 209–27.

YAMAMURA, Y., OGURA, T., YOSHIMOTO, T., NISHIKAWA, H., SAKATANI, M., ITOH, M., MASUMO, T., NAWBA, M., YASAKI, H., HIRAO, F. & AZUMA, I. (1976b). Successful treatment of the patients with malignant pleural effusion with BCG cell wall skeleton. *Gann*, **67**, 669–77.

YAMAMURA, Y., YOSHIZAKI, K., AZUMA, I., YAGURA, T. & WATANABE, T. (1975). Immunotherapy of human malignant melanoma with oil-attached BCG cell wall skeleton. *Gann*, **66**, 355–63.

YARKONI, E. & BEKIERKUNST, A. (1976). Non-specific resistance against infection with *Salmonella typhi* and *Salmonella typhimurium* induced in mice by cord factor (trehalose-6,6′-dimycolate) and its analogues. *Infection and Immunity*, **14**, 1125–9.

YARKONI, E., BEKIERKUNST, A., ASSELINEAU, J., TOUBIANA, B., TOUBIANA, M. J. & LEDERER, E. (1973). Suppression of growth of Ehrlich ascites cells in mice pretreated with synthetic analogs of trehalose-6,6′-dimycolate (cord factor). *Journal of the National Cancer Institute*, **51**, 717–20.

YARKONI, E., WANG, L. & BEKIERKUNST, A. (1974). Suppression of growth of Ehrlich ascites tumor cells in mice by trehalose-6,6′-dimycolate (cord factor) and BCG. *Infection and Immunity*, **9**, 977–84.

YRON, I., WEISS, D. W., ROBINSON, E., COHEN, D., ADELBERG, M. G., MEKORI, T. & HABER, M. (1973). Immunotherapeutic studies in mice with the methanol extraction residue (MER) fraction of BCG: studies with solid tumors. *National Cancer Institute, Monographs*, **39**, 33–54.

ZBAR, B., BERNSTEIN, I. D. & RAPP, H. J. (1971). Suppression of tumor growth at the site of infection with living bacillus Calmette-Guérin. *Journal of The National Cancer Institute*, **46**, 831–9.

ZBAR, B., RAPP, H. J. & RIBI, E. (1972). Tumor suppression by cell walls of *Mycobacterium bovis* attached to oil droplets. *Journal of the National Cancer Institute*, **48**, 831–5.

ZBAR, B., RIBI, E., MEYER, T., AZUMA, I. & RAPP, H. J. (1974). Immunotherapy of cancer: regression of established intradermal tumors after intralesional-injection of mycobacterial cell walls attached to oil droplets. *Journal of the National Cancer Institute*, **52**, 1571–7.

STRUCTURE AND FUNCTION OF BACTERIAL PLASMA MEMBRANES

MILTON R. J. SALTON

Department of Microbiology,
New York University School of Medicine,
New York, NY 10016, USA

INTRODUCTION

The importance of a cell surface boundary separating cellular contents from their external environment was clearly recognized by Antoine van Leeuwenhoek from his first remarkable observations of the microscopic morphology of bacteria and other microorganisms. Indeed, van Leeuwenhoek obviously understood the essence of compartmentalization of cells and that the cell surface structures were responsible for the characteristic shapes of the organisms. Mitchell (1970) has succinctly emphasized the biological significance of membranes in that 'the natural membranes occupy a special position in the organization of living cells because they define, structurally and functionally, the boundary between the cell and its environment and the boundaries between intracellular compartments'. The cell surface envelope structures are of special importance for free-living bacteria because of the enormous range of environments they may encounter. Their surface structures have to be totally self-sufficient since the prokaryotic cell lacks the supportive advantages (or disadvantages) of forming a tissue or aggregate of cells in intimate contact with one another. Moreover, the plasma membrane of the bacterial cell occupies a central functional role for the prokaryotic cell, characteristically devoid of all the specialized membranous organelles (e.g. endoplasmic reticulum, mitochondria, chloroplasts, nuclear membranes, golgi, and lysosomes) typical of the functional compartments of the eukaryotic cell (Stanier, 1970). Although the plasma membrane is not the sole surface structure of the bacterial cell (with the notable exceptions of the mycoplasma group and certain L-forms), it is unique in that it constitutes a single membranous organelle performing transport, mitochondrial, biosynthetic and secretory functions of the cell. Apart from the specialized intracellular membranes in certain bacteria (e.g. spore membranes in sporulating organisms, chromatophores of photosynthetic species, cytomembranes of nitrifying organisms, etc.), the mesosomal membrane vesicles are the only other

major membranous organelles of bacteria (Salton, 1971). These structures which still appear to lack a unique function, will be discussed in
the concluding section of this contribution.

The term 'membrane' as applied to the bacterial cell is often used
loosely to encompass the wall, outer membranous layers and the
envelopes (outer+inner membranes with or without the rigid peptidoglycan layer) of Gram-negative organisms. In certain species of Gram-
positive and Gram-negative organisms, outer layers of 'crystalline'
protein arrays may be found external to the cell wall or outer membrane
respectively (Glauert & Thornley, 1969). The functions of these outermost, fine-structured layers are not clearly apparent at present. Both
these outer 'crystalline' protein layers and the outer membranes of
Gram-negative bacteria could provide the bacterial cell with additional
peripheral cellular compartments. Indeed, the outer membrane of
Gram-negative organisms clearly constitutes the external boundary for
the periplasmic space (Costerton, Ingram & Cheng, 1974). Although
this contribution will deal primarily with the plasma membranes of
both Gram-positive and Gram-negative bacteria we do not wish to
infer that the external periodic protein layers, nor the outer membranes
of Gram-negative organisms are unimportant. On the contrary, much
is now known about the properties and functions of the outer membranes of Gram-negative bacteria and their major protein components
have been extensively characterized (Bragg & Hou, 1972; Braun, 1975,
1977; Haller, Hoehn & Henning, 1975; Henning & Haller, 1975; Nakae
& Nikaido, 1975; Rosenbusch, 1974; Schnaitman, 1970, 1974; Smit,
Kamio & Nikaido, 1975). Apart from their role in iron transport,
bacteriophage receptor activity and providing hydrophobic and
hydrophilic pores with selective permeability barrier properties (Braun,
1977; Nakae & Nikaido, 1975; Schnaitman, Smith & Forn de Salsas,
1975), the outer membranes do not appear to possess major enzymatic
functions. The only enzyme so far found to be located almost exclusively
in the outer membrane of various Gram-negative species is phospholipase A (Bell, Mavis, Osborn & Vagelos, 1971; White, Albright, Lennarz
& Schnaitman, 1971). However, its role in phospholipid metabolism in
the bacterial cell is still obscure. All other membrane-bound activities,
respiratory components and energized transport systems in the bacterial
cell appear to reside solely in the plasma membranes with of course the
exceptions noted above for specialized intracellular membranes. Our
remarks on the structure and function of bacterial membranes will
accordingly be restricted to the plasma membranes and the mesosomal
vesicles.

ULTRASTRUCTURAL ANALYSIS OF
BACTERIAL PLASMA MEMBRANES

The ultrastructural features of the bacterial cell surface components including the plasma membranes have been established by all of the basic electron microscopy procedures. From the profiles of the cell surface as seen in thin sections of bacteria, the plasma membranes exhibited the typical 'double-track' appearance of all 'unit' membrane structures (Glauert & Thornley, 1969). In contrast with the sharply defined profiles of plasma membranes, the cell walls of Gram-positive organisms are generally seen as rather thick amorphous structures and evidence for layers of different electron densities is found only in certain species (Murray, 1968; Barksdale & Kim, 1977). Initial studies of profiles through the surface of *Escherichia coli* showed a double membrane structure in which the outer membrane profile was indistinguishable from the inner (plasma) membrane in thickness and electron density layering (Kellenberger & Ryter, 1968). The peptidoglycan layer was not resolved in such profiles and the demonstration of its location between the plasma and outer membranes awaited the use of special staining procedures (Murray, Steed & Elson, 1965). Moreover, the profiles of the outermost periodic protein layers of organisms such as the *Spirillum* species studied by Murray and his colleagues were readily distinguishable from those of the plasma (inner) and outer membranes (Buckmire & Murray, 1970; Murray, 1968). Electron microscopy of bacterial cells in which the 'stratagraphical' relationships of the surface structures were undisturbed except by fixation and embedding procedures, established the basic structural similarity of bacterial plasma membranes to those of other cells. Thin sectioning combined with procedures such as labelling with specific ferritin–antibody conjugates (Singer & Schick, 1961) has further extended our knowledge of the topography of bacterial membranes (see Oppenheim & Nachbar, 1977).

With the development of the techniques of freeze-fracturing and freeze-etching it has been possible to examine cellular structures in a near normal state of preservation. Such studies have been particularly valuable in elucidating the ultrastructure of cell surfaces and biomembranes (Branton, 1966; Branton & Deamer, 1972). Freeze-fracture studies have provided strong supportive evidence for the existence of substantial regions of bilayer lipid in cell membranes. There is now general acceptance that the fracture plane of membrane structures occurs through the hydrophobic centre, thus revealing the internal surfaces of the cleaved membranes. Freeze-etching can be used to reveal the outer surface following fracture through the plane of the membrane (Branton, 1966). The application of these techniques to studies of the bacterial surface structures has been especially valuable in further substantiating the ultrastructure of their surface membranes (Holt & Leadbetter, 1969). A more detailed discussion of the results of bacterial cell surface architecture studied by freeze-fracturing and freeze-etching has recently appeared (Salton & Owen, 1976). Accordingly, our remarks will be largely confined to the evidence obtained with the plasma membranes.

If one excludes the fractures through the cytoplasm, then the number of fracture surfaces commonly seen in Gram-positive bacteria subjected to freeze-etching is two, whereas four are generally observed for Gram-negative organisms. The analysis of complementary replicas has established that the fracture faces observed for Gram-positive bacteria result from one distinct fracture plane and those for Gram-negative species from two (for discussion of original papers, see Salton & Owen, 1976). The evidence from the complementary replicas and thin-sectioning of freeze-fractured cells (Nanninga, 1971) is in accord with the conclusion that the main fracture plane of Gram-positive and of most Gram-negative bacteria is through the hydrophobic centre of the plasma membrane. As with other plasma membranes, the convex fracture surface of the bacterial membrane is characterized by a dense covering of particles (about 5–10 nm in diameter) and the corresponding concave fracture faces possess fewer particles and have occasional depressions (Holt & Leadbetter, 1969). Thus, the convex and concave fracture surfaces revealed by the fracture through the internal bilayer centre of the bacterial plasma membrane present very similar appearances to those observed for plasma membranes of other types of cells (Branton, 1966; Branton & Deamer, 1972). Moreover, aggregation of intercalated particles also occurs in the bacterial plasma membrane upon transition of the lipids from a liquid crystalline state to a gel phase. Examples of this behaviour have been documented for plasma membranes of *E. coli* (Haest *et al.*, 1974; Kleeman & McConnell, 1974; Schechter, Letellier & Gulik-Krzywicki, 1974; Verkleij & Ververgaert, 1975), *Acholeplasma laidlawii* (Verkleij, Ververgaert, van Deenen & Elbers, 1972) and *Streptococcus faecalis* (Tsien & Higgins, 1974). Although the normal distribution of intramembrane particles on the convex fracture faces is usually random, clustering and aggregation has been correlated with age of culture, membrane growth, and divalent cation concentration (see Salton and Owen, 1976, for further discussion).

The second fracture plane observed in Gram-negative bacteria is generally much weaker than that of the plasma membrane. The location of this particular fracture plane has been the subject of much controversy but the evidence now available strongly suggests that it occurs through the plane of the outer membrane. Many studies have been devoted to investigating the effects of divalent cations, EDTA treatment, mutations in lipopolysaccharide biosynthesis, and phospholipid and outer membrane protein alterations on the fracture faces of the outer membrane but a discussion of the details of this aspect of bacterial outer membrane structure is beyond the scope of this contribution.

The final technique of electron microscopy worthy of mention for its use in establishing membrane ultrastructure is that of negative staining. Although of rather limited value in studies of the topography and architecture of cell membranes, it has been of value in 'highlighting' the mesosomal vesicles in intact cells, in distinguishing between plasma membranes and mesosomal vesicles of Gram-positive bacteria and the inner (plasma) and outer membranes from Gram-negative bacteria. It has also been valuable in establishing the identity of the ATPase 'particles' (10 nm diameter structures) of the bacterial plasma membrane (Oppenheim & Salton, 1973). Because of the simplicity of the negative-staining procedure, it has provided a rapid and ideal method for monitoring membrane isolation, dissociation and reassociation studies and membrane vesicle preparation. Moreover, contamination of isolated membrane fractions with other structural components (e.g. pili) can be readily assessed by negative staining (Pollock et al., 1974).

The distinctive appearance of plasma membranes and mesosomal vesicles is clearly illustrated in Plate 1. Negatively stained preparations of membrane fractions from *Micrococcus lysodeikticus* show the typical appearance of plasma membrane sheets and fragments with the uniform 10 nm ATPase particles (Muñoz, Freer, Ellar & Salton, 1968) randomly distributed on the membrane surface, contrasted with the smooth vesicular nature of the mesosomal components. Moreover, the identity of the particles as the ATPase complex has been demonstrated in a variety of ways, including negative staining of the released particles and membrane residues following selective 'shock' wash release of ATPase activity (Nachbar & Salton, 1970) and by examination of negatively stained preparations labelled with ferritin-anti-ATPase conjugates (Oppenheim & Salton, 1973).

The characteristic appearance of the plasma membranes of *E. coli* K12 is also readily distinguishable from that of the outer membrane structures (Plate 2). As in Gram-positive bacteria, the isolated plasma membrane structures appear as sheets, fragments and occasionally as small vesicles, all having an abundance of the ATPase particles (10 nm) distributed randomly over the membrane surface. The isolated outer membranes (Plate 2b) are generally rounded, vesicular structures, and are much more uniform in both the appearance of surface texture and in overall size of the vesicles. The negative-staining technique is also useful in distinguishing between sealed and unsealed vesicle structures irrespective of their origins (e.g. mesosomal membranes, outer membranes of Gram-negative organisms or the less frequently seen plasma membrane vesicles in isolated plasma membrane preparations). Thus, where the vesicles are sealed, the negative stain fails to penetrate.

MAJOR CHEMICAL COMPONENTS OF BACTERIAL MEMBRANES

In contrast to their external cell-wall neighbour, the plasma membranes of both Gram-positive and Gram-negative bacteria are essentially lipid, protein structures. They, thus, conform to the general chemical features of all types of biological membranes (Wallach, 1972). However, the range of protein to lipid ratios for bacterial membranes (2.0–4.0) tend to be somewhat higher than those observed for many other types of membranes, with one of the notable exceptions being that of the inner mitochondrial membrane (Wallach, 1972). The proteins of bacterial plasma membranes have been studied in a variety of ways and this topic will be discussed in more detail in a subsequent section of this contribution. Suffice it to say that sodium dodecyl sulphate (SDS)–polyacrylamide gel electrophoresis gives an approximation of the array of polypeptide constituents and the number of individual bands may range from some 20–50 components in a variety of plasma membrane fractions (Salton, 1971; Steck & Fox, 1972). The significance of such figures is difficult to assess at this stage since very few membrane enzymes have been purified to homogeneity and their subunit polypeptide structure determined. One notable exception is the bacterial ATPase with its five (α, β, γ, δ, and ϵ) distinct polypeptide subunits (Abrams & Smith, 1974) all of which contribute to the complex banding patterns seen in single dimension SDS–polyacrylamide gel electrophoresis. Higher resolution, two dimensional studies separating polypeptides on the basis of charge and molecular weight would be needed to gain a more accurate estimate of the total variety of polypeptide constituents. In turn, they would ultimately have to be interpreted in terms of the subunit composition of the variety of membrane enzymes and cytochromes, a truly formidable task.

Glycoproteins appear to be much less abundant in bacterial plasma membranes than in animal cell membranes (Steck & Fox, 1972; Wallach, 1972). Evidence suggesting the existence of several glycoprotein antigens in *M. lysodeikticus* membranes has been based upon their reactions with lectins (Owen & Salton, 1975a; P. Owen & M. R. J. Salton, unpublished results). However, the isolation and chemical characterization of a bacterial plasma membrane glycoprotein has to the author's knowledge not yet been achieved. Such studies would of course be essential to establish the existence of the glycoprotein class of components in bacterial plasma membranes. Their presence in isolated inner (plasma) membranes of Gram-negative bacteria has not so far been reported.

The lipids of bacterial membranes have been investigated quite extensively and were much more fully characterized long before efforts were made to determine the protein constitution of the membranes. It has long been known that the lipids of the bacterial cell are almost exclusively localized in the membrane systems (i.e. plasma and mesosome membranes of Gram-positive bacteria and outer and inner membranes of Gram-negative organisms). Studies of the distribution and classes of lipids in bacterial membranes and their biosynthesis, have been extensively reviewed and need very little further comment (Ambron & Pieringer, 1973; Cronan & Vagelos, 1972; Finnerty & Makula, 1975). As in other types of cell membranes, phospholipids are the major class of bacterial plasma membrane lipid and they may account for up to 90–95% of the total lipid (Law, 1967). Phosphatidylethanolamine (PE) is the principal phospholipid in the membranes of many Gram-negative and Gram-positive organisms. It is not universally found in plasma membranes of all bacteria since the phospholipid composition is often a species characteristic (Salton, 1971; Whiteside, De Siervo & Salton, 1971). Thus certain pigmented micrococci contain predominantly cardiolipin (diphosphatidylglycerol) and phosphatidylglycerol together with smaller amounts of phosphatidylinositol (Whiteside *et al.*, 1971).

Glycolipids are the other major class of lipids in bacterial membranes. They are quite widely distributed in Gram-positive organisms and the mycoplasma group and are seen much less frequently, if at all, in many Gram-negative bacteria, all of the latter possessing their own unique class of lipopolysaccharide (Lüderitz, Westphal, Staub & Nikaido, 1971). Other minor lipid-soluble constituents include the carotenoids, hydrocarbons, menaquinones and ubiquinones found characteristically in certain species (Salton, 1971).

Much of the lipid of the bacterial membrane occurs in bilayer form (Reinert & Steim, 1970) and thus shares this arrangement of the phospholipids with other membrane systems (Wallach, 1972). Moreover, the asymmetry of the bilayer phospholipids first discovered in erythrocyte membranes by Bretscher (1972) has also been found to be a feature of several bacterial membranes (Barsukov, Kulikov & Bergelson, 1976; Rothman & Kennedy, 1977). From a study of M. lysodeikticus membranes it was concluded that diphosphatidylglycerol distributes almost evenly between the inner and outer monolayers, phosphatidylglycerol predominantly on the outer surface and phosphatidylinositol in the inner monolayer (Barsukov et al., 1976). In Bacillus megaterium membranes, the inner (cytoplasmic) monolayer contained twice as much phosphatidylethanolamine as the outer monolayer (Rothman & Kennedy, 1977). Thus, the bilayer arrangement and asymmetrical distribution of individual phospholipids within the bilayer is an intrinsic feature of biological membranes including bacterial membranes.

Although plasma membranes of bacteria have so many structural and chemical properties in common with other membranes, the discovery of a membrane-bound form of teichoic acid (lipoteichoic acid) in Gram-positive organisms gives them a unique feature (Coley, Duckworth & Baddiley, 1972; Hay, Wicken & Baddiley, 1963; Shockman & Slade, 1964; Wicken & Knox, 1975). These lipid-linked polymers carrying a strong negative charge could function in the regulation of divalent cation passage across the cell surface through their ability to sequester these ions (Heptinstall, Archibald & Baddiley, 1970). The biochemical significance of the lipoteichoic acids believed to be largely localized in the plasma membranes and mesosomal structures of Gram-positive organisms has yet to be elucidated (Theodore, Cole & Huff, 1974). They have been found to have an inhibitory effect on the cell wall growth and extension (Cleveland et al., 1975). Not all Gram-positive bacteria possesss lipoteichoic acids but a rather analogous type of structure, a succinylated lipomannan (also membrane bound) has been found in M. lysodeikticus and other micrococci devoid of lipoteichoic acids (Owen & Salton, 1975b, c; Pless, Schmit & Lennarz, 1975; Powell, Duckworth & Baddiley, 1975a, b). Such a lipid-linked, negatively charged membrane component could serve as an 'analogue' of the lipoteichoic acids in these species. It will thus be of considerable interest to determine the precise function of this interesting class of membrane constituent in the Gram-positive cell. The succinylated lipomannan is expressed as a major antigen on the protoplast surface and it is conceivable that amphipathic molecules of this kind and the lipoteichoic acids may be involved in determining yet another aspect of the asymmetry of the plasma membrane.

ANTIGENIC AND ENZYMIC ARCHITECTURE OF BACTERIAL PLASMA MEMBRANES

The variety of biochemical functions performed by the plasma membrane has been widely recognized from the time Weibull (1953a, b) succeeded in releasing functional protoplasts from the bacterial cell and isolating and characterizing the protoplast (plasma) membrane. In the prokaryotic cell, the plasma membrane performs many of the various functions organized in the individual membranous organelles of the eukaryotic cell. Thus, the bacterial plasma membrane provides the 'mitochondrial' functions for the cell, it is the site of energized transport systems, it is the organelle involved in secretion of exocellular enzymes and protein toxins, it is the site of biosynthesis of its own phospholipids, glycolipids and accessory lipid molecules, and it is the site of a major stage in the biosynthesis and assembly of the external cell-wall-supporting structure, membrane and wall polymers and the external capsular polysaccharides and outer membrane lipopolysaccharides of Gram-negative bacteria. It is therefore involved in the biosynthesis of a diverse variety of the major chemical components of walls, membranes and capsules, including peptidoglycan, teichoic and lipoteichoic acids, teichuronic acids, polysaccharides, lipopolysaccharides, capsular γ-glutamyl polypeptides and the various types of lipids and lipid intermediates (for reviews seee Ellar, 1970; Salton, 1974). The relationships of the plasma membranes to other important biochemical events in the bacterial cell such as protein biosynthesis and DNA replication and anchoring are much less well defined. The functional significance of membrane-bound ribosomes (Hendler, 1968) has yet to be clearly established, although a plasma membrane role in post-transcriptional processing of proteins destined for export (e.g. β-lactamase) is evident from the work of Lampen and his colleagues (Yamamoto & Lampen, 1976). A specific membrane–DNA complex has been investigated (Tremblay, Daniels & Schaechter, 1969) but the nature of DNA anchoring or association is still a topic of contention with claims being proposed for a specific mesosomal role in nucleoid association and the division cycle in *B. subtilis* (van Iterson & Aten, 1976). Even neglecting these latter unresolved functions, it is clearly evident that the bacterial plasma membrane is a multifunctional structure. The organization and resolution of the multiplicity of enzymes in the bacterial plasma membrane is therefore at the heart of the problem of elucidating the structure–function relationships in the prokaryotic membrane.

One of the many possible approaches to this problem has been to determine the antigenic and enzymic architecture of the plasma membranes. Investigations in our laboratory have been directed towards this goal and the lysozyme-sensitive organism *M. lysodeikticus* has been particularly suitable for this purpose. The expression of antigenic components exposed on the surface can be determined on the intact organisms and on stable, intact protoplasts after selective removal of the wall by digestion with lysozyme. Plasma membranes and mesosomal vesicles can be separated and isolated as homogeneous structural entities of *M. lysodeikticus*. By carrying out absorption experiments with antisera generated to the various components, it is therefore possible to deduce their expression on the surface structures and the asymmetry of the disposition of membrane antigens including specific enzymes.

An important prerequisite of the approach to this work was the availability of (*a*) suitable procedures for membrane solubilization with retention of biological activities and (*b*) a high resolution procedure in which complex mixtures of antigens such as those derived from membranes could be reacted with antibodies for the specific identification of individual antigenic components (including enzymes where possible). These prerequisites were met, in part, by using Triton X-100 'solubilization' of isolated membranes and the resolution of these fractions by the two dimensional, crossed immunoelectrophoresis (CIE) technique first introduced by Laurell (1965) and developed into a highly valuable analytical procedure, complete with enzyme identification by zymogram staining of immunoprecipitates (see Axelsen, Krøll & Weeke, 1973; Uriel, 1971).

Earlier attempts in our laboratory (Fukui, Nachbar & Salton, 1971) and by other investigators (Freimer, 1963; Kahane & Razin, 1969) gave poor resolution of membrane fractions examined by gel diffusion and simple immunoelectrophoresis. The limited number of immunoprecipitates (3–4) detected by these 'classical' methods was contrary to the expectations from the known complexity of the membranes and from the variety of specific enzymes present in bacterial membranes. In contrast the crossed immunoelectrophoresis of Triton X-100 solubilized plasma membranes of *M. lysodeikticus* against membrane antibodies has established the presence of twenty-seven discrete antigens (P. Owen & M. R. J. Salton, *J. Bact*, in press). The resolution of membrane antigens of *M. lysodeikticus* is illustrated in the two dimensional (CIE) display of the immunoprecipitates stained with Coomassie brilliant blue (Plate 3*a*, *b*). Not all twenty-seven antigens are visible in a single immunoplate, since it is necessary to examine the patterns at different antigen and antibody concentrations to optimize the detection of some immunoprecipitates as previously reported by Bjerrum and Bøg-Hansen (1976). The total number of individual antigens detectable by this high resolution procedure is, therefore, much more in accord with the complexity of the membranes expected from biochemical studies and SDS-polyacrylamide gel electrophoresis.

Individual antigens of the *M. lysodeikticus* plasma membranes have also been identified by their reactivity to the lectins, concanavalin A and soybean agglutinin, in the crossed immunoaffinoelectrophoresis system; with the lectins in intermediate gels and by zymogram staining for antigens possessing enzymic activity. The succinylated lipomannan has been identified by its reactivity with concanavalin A and coelectrophoresis of the purified lipomannan isolated from plasma or mesosomal membranes (Owen & Salton, 1975, *a, b, c*). Five of the antigens have been identified as enzymes by zymogram staining of the immunoplates. These included the ATPase complex, two antigenically distinct NADH dehydrogenases, succinate dehydrogenase and malate dehydrogenase (Owen & Salton, 1975*a*). Attempts to identify additional enzymes by a selection of procedures for staining other dehydrogenases, phosphatases, esterases and proteases have not succeeded. Nor did we succeed in an effort to identify cardiolipin synthetase (De Siervo & Salton, 1971) with either ^{32}P-labelled phosphatidylglycerol (PG) or anti-phosphatidylglycerol antibodies (L. Trepo-Vitvitsky & M. R. J. Salton, unpublished results). The identification of other enzymes and the cytochromes of this organism must await the perfection of suitable, sensitive detection systems.

The asymmetric distribution of the plasma membrane antigens (and enzymes) of *M. lysodeikticus* has been established by absorption experiments with intact, stable protoplasts and isolated plasma membranes. Of the total of 27 antigens detected in solubilized membranes by CIE, 12 of the antigens were exposed on the protoplast surface and the succinylated lipomannan was identified as a major surface antigen (P. Owen & M. R. J. Salton, *J. Bact.*, in press). The removal of antibodies to the surface-expressed antigens by absorption with protoplasts is illustrated in Plate 4 and shows that immunoprecipitates of certain antigens are virtually unaffected by absorption and are therefore inaccessible to antibodies. Fourteen antigens possessed determinants that were expressed solely on the cytoplasmic face of the membrane and of these, five were identifiable as succinate and malate dehydrogenases, together with the two NADH dehydrogenases and ATPase. Thus, these enzymes and a component reacting with soybean agglutinin (primary sugar specificity for N-acetyl-D-galactosaminyl residues) occur on the cytoplasmic side of the plasma membrane.

Absorption experiments with isolated plasma membranes having both the outer face and cytoplasmic side of the membrane freely accessible to antibodies showed that all antigens detected by crossed immunoelectrophoresis (CIE) were fully expressed. Such absorption-CIE studies do not, however, permit the identification of antigens which may span the membrane (i.e. transmembrane components). Special labelling techniques or studies with proteolytic enzymes acting on components accessible on the protoplast surface would be needed to establish the presence of such antigens.

The membranes of Gram-negative bacteria have been much more difficult to analyse in a comparable fashion to those described above for Gram-positive organisms. The outer membrane structure of the Gram-negative cell envelope is molecularly heterogeneous and no single enzyme is capable of selectively removing this external structure analogous to the action of muramidases on cell-wall peptidoglycans of Gram-positive bacteria. Uniform protoplasts of Gram-negative bacteria are generally not available for the determination of the asymmetry of the plasma membranes of Gram-negative bacteria. Only in a few instances has it been possible to detach or dissociate the outer membrane and remove the peptidoglycan layer so that true protoplasts of Gram-negative species could be prepared for studies similar to those performed with protoplasts of Gram-positive organisms. The marine pseudomonad investigated by MacLeod and his colleagues (De Voe, Thompson, Costerton & MacLeod, 1970; Martin & MacLeod, 1971) was the first Gram-negative organism amenable to this type of study. The outer membrane could be dissociated to yield rod-shaped 'mureinoplasts' which could then be converted with lysozyme to form true, wall-less protoplasts (Costerton et al., 1967). Such protoplasts would be fully amenable to the approaches described above for the structural analysis of Gram-positive organisms. Although Weiss (1976) has succeeded in preparing protoplasts of E. coli, it is not yet clear whether the procedure is generally applicable to various strains of this organism and other Gram-negative species. Moreover, it has not yet been established whether the protoplasts are sufficiently uniform (i.e. quantitative conversion) to permit a definitive analysis of the surface expression of plasma membrane antigens. Further efforts will be needed to develop reproducible and quantitative methods to permit definitive structural analysis of the asymmetry of the outer and inner membranes of Gram-negative bacteria. However, the separation of relatively homogeneous outer and inner membranes of Gram-negative by the procedures described by Schnaitman (1970) and Osborn, Gander, Parisi & Carson (1972) have represented significant advances in the characterization of the proteins, enzymes and antigens of these structures. Studies of the distribution of enzymes between outer and inner (plasma) membranes have already clearly indicated that the inner membranes, as anticipated, possess all of the major functions known to be located in the plasma membranes of Gram-positive organisms as well as the enzymes involved in the biosynthesis of the outer membrane lipopolysaccharides (Osborn & Munson, 1974; Salton, 1976).

There have been numerous studies on the SDS–polyacrylamide gel electrophoresis of both outer and inner membranes of Gram-negative organisms, but apart from direct isolation of particular enzymes very little has been done on the identification of individual antigenic and/or enzyme components in solubilized membranes. Resolution of the complex Gram-negative cell envelopes by the conventional immunochemical methods applied to detergent-solubilized membrane preparations has been poor. Two recent studies with detergent-dissociated membranes of *A. laidlawii* (Johansson & Hjertén, 1974) and envelopes of *Neisseria gonorrhoeae* (Smyth, Friedman-Kien & Salton, 1976) by crossed immunoelectrophoresis has established the complexity of these systems by a high degree of resolution of the membrane antigens. In the latter study zymogram staining was used to identity specific antigens as enzymes. This immunochemical approach to the investigation of membrane architecture and cell surface organization is thus equally applicable to surface membranes such as those of the *A. laidlawii* and envelopes of the Gram-negative gonococcus. The application of these methods to an organism such as *E. coli* in which the outer and inner membranes of the envelope can be separated more readily than in the gonococcus would provide a valuable understanding of the antigenic structure of the surfaces of Gram-negative organisms. Moreover, as we have demonstrated with the plasma membranes of the Gram-positive organism, *M. lysodeikticus*, enough enzyme activities are preserved on detergent solubilization and crossed immunoelectrophoresis analysis to permit the identification of enzymes and/or enzyme complexes. Such studies with *E. coli* have thus provided a more 'functional' approach to establishing the unique differences between plasma (inner) membranes and outer membranes of this Gram-negative species.

8

Investigations in our laboratory (C. J. Smyth, J. Siegel, M. R. J. Salton & P. Owen, *J. Bact.*, in press) have resolved 46 antigens in the immunoprecipitate patterns of the inner (plasma) membranes of *E. coli* K-12, strain 44 (a β-lactamase negative mutant used by Pollock *et al.* (1974). Of the 46 antigens detected in the Triton X-100 extract of the inner membranes, eleven have been identified as enzymes or enzyme complexes by zymogram staining and one additional immunoprecipitate specifically identified as Braun's lipoprotein with antisera to this component. Immunoprecipitates possessing the following enzyme activities were identified: ATPase, 6-phosphogluconate dehydrogenase, glutamate dehydrogenase (two separate components), malate, dihydro-orotate, succinate and lactate dehydrogenases, NADH dehydrogenase, glycerol-3-phosphate dehydrogenase and a protease. The corresponding immunoprecipitate pattern for the isolated outer membranes was strikingly different and consisted of at least 25 discrete antigens. Two of the major immunogens of the outer membranes have been identified as lipopolysaccharide and Braun's lipoprotein and an immunoprecipitate possessing protease activity was also detected. Thus the distinction between inner and outer membranes can be readily determined by crossed immunoelectrophoresis and functional components identified by detection of specific enzymic activities. It is significant that neither with the plasma membranes of *M. lysodeikticus* nor those of *E. coli* did we see any evidence of multiple enzyme staining. A given immunoprecipitate only stained specifically for one enzyme, although heterogeneity of the components was evident from the morphology of the immunoprecipitates. When isolated envelopes (outer+inner membranes) were examined in a similar fashion, the immuno-precipitate pattern contained many of the individual precipitates seen in separated inner and outer membrane fractions, thus indicating that envelopes from disrupted cells possess components from both membranes. The individual patterns are illustrated in Plate 3*c*.

In addition to the investigations of the use of membrane antibody absorption experiments for the determination of the structural asymmetry of the bacterial plasma membranes, two other basic approaches (lactoperoxidase ^{125}I-labelling and ferritin–antibody labelling) have yielded very similar results. Lactoperoxidase-^{125}I labelling of cell surface proteins (Phillips & Morrison, 1971) established that the *M. lysodeikticus* ATPase was not accessible to iodination when intact protoplasts were used (Salton, Schor & Ng, 1972). However, treatment of isolated plasma membrane fractions by either the lactoperoxidase method or by reaction with ^{125}ICl iodinated polypeptides identifiable as the ATPase subunits upon electrophoresis in SDS–polyacrylamide gels (Salton *et al.*, 1972). These results thus indicated that the ATPase could only be labelled when the inner face of the membrane was also accessible to the reagents. Ferritin-labelled antibody conjugates specific for cell membrane components have provided an elegant way of determining distribution and location of a particular antigen (Singer & Schick, 1961). This procedure was used by Oppenheim & Salton (1973), in identifying the ATPase 'particles' (approximately 10 nm diameter) on the plasma membrane of *M. lysodeikticus* by reaction with ferritin conjugated to monospecific antibodies to the purified enzyme. These studies, moreover, established the asymmetry of distribution of the ATPase, since no labelling occurred on the surface of intact protoplasts but was clearly evident on membrane sheets and fragments and only on one side of membrane vesicles (inside-out orientation). The ferritin-antibody-labelling procedure was also used to confirm the identity of the ATPase by the selective shock wash (Muñoz, Salton, Ng & Schor, 1969) which released the ferritin–antibody–enzyme conjugate from the membranes (Oppenheim & Salton, 1973). Thus the ferritin-labelling is not only superior to heavy metal-cytochemical staining of membranes for the detection of such enzymes but has the added advantage of permitting definitive identification of a membrane enzyme as a ferritin–antibody–enzyme complex.

MEMBRANE DISSOCIATION AND RECONSTITUTION

Much effort in the past has gone into studies of membrane dissociation and reconstitution, a topic which has been admirably reviewed by Razin (1972). Solubilization procedures have often been too drastic to permit reconstitution of fully functional membranes with a few notable exceptions discussed in detail by Razin (1972). The formation of membranes following removal or depletion of SDS from fully dissociated membranes has been well documented for many membrane systems but is clearly of less value in establishing structure–function relationships. One of the major goals of membrane dissociation studies has been to provide the necessary procedures for the selective release of membrane-bound enzymes for purification and characterization studies.

Membrane dissociation studies combined with electron microscopy has added a limited amount of information on structure–function relationships of bacterial plasma membranes. Thus the release of ATPase particles from the membranes and the stripped appearance of the membrane residues correlate well with the distributions of enzyme activities. Moreover, the conditions for reassociation of ATPase with the membrane have been determined although there may be uncertainties as to which ATPase subunits are critical for this process (Abrams & Smith, 1974).

In general, detergent 'solubilized' fractions of bacterial membranes examined by electron microscopy have yielded little information of structural value. The residues after deoxycholate extraction of *M. lysodeikticus* membranes do possess well defined structural features. They exist as folded, lipid-depleted membrane sheets or vesicles highly enriched in succinate dehydrogenase and cytochromes (Salton, Freer & Ellar, 1968). Such membranous structures could represent localized regions containing the electron transport components or may simply reflect the coalescence of hydrophobic proteins following the removal of the major portion of the membrane lipid. The random distribution of ATPase over the whole inner surface of the membrane may suggest that the latter interpretation of such structures is more likely rather than localization of the electron transport chain components as relatively large regions of the membrane. Ferritin-labelling of dehydrogenases or cytochromes would clearly be needed to settle such problems.

Reconstitution studies such as those recently reported by Kagawa, Sone & Yoshida (1976) with the proton translocating ATPase of a thermophilic bacterium will add much to our understanding of the organization of bacterial membranes at the structural and functional levels.

THE ENIGMA OF MESOSOMAL MEMBRANE VESICLES

The structural and biochemical significance of mesosomes has been one of the most recalcitrant problems in bacterial anatomy and physiology. Much of the literature on these entities has been extensively reviewed in recent years by Reusch & Burger (1973), Ghosh (1974) and Greenawalt & Whiteside (1975) and we still appear to be no closer to a definitive answer as to what these enigmatic structures are and as to what they do (Salton & Owen, 1976). The almost complete absence of the enzymes so prominent in the plasma membrane fractions and found on the cytoplasmic face of these membranes adds further to the puzzle (Salton & Owen, 1976). Enzyme(s) characteristic of the mesosome fraction have so far not been found or authenticated. Apart from its enrichment in the lipomannan and the presence of the antigens expressed on the outer surface of the membrane, the only unique property we have been able to detect is an inhibitory action on the cardiolipin synthetase of the plasma membranes (L. Trepo-Vitvitsky & M. R. J. Salton, unpublished results). The nature and significance of this inhibitory activity of the mesosomal vesicles has yet to be determined.

The outer surface antigens found in the mesosomes together with the absence or low levels of plasma membrane enzymes has prompted us to suggest that they may represent the peeling off or rearrangement of the outer leaflet of the plasma membrane due to perturbation of the latter structure. This possibility should be amenable to experimental testing and may help us to make a decision as to whether they are 'fact or fiction'.

The author is grateful to the National Science Foundation for a grant (PCM 76-83074) supporting the investigations on the biochemistry of bacterial membranes in his laboratory and wishes to thank Dr Kwang S. Kim for the electron micrographs.

REFERENCES

ABRAMS, W. & SMITH, J. B. (1974). Bacterial membrane ATPase. In *The Enzymes*, ed. P. D. Boyer, 3rd edn, pp. 395–429. New York: Academic Press.

AMBRON, R. T. & PIERINGER, R. A. (1973). Phospholipids in micro-organisms. In *Form and Function of Phospholipids*, ed. G. B. Ansell, R. M. C. Dawson & J. N. Hawthorne, 2nd revised edn, BBA Library, 3, pp. 289–331. Amsterdam: Elsevier.

AXELSEN, N. H., KRØLL, J. & WEEKE, B. (eds.) (1973). A Manual of Quantitative Immunoelectrophoresis. Methods and Applications. *Scandinavian Journal of Immunology*, 2 (Suppl. 1), 15–169.

BARKSDALE, L. & KIM, K.-S. (1977). *Mycobacterium. Bacteriological Reviews*, 41, 217–372.

BARSUKOV, L. I., KULIKOV, V. I. & BERGELSON, L. D. (1976). Lipid transfer proteins as a tool in the phospholipids in the protoplasmic membrane of *Micrococcus lysodeikticus. Biochemical and Biophysical Research Communications*, 71, 704–11.

BELL, R. M., MAVIS, R. D., OSBORN, M. J. & VAGELOS, P. R. (1971). Enzymes of phospholipid metabolism: Localization in the cytoplasmic and outer membrane of the cell envelope of *Escherichia coli* and *Salmonella typhimurium. Biochimica et Biophysica Acta*, 249, 628–35.

BJERRUM, O. J. & BØG-HANSEN, T. C. (1976). Analysis of partially degraded proteins by quantitative immunoelectrophoresis. *Scandinavian Journal of Immunology*, 4 (Suppl. 2), 89–99.

BRAGG, P. D. & HOU, C. (1972). Organization of proteins in the native and reformed outer membrane of *Escherichia coli. Biochimica et Biophysica Acta*, 274, 478–88.

BRANTON, D. (1966). Fracture faces of frozen membranes. *Proceedings of the National Academy, USA*, 55, 1048–56.

BRANTON, D. & DEAMER, D. W. (1972). *Membrane Structure*, 2. Vienna: Springer-Verlag.

BRAUN, V. (1975). Covalent lipoprotein from the outer membrane of *Escherichia coli. Biochimica et Biophysica Acta*, 415, 335–77.

BRAUN, V. (1977). Membranpermeation und Antibiotika-Resistenz bein Bakterien. *Naturwissenschaften*, **64**, 126–32.

BRETSCHER, M. S. (1972). Asymmetrical lipid bilayer structure for biological membranes. *Nature New Biology*, **236**, 11–12.

BUCKMIRE, F. L. A. & MURRAY, R. G. E. (1970). Studies on the cell wall of *Spirillum serpens*. I. Isolation and partial purification of the outermost cell wall layer. *Canadian Journal of Microbiology*, **16**, 1011–22.

CLEVELAND, R. F., HOLTJE, J.-W., WICKEN, A. J., TOMASZ, A., DANEO-MOORE, L. & SHOCKMAN, G. D. (1975). Inhibition of bacterial wall lysins by lipoteichoic acids and related compounds. *Biochemical and Biophysical Research Communications*, **67**, 1128–35.

COLEY, J., DUCKWORTH, M. & BADDILEY, J. (1972). The occurrence of lipoteichoic acids in the membranes of Gram-positive bacteria. *Journal of General Microbiology*, **73**, 587–91.

COSTERTON, J. W., FORSBERG, C. W., MATULA, T. I., BUCKMIRE, F. L. A. & MacLEOD, R. A. (1967). Nutrition and metabolism of marine bacteria. XVI. Formation of protoplasts, spheroplasts, and related forms from a gram-negative marine bacterium. *Journal of Bacteriology*, **94**, 1764–77.

COSTERTON, J. W., INGRAM, J. M. & CHENG, K. J. (1974). Structure and function of the cell envelope of gram-negative bacteria. *Bacteriological Reviews*, **38**, 87–110.

CRONAN, J. E., JR. & VAGELOS, P. R. (1972). Metabolism and function of membrane phospholipids of *E. coli*. *Biochimica et Biophysica Acta*, **265**, 25–60.

DE SIERVO, A. J. & SALTON, M. R. J. (1971). Biosynthesis of cardiolipin in the membranes of *Micrococcus lysodeikticus*. *Biochimica et Biophysica Acta*, **239**, 280–92.

DE VOE, I. W., THOMPSON, J., COSTERTON, J. W. & MacLEOD, R. A. (1970). Stability and comparative transport capacity of cells, mureinoplasts, and true protoplasts of a gram-negative bacterium. *Journal of Bacteriology*, **101**, 1014–26.

ELLAR, D. J. (1970). The biosynthesis of protective surface structures of prokaryotic and eukaryotic cells. In *Organization and Control in Prokaryotic and Eukaryotic Cells*; *Symposia of the Society for General Microbiology*, **20**, ed. H. P. Charles & B. C. J. G. Knight, pp. 167–202. Cambridge University Press.

FINNERTY, W. R. & MAKULA, R. A. (1975). Microbial lipid metabolism. *Critical Review of Microbiology*, **4**, 1–40.

FREIMER, E. H. (1963). Studies of L forms and protoplasts of group A. *Streptococci*. II. Chemical and immunological properties of the cell membrane. *Journal of Experimental Medicine*, **117**, 377–99.

FUKUI, Y., NACHBAR, M. S. & SALTON, M. R. J. (1971). Immunological properties of *Micrococcus lysodeikticus* membranes. *Journal of Bacteriology*, **105**, 86–92.

GHOSH, B. K. (1974). The mesosome-A clue to the evolution of the plasma membrane. *Sub-cellular Biochemistry*, **3**, 311–67.

GLAUERT, A. M. & THORNLEY, M. J. (1969). The topography of the bacterial cell wall. *Annual Review of Microbiology*, **23**, 159–98.

GREENAWALT, J. W. & WHITESIDE, T. L. (1975). Mesosomes: Membranous bacterial organelles. *Bacteriological Reviews*, **39**, 405–63.

HAEST, C. W. M., VERKLEIJ, A. J., DE GIER, J., SCHEEK, R., VERVERGAERT, P. H. J. & VAN DEENAN, L. L. M. (1974). The effect of lipid phase transitions on the architecture of bacterial membranes. *Biochimica et Biophysica Acta*, **356**, 17–26.

HALLER, I., HOEHN, B. & HENNING, U. (1975). Apparent high degree of asymmetry of protein arrangement in the *Escherichia coli* outer cell envelope membrane. *Biochemistry*, **14**, 478–84.

HAY, J. B., WICKEN, A. J. & BADDILEY, J. (1963). The location of intracellular teichoic acids. *Biochimica et Biophysica Acta*, **71**, 188–90.

HENDLER, R. W. (1968). *Protein Biosynthesis and Membrane Biochemistry*, p. 296. New York: Wiley.

HENNING, U. & HALLER, I. (1975). Mutants of *Escherichia coli* K 12 lacking all 'major' proteins of the outer cell envelope membrane. *FEBS Letters*, **45**, 161–4.

HEPTINSTALL, S., ARCHIBALD, A. R. & BADDILEY, J. (1970). Teichoic acids and membrane function in bacteria. *Nature, London*, **225**, 519–21.

HOLT, S. C. & LEADBETTER, E. R. (1969). Comparative ultrastructure of selected aerobic spore-forming bacteria: a freeze-etching study. *Bacteriological Reviews*, **33**, 346–78.

JOHANSSON, K.-E. & HJERTÉN, S. (1974). Localization of the Tween 20-soluble membrane proteins of *Acholeplasma laidlawii* by crossed immunoelectrophoresis. *Journal of Molecular Biology*, **86**, 341–8.

KAGAWA, L. Y., SONE, N. & YOSHIDA, M. (1976). Proton translocating ATPase of a thermophilic bacterium. Morphology, subunits and chemical composition. *Journal of Biochemistry*, **80**, 141–51.

KAHANE, I. & RAZIN, S. (1969). Immunological analysis of *Mycoplasma* membranes. *Journal of Bacteriology*, **100**, 187–94.

KELLENBERGER, E. & RYTER, A. (1958). Cell wall and cytoplasmic membrane of *Escherichia coli*. *Journal of Biophysical and Biochemical Cytology*, **4**, 323–6.

KLEEMAN, W. & MCCONNELL, H. M. (1974). Lateral phase separation in *Escherichia coli* membranes. *Biochimica et Biophysica Acta*, **345**, 220–30.

LAURELL, C. (1965). Antigen-antibody crossed immunoelectrophoresis. *Analytical Biochemistry*, **10**, 358–61.

LAW, J. H. (1967). Bacterial lipids. In *The Specificity of Cell Surfaces*, ed. B. D. Davis & L. Warren, pp. 87–105. Englewood Cliffs, New Jersey: Prentice Hall.

LÜDERTIZ, O., WESTPHAL, O., STAUB, M. A. & NIKAIDO, H. (1971). Isolation and chemical and immunological characterization of bacterial lipopolysaccharides. In *Microbial Toxins*, **4**, ed. G. Weinbaum, S. Kadis & S. J. Ajl, pp. 145–233. New York: Academic Press.

MARTIN, E. L. & MACLEOD, R. A. (1971). Isolation and chemical composition of the cytoplasmic membrane of a gram-negative bacterium. *Journal of Bacteriology*, **105**, 1160–7.

MITCHELL, P. (1970). Membranes of cells and organelles: Morphology, transport and metabolism. In *Organization and Control in Prokaryotic and Eukaryotic Cells*; *Symposia of the Society for General Microbiology*, **20**, ed. H. P. Charles & B. C. J. G. Knight, pp. 121–66. Cambridge University Press.

MUÑOZ, E., FREER, J. H., ELLAR, D. J. & SALTON, M. R. J. (1968). Membrane-associated ATPase activity from *Micrococcus lysodeikticus*. *Biochimica et Biophysica Acta*, **150**, 531–3.

MUÑOZ, E., SALTON, M. R. J., NG, M. H. & SCHOR, M. T. (1969). Membrane adenosine triphosphatase of *Micrococcus lysodeikticus*. Purification, properties of the 'soluble' enzyme and properties of the membrane-bound enzyme. *European Journal of Biochemistry*, **7**, 490–501.

MURRAY, R. G. E. (1968). Bacterial cell wall anatomy in relation to the formation of spheroplasts and protoplasts. In *Microbial Protoplasts, and L-forms*, ed. L. B. Guze, pp. 1–18. Baltimore: Williams and Wilkins.

MURRAY, R. G. E., STEED, P. & ELSON, H. E. (1965). The location of the mucopeptide in sections of the cell wall of *Escherichia coli* and other gram-negative bacteria. *Canadian Journal of Microbiology*, **11**, 547–60.

NACHBAR, M. S. & SALTON, M. R. J. (1970). Dissociation of functional markers in bacterial membranes. In *Surface Chemistry of Biological Systems*, pp. 175–90. New York: Plenum.

NAKAE, T. & NIKAIDO, H. (1975). Outer membrane as a diffusion barrier in *Salmonella typhimurium*. Penetration of oligo- and poly-saccharides into isolated outer membrane vesicles and cells with degraded peptidoglycan layer. *Journal of Biological Chemistry*, **250**, 7359–65.

NANNINGA, N. (1971). Uniqueness and location of the fracture plane in the plasma membrane of *Bacillus subtilis*. *Journal of Cell Biology*, **49**, 564–70.

OPPENHEIM, J. D. & NACHBAR, M. S. (1977). Immunochemistry of bacterial ATPases. In *Immunochemistry of Enzymes and Their Antibodies*, ed. M. R. J. Salton, pp. 89–124. New York: Wiley.

OPPENHEIM, J. D. & SALTON, M. R. J. (1973). Localization and distribution of *Micrococcus lysodeikticus* membrane ATPase determined by ferritin labeling. *Biochimica et Biophysica Acta*, **298**, 297–322.

OSBORN, M. J., GANDER, J. E., PARISI, E. & CARSON, J. (1972). Mechanism of assembly of the outer membrane of *Salmonella typhimurium*. *Journal of Biological Chemistry*, **247**, 3962–72.

OSBORN, M. J. & MUNSON, R. (1974). Separation of the inner (cytoplasmic) and outer membranes of gram-negative bacteria. In *Methods of Enzymology*, **31**, ed. S. P. Colowick & N. O. Kaplan, pp. 642–53. New York: Academic Press.

OWEN, P. & SALTON, M. R. J. (1975*a*). Antigenic and enzymatic architecture of *Micrococcus lysodeikticus* membranes established by crossed immunoelectrophoresis. *Proceedings of the National Academy of Sciences, USA*, **72**, 3711–5.

OWEN, P. & SALTON, M. R. J. (1975*b*). A succinylated mannan in the membrane system of *Micrococcus lysodeikticus*. *Biochemical and Biophysical Research Communications*, **63**, 875–80.

OWEN, P. & SALTON, M. R. J. (1975*c*). Isolation and characterization of a mannan from mesosomal membrane vesicles of *Micrococcus lysodeikticus*. *Biochimica et Biophysica Acta*, **406**, 214–34.

PHILLIPS, D. R. & MORRISON, M. (1971). Exposed protein on the intact human erythrocyte. *Biochemistry*, **10**, 1766–71.

PLESS, D. D., SCHMIT, A. S. & LENNARZ, W. J. (1975). The characterization of mannan of *Micrococcus lysodeikticus* as an acidic lipopolysaccharide. *Journal of Biological Chemistry*, **250**, 1319–27.

POLLOCK, J. J., NGUYEN-DISTÈCHE, M., GHUYSEN, J.-M., COYETTE, J., LINDER, R., SALTON, M. R. J., KIM, K. S., PERKINS, H. R. & REYNOLDS, P. (1974). Fractionation of the DD-carboxypeptidase activities solubilized from membranes of *Escherichia coli* K 12, strain 44. *European Journal of Biochemistry*, **41**, 439–46.

POWELL, D. A., DUCKWORTH, M. & BADDILEY, J. (1975*a*). A membrane-associated lipomannan in micrococci. *Biochemical Journal*, **151**, 387–97.

POWELL, D. A., DUCKWORTH, M. & BADDILEY, J. (1975*b*). The presence in micrococci of a membrane-associated lipomannan. *Proceedings of the Society for General Microbiology, Great Britin*, **3** (1), 28.

RAZIN, S. (1972). Reconstitution of biological membranes. *Biochimica et Biophysica Acta*, **265**, 241–96.

REINERT, J. C. & STEIM, J. M. (1970). Calorimetric detection of a membrane-lipid phase transition in living cells. *Science*, **168**, 1580–2.

REUSCH, V. M., JR., & BURGER, M. M. (1973). The bacterial mesosome. *Biochimica et Biophysica Acta*, **300**, 79–104.

ROSENBUSCH, J. P. (1974). Characterization of the major envelope protein from *Escherichia coli*. Regular arrangement on the peptidoglycan and unusual dodecyl sulfate binding. *Journal of Biological Chemistry*, **249**, 8019–29.

ROTHMAN, J. E. & KENNEDY, E. P. (1977). Asymmetrical distribution of phospho-lipids in the membrane of *Bacillus megaterium*. *Journal of Molecular Biology*, **110**, 603–18.

SALTON, M. R. J. (1971). The bacterial membrane. In *Biomembranes*, **1**, ed. L. A. Manson, pp. 1–65. New York: Plenum.

SALTON, M. R. J. (1974). Membrane associated enzymes in bacteria. In *Advances in Microbial Physiology*, **11**, ed. A. H. Rose & D. W. Tempest, pp. 213–83. London: Academic Press.

SALTON, M. R. J. (1976). Methods of isolation and characterization of bacterial membranes. In *Methods in Membrane Biology*, **6**, ed. E. D. Korn, pp. 101–50. New York: Plenum.

SALTON, M. R. J., FREER, J. H. & ELLAR, D. J. (1968). Electron transport compo-nents localized in a lipid-depleted sheet isolated from *Micrococcus lysodeikticus* membranes by deoxycholate extraction. *Biochemical and Biophysical Research Communications*, **33**, 909–15.

SALTON, M. R. J. & OWEN, P. (1976). Bacterial membrane structure. *Annual Re-views of Microbiology*, **30**, 451–82.

SALTON, M. R. J., SCHOR, M. & NG, M. H. (1972). Internal localization of *Micro-coccus lysodeikticus* membrane ATPase by iodination with ^{125}I. *Biochimica et Biophysica Acta*, **290**, 408–13.

SCHECHTER, E., LETELLIER, L. & GULIK-KRZYWICKI, T. (1974). Relations between structure and function in cytoplasmic membrane vesicles isolated from an *Escherichia coli* fatty-acid auxotroph, high-angle X-ray diffraction, freeze-etch electron microscopy and transport studies. *European Journal of Biochem-istry*, **49**, 61–76.

SCHNAITMAN, C. A. (1970). Protein composition of the cell wall and cytoplasmic membrane of *Escherichia coli*. *Journal of Bacteriology*, **104**, 890–901.

SCHNAITMAN, C. A. (1974). Outer membrane proteins of *Escherichia coli*. III. Evidence that the major protein of *Escherichia coli* 0111 outer membrane consists of four distinct polypeptide species. *Journal of Bacteriology*, **118**, 442–53.

SCHNAITMAN, C. A., SMITH, D., & FORN DE SALSAS, M. (1975). Temperate bacterio-phage which causes the production of a new major outer membrane protein by *Escherichia coli*. *Journal of Virology*, **15**, 1121–30.

SHOCKMAN, G. D. & SLADE, H. D. (1964). The cellular location of the streptococcal group D antigen. *Journal of General Microbiology*, **37**, 297–305.

SINGER, S. J. & SCHICK, A. F. (1961). The properties of specific stains for electron microscopy prepared by the conjugation of antibody molecules with ferritin. *Journal of Biophysical and Biochemical Cytology*, **9**, 519–37.

SMIT, J., KAMIO, Y. & NIKAIDO, H. (1975). Outer membrane of *Salmonella typhi-murium*: Chemical analysis and freeze-fracture studies with lipopolysaccharide mutants. *Journal of Bacteriology*, **124**, 942–58.

SMYTH, C. J., FRIEDMAN-KIEN, A. E. & SALTON, M. R. J. (1976). Antigenic analysis of *Neisseria gonorrhoeae* by crossed immuno-electrophoresis. *Infection and Immunity*, **13**, 1273–88.

STANIER, R. Y. (1970). Some aspects of the biology of cells and their possible evo-lutionary significance. In *Organization and Control in Prokaryotic and Eukary-otic Cells*; *Symposia of the Society for General Microbiology*, **20**, ed. H. P. Charles & B. C. J. G. Knight, pp. 1–38. Cambridge University Press.

STECK, T. L. & FOX, C. F. (1972). Membrane proteins. In *Membrane Molecular Biology*, ed. C. F. Fox & A. Keith, pp. 27–75. Stamford, Conn.: Sinauer As-sociates Inc.

THEODORE, T. S., COLE, R. M. & HUFF, E. (1974). Localization of glycerol phosphate in mesosomal vesicles of *Staphylococcus aureus*. *Biochemical and Biophysical Research Communications*, **59**, 215–20.

TREMBLAY, G. Y., DANIELS, M. J. & SCHAECHTER, M. (1969). Isolation of a cell membrane-DNA-nascent RNA complex from bacteria. *Journal of Molecular Biology*, **40**, 65–76.

TSIEN, H. C. & HIGGINS, M. L. (1974). Effect of temperature on the distribution of membrane particles in *Streptococcus faecalis* as seen by the freeze-fracture technique. *Journal of Bacteriology*, **118**, 725–34.

URIEL, J. (1971). Characterization of precipitates in gels. I. Color reactions for identification of antigen-antibody precipitates in gels. In *Methods in Immunology and Immunochemistry*, **3**, ed. C. A. Williams & M. W. Chase, pp. 294–321. New York: Academic Press.

VAN ITERSON, W. & ALTEN, J. A. (1976). Nuclear and cell division in *Bacillus subtilis*: Cell development from spore germination. *Journal of Bacteriology*, **26**, 384–99.

VERKLEIJ, A. J. & VERVERGAERT, P. H. J. (1975). The architecture of biological and artificial membranes as visualized by freeze-etching. *Annual Review of Physical Chemistry*, **26**, 101–22.

VERKLEIJ, A. J., VERVERGAERT, P. H. J., VAN DEENEN, L. L. M. & ELBERS, P. F. (1972). Phase transitions of phospholipid bilayers and membranes of *Acholeplasma laidlawii* B visualized by freeze fracturing electron microscopy. *Biochimica et Biophysica Acta*, **288**, 326–32.

WALLACH, D. F. H. (1972). *The Plasma Membrane: Dynamic Perspectives, Genetics and Pathology*. New York: Springer-Verlag.

WEIBULL, C. (1953a). The isolation of protoplasts from *Bacillus megaterium* by controlled treatment with lysozyme. *Journal of Bacteriology*, **66**, 688–95.

WEIBULL, C. (1953b). Characterization of the protoplasmic constituents of *Bacillus megaterium*. *Journal of Bacteriology*, **66**, 696–702.

WEISS, R. L. (1976). Protoplast formation in *Escherichia coli*. *Journal of Bacteriology*, **128**, 668–70.

WHITE, D. A., ALBRIGHT, F. R., LENNARZ, W. J. & SCHNAITMAN, C. A. (1971). Distribution of phospholipid-synthesizing enzymes in the wall and membrane subfractions of the envelope of *Escherichia coli*. *Biochimica et Biophysica Acta*, **249**, 636–42.

WHITESIDE, T. L., DE SIERVO, A. J. & SALTON, M. R. J. (1971). Use of antibody to membrane adenosine triphosphatase in the study of bacterial relationships. *Journal of Bacteriology*, **105**, 957–67.

WICKEN, A. J. & KNOX, K. W. (1975). Lipoteichoic acids: A new class of bacterial antigen. *Science*, **187**, 1161–67.

YAMAMOTO, S. & LAMPEN, J. O. (1976). Membrane penicillinase of *Bacillus licheniformis* 749/C. Sequence and possible repeated tetrapeptide structure of the phospholipopeptide region. *Proceedings of the National Academy of Sciences, USA*, **73**, 1457–61.

EXPLANATION OF PLATES

PLATE 1

(a) Electron micrograph showing the typical appearance of negatively stained (2% ammonium molybdate) plasma membrane fragments of *M. lysodeikticus*. The small particles (approximately 10 nm diameter) have been identified as the ATPase; (b) negatively stained preparation of isolated mesosomal vesicles of *M. lysodeikticus*. Note the absence of the ATPase particles (L. Trepo-Vitvitsky & M. R. J. Salton, unpublished results).

PLATE 1

(a)

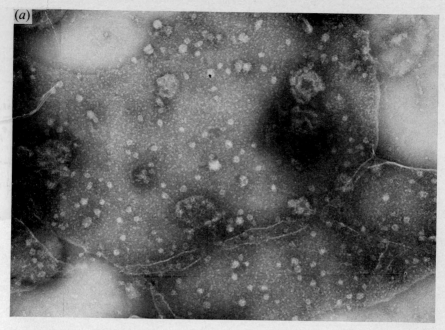

(b)

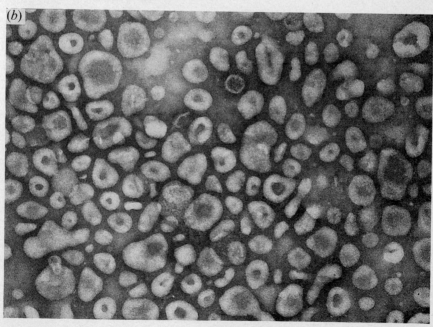

PLATE 2

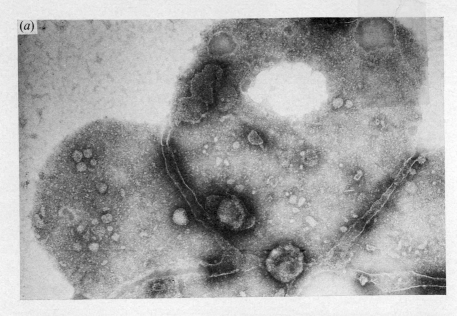

(a)

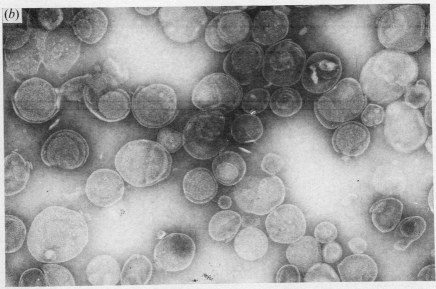

(b)

PLATE 3

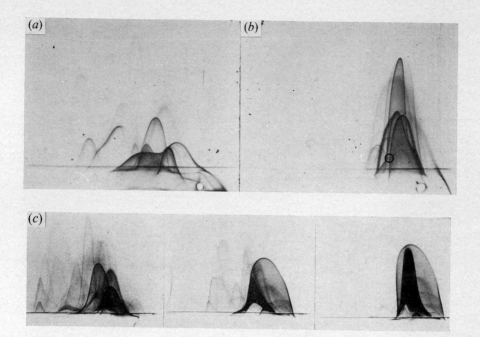

PLATE 4

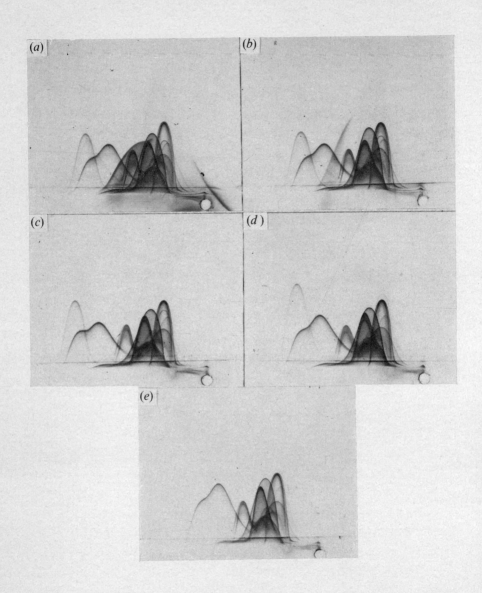

PLATE 2

(*a*) Electron micrograph of negatively stained (2% ammonium molybdate) isolated plasma membrane fraction of *E. coli*, K-12. The plasma membrane fragments have a very similar appearance to those of *M. lysodeikticus*; (*b*) The characteristic appearance of the isolated outer membrane fraction of *E. coli* K-12 as seen by electron microscopy of negatively stained preparations. (J. Siegel & M. R. J. Salton, unpublished results).

PLATE 3

Resolution of Triton X-100 soluble fraction of *M. lysodeikticus* plasma membranes by crossed immunoelectrophoresis under conditions of (*a*) relatively low antigen loading and (*b*) shortened electrophoresis in the first direction to resolve additional immunoprecipitates. Immunoplates were stained with Coomassie brilliant blue. (From P. Owen & M. R. J. Salton, unpublished results). (*c*) Crossed immunoelectrophoresis of Triton X-100 extracts of *E. coli* K-12 membrane structures. Left to right, isolated inner (plasma) membranes, envelopes prepared by mechanical disruption of cells; isolated outer membranes. Note that the envelopes represent a partial composite pattern of the inner and outer membranes (C. J. Smyth, J. Siegel, M. R. J. Salton & P. Owen, *J. Bacteriology*, in press).

PLATE 4

Absorption experiments with either intact cells or protoplasts remove antibodies to the surface-expressed antigens of the protoplast surface and yield identical results. The progressive removal of antibodies to such components is illustrated. Note that the pattern of antigens expressed on the inner membrane remain unaltered. Absorbed (*b–e*) and unabsorbed sera (*a*) were tested against Triton X-100 soluble membrane fraction of *M. lysodeikticus*. (From P. Owen & M. R. J. Salton, *J. Bacteriology*, in press).

THE PURPLE MEMBRANE OF HALOBACTERIA

RICHARD HENDERSON

Medical Research Council Laboratory of Molecular Biology,
Hills Road, Cambridge CB2 2QH, UK

Since its discovery as part of the membrane of *Halobacterium halobium* (Stoeckenius & Rowen, 1967; Stoeckenius & Kunau, 1968), the purple membrane has been characterised as a uniquely simple and interesting structure (Oesterhelt & Stoeckenius, 1971; Blaurock & Stoeckenius, 1971) and shown to function as a light-driven proton pump creating an electrochemical gradient of hydrogen ions across the membrane (Oesterhelt & Stoeckenius, 1973). There are several recent reviews for those who would like to read more thoroughly (Stoeckenius, 1976; Oesterhelt, 1976; Henderson, 1977). I shall present a summary of some of the work which bears most directly on the structure and molecular function of the purple membrane and the bacteriorhodopsin molecules of which it is composed, without paying too much attention to its role in cellular function, since this would involve extensive discussion of coupling of hydrogen ion movements to the movements of other ions and small molecules across the cell membrane.

Purple membrane forms specialised patches which are integrated within the cytoplasmic membrane of halobacteria. They can be seen clearly in freeze-fracture pictures (Plate 1) of whole bacteria where the characteristic hexagonal lattice structure of the purple membrane sets it apart from the rest of the cell membrane which is less regularly arranged and on isolation has a dark red colour due to the pigment bacterioruberin. Figure 1 represents schematically the relationship between the red and purple membrane fractions and the cell wall, which is composed of glycoproteins (Oesterhelt & Stoeckenius, 1974; Mescher & Strominger, 1976). The purple membrane is composed (Oesterhelt & Stoeckenius, 1971) of identical protein molecules of 26000 molecular weight which make up 75% of the mass of the membrane and several different types of lipid which make up the remaining 25%. Retinal, the chromophore responsible for the characteristic purple colour, is found bound to each protein molecule in a 1:1 ratio. All three components in the membrane are arranged in a two-dimensional hexagonal array, properly described (Blaurock, 1975; Henderson, 1975) as a crystal of space group P3, having a thickness of only one molecule. The

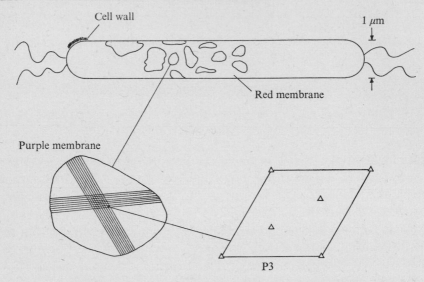

Fig. 1. Schematic diagram of *H. halobium* showing the relationship between the red membrane, the purple membrane, the cell wall and the flagellae in whole bacteria.

retinal is bound via a Schiff base to a lysine(*) residue in the protein (Oesterhelt & Stoeckenius, 9171). This residue is found in the sequence (Met)–Gly–Val–Ser–Asp–Pro–Asp—Lys–Lys(*)–Phe–Tyr–Ala–Ileu–Met (Bridgen & Walker, 1976). The protein–retinal complex has been named bacteriorhodopsin because of its similarity to rhodopsin, the visual pigment. Both molecules are light receptors and both contain retinal as their chromophores.

Purple membrane has been shown to function as a light-driven proton pump (Oesterhelt & Stokeckenius, 1973). Each bacteriorhodopsin molecule appears to undergo a photochemical cycle (Fig. 2) which is initiated by the light-driven transition $bR_{570} \rightarrow K_{590}$. The state K_{590} is a high energy state which then decays through a cycle of intermediate structures with no additional energy input. The rough time scale of these events, derived from flash spectroscopy experiments (Stoeckenius & Lozier, 1974; Dencher & Wilms, 1975; Sherman, Slifkin & Caplan, 1976; Chance, Porte, Hess & Oesterhelt, 1975; Kung, Devault, Hess & Oesterhelt, 1975; Lozier & Niederberger, unpublished observations), is shown in Fig. 2.

During this photochemical cycle, approximately one proton is released and later taken up again, per molecule cycling. The released proton appears on the outside of the membrane and the subsequent uptake is from the intracellular surface of the membrane (Lozier *et al.*, 1976). Hence the net effect of these molecular events is to create an

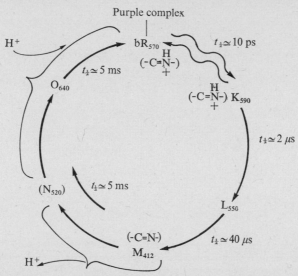

Fig. 2. A current model of the intermediates in the photochemical cycle of the purple membrane mainly derived from flash spectroscopy (see text). A hydrogen ion is released and taken up again, near the positions shown. The drawing is slightly adapted from Lozier, Bogomolni & Stoeckenius (1975).

energy store in the form of an electrochemical gradient across the membrane. The pumping of protons eventually acidifies the extracellular solution particularly when usage of the gradient by other molecules in the membrane is blocked (Oesterhelt, 1975). However, experimental observation of pH changes depends on the transfer of other counterions, making the interpretation of some of the pH changes complex (Oesterhelt *et al.*, 1975; Bogomolini, Baker, Lozier & Stoeckenius, 1976).

Our approach to understanding the nature of the purple membrane has been a straightforward structural one (Unwin & Henderson, 1975; Henderson & Unwin, 1975). Using electron microscopy at low electron doses and without the use of heavy atom stains, we have obtained a 0.7 nm resolution three-dimensional map of the structure of the membrane. The map shows that each protein molecule is composed of seven α-helical segments. Each helical segment is approximately 4 nm long and traverses the membrane roughly perpendicular to its plane. The lipid forms bilayer regions which fill up the spaces between the protein molecules. A schematic diagram of one molecule in the membrane is shown in Fig. 3. The position of the retinal is not yet known, and a more detailed idea of how the molecule functions as a proton pump depends on getting a higher resolution structure, of at least 0.35 nm, so that individual amino acid side chains can be discerned.

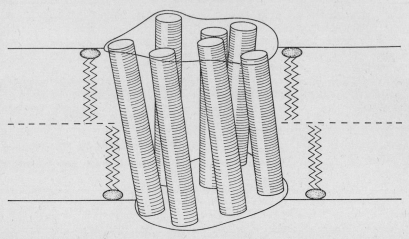

Fig. 3. Schematic diagram of the arrangement of α-helices in one protein molecule of the purple membrane. The lipid molecules are probably close to the positions shown. The model was derived by electron microscopy (Henderson & Unwin, 1975).

The chemical nature of the membrane is quite different on the two sides. Plate 2 shows the result of labelling purple membrane with the lysine reagent biotin-N-hydroxy-succinimide ester (Henderson, unpublished experiments). After attachment of membranes to the carbon film used for electron microscopy, subsequent application of the reagent made by conjugation of avidin with ferritin (Heitzmann & Richards, 1974) shows that only half the membranes react with ferritin, the other half being unreactive. This is because (probably only) one lysine is labelled on one side of the membrane and none are labelled on the other. Membranes with their labelled lysine in contact with the carbon cannot react with the avidin–ferritin reagent. This demonstrates quite strikingly the chemical difference between the two sides of the membrane. It can be shown, by electron diffraction of ferritin-labelled membranes, that the labelled lysine(s) is on the bottom surface of the model shown in Fig. 3, and, by further experiments using labelled cell envelopes in which the two compartments representing the inside and outside of the cell are separately accessible, that the lysine is on the outside surface of the cell membrane (Henderson, unpublished experiments). That is, the bottom of the model corresponds to the outside surface of the cell membrane. Similar kinds of labelling of other chemical sites, together with higher resolution crystallographic analysis should eventually provide the framework for an interpretation of the molecular mechanism of this light-driven proton pump.

REFERENCES

BLAUROCK, A. E. (1975). Bacteriorhodopsin: a trans-membrane pump containing α-helix. *Journal of Molecular Biology*, **93**, 139–58.

BLAUROCK, A. E. & STOECKENIUS, W. (1971). Structure of the purple membrane. *Nature New Biology*, **233**, 152–5.

BOGOMOLNI, R. A., BAKER, R. A., LOZIER, R. H. & STOECKENIUS, W. (1976). Light-driven proton translocations in *Halobacterium halobium*. *Biochemica et Biophysica Acta*, **440**, 68–88.

BRIDGEN, J. & WALKER, I. D. (1976). Photoreceptor protein from the purple membrane of *Halobacterium halobium*: molecular weight and retinal binding site. *Biochemistry*, **15**, 792–8.

CHANCE, B., PORTE, M., HESS, B. & OESTERHELT, D. (1975). Low temperature kinetics of H^+ changes of bacterial rhodopsin. *Biophysical Journal*, **15**, 913–17.

DENCHER, N. & WILMS, M. (1975). Flash-photometric experiments in the photochemical cycle of bacteriorhodopsin. *Biophysics of Structure & Mechanism*, **1**, 259–71.

HEITZMANN, H. & RICHARDS, F. M. (1974). Use of the Avidin–Biotin complex for specific staining of biological membranes in electron microscopy. *Proceedings of the National Academy of Sciences, USA*, **71**, 3537–41.

HENDERSON, R. (1975). The structure of the purple membrane from *Halobacterium halobium*: analysis of the X-ray diffraction pattern. *Journal of Molecular Biology*, **93**, 123–38.

HENDERSON, R. (1977). The purple membrane from *Halobacterium halobium*. *Annual Review of Biophysics and Bioengineering*, **6**, 87–109.

HENDERSON, R. & UNWIN, P. N. T. (1975). Three-dimensional model of purple membrane obtained by electron microscopy. *Nature, London*, **257**, 28–32.

KUNG, M. C., DEVAULT, D., HESS, B. & OESTERHELT, D. (1975). Photolysis of bacterial rhodopsin. *Biophysical Journal*, **15**, 907–11.

LOZIER, R. H., BOGOMOLNI, R. A. & STOECKENIUS, W. (1975). Bacteriorhodopsin: a light-driven proton pump in *Halobacterium halobium*. *Biophysical Journal*, **15**, 955–62.

LOZIER, R. H., NIEDERBERGER, W., BOGOMOLNI, R. A., HWANG, S. & STOECKENIUS, W. (1976). Kinetics and stoichiometry of light-induced proton release and uptake from purple membrane fragments, *Halobacterium halobium* cell envelopes, and phospholipid vesicles containing oriented purple membrane. *Biochimica et Biophysica Acta*, **440**, 545–56.

MESCHER, M. F. & STROMINGER, J. L. (1976). Purification and characterization of a prokaryotic glycoprotein from the cell envelope of *Halobacterium salinarium*. *Journal of Biological Chemistry*, **251**, 2005–14.

OESTERHELT, D. (1975). The purple membrane of *Halobacterium halobium*: a new system for light energy conversion. In *Energy Transformation in Biological Systems; Ciba Foundation Symposium*, **31**, 147–67.

OESTERHELT, D. (1976). Bacteriorhodopsin as an example of a light-driven proton pump. *Angewandte Chemie*, **15**, 17–24.

OESTERHELT, D., HARTMANN, R., FISCHER, U., MICHEL, H. & SCHRECKENBACH, T. (1975). The biochemistry of a light-driven proton pump: bacteriorhodopsin. *Proceedings of the 10th FEBS Meeting*, **40**, 239–51.

OESTERHELT, D. & STOECKENIUS, W. (1971). Rhodopsin-like protein from the purple membrane of *Halobacterium halobium*. *Nature New Biology*, **233**, 149–52.

OESTERHELT, D. & STOECKENIUS, W. (1973). Functions of a new photoreceptor membrane. *Proceedings of the National Academy of Sciences, USA*, **70**, 2853–7.

OESTERHELT, D. & STOECKENIUS, W. (1974). Isolation of the cell membrane of *Halobacterium halobium* and its fractionation into red and purple membranes. In *Methods in Enzymology*, **31**, ed. S. Fleischer & L. Packer, pp. 667–78. New York: Academic Press.

SHERMAN, W. V., SLIFKIN, M. A. & CAPLAN, S. R. (1976). Kinetic studies of photo-transients in bacteriorhodopsin. *Biochimica et Biophysica Acta*, **423**, 238–48.

STOECKENIUS, W. (1976). The purple membrane of salt-loving bacteria. *Scientific American*, **234**, (6), 38–46.

STOECKENIUS, W. & KUNAU, W. H. (1968). Further characterization of particulate fractions from lysed cell envelopes of *Halobacterium halobium* and isolation of gas vacuole membranes. *Journal of Cell Biology*, **38**, 337–57.

STOECKENIUS, W. & LOZIER, R. H. (1974). Light energy conversion in *Halobacterium halobium*. *Journal of Supramolecular Structure*, **2**, 769–74.

STOECKENIUS, W. & ROWEN, R. (1967). A morphological study of *Halobacterium halobium* and its lysis in media of low salt concentration. *Journal of Cell Biology*, **34**, 365–93.

UNWIN, P. N. T. & HENDERSON, R. (1975). Molecular structure determination by electron microscopy of unstained crystalline specimens. *Journal of Molecular Biology*, **94**, 425–40.

EXPLANATION OF PLATES

PLATE 1

Freeze-fracture picture of whole cells of *H. halobium*. The characteristic hexagonal lattice can be seen in the areas which represent the purple membrane patches. The protein molecules appear to stick to the cytoplasmic fracture face, whereas the extracellular fracture face is smooth or contains only very faint striations. The rest of the cell membrane has the normal irregular structure found in many other fractured membranes. (Picture courtesy of Dr W. Stoeckenius.)

PLATE 2

Electron micrograph of purple membrane coupled to biotin. The biotin groups (linked covalently to lysine residues) appear to react with only one side of the membrane. They are visualised by reaction with avidin–ferritin conjugates after being deposited on the carbon film used for the electron microscopy. Ferritin binding occurs only to those membranes with their biotinylated surface directed away from the carbon film. Other experiments (see text) have shown that the side of the membrane which contains the reactive lysine(s) corresponds to the bottom of the model in Fig. 3 and the outside surface of the cell membrane. The two surfaces of the membrane are quite clearly different.

PLATE 1

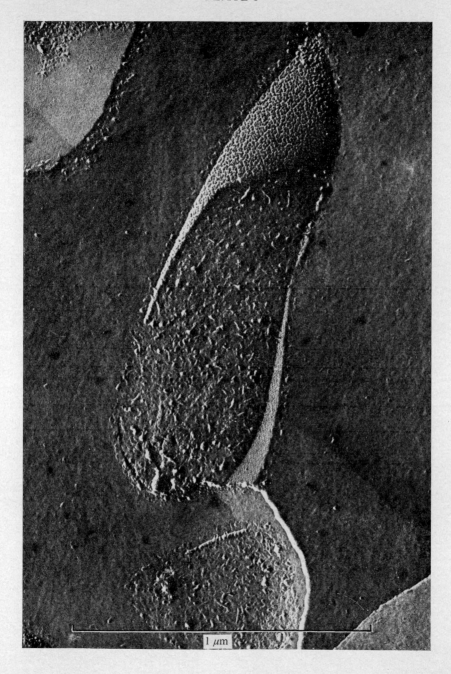

1 μm

PLATE 2

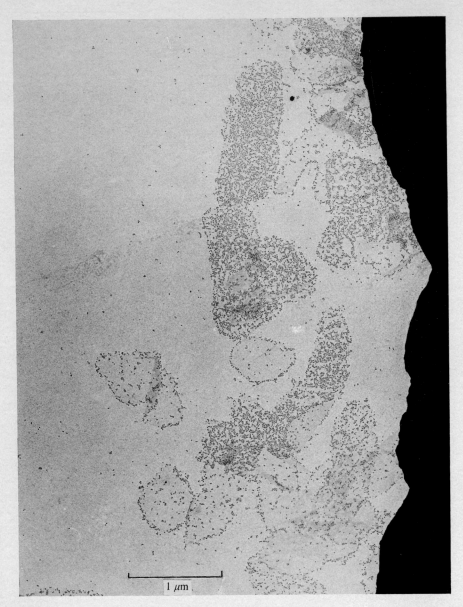

1 μm

PHOSPHOLIPOPROTEINS IN ENZYME EXCRETION BY BACTERIA

J. OLIVER LAMPEN

Waksman Institute of Microbiology, Rutgers – The State University of New Jersey, PO Box 759, Piscataway, NJ 08854, USA

INTRODUCTION

The excretion of protein molecules (enzymes, cell wall components, hormones, toxins, etc.) that are too large or hydrophilic to pass readily through the lipid diffusion barrier of the plasma membrane is an important physiological feature of virtually all cells. A substantial body of information has been assembled concerning the nature of the process in mammalian cells, but it is only in the last 10–15 years that the characteristics of excretion in microorganisms have been intensively investigated. The general view of the process in mammalian cells (recently reviewed by Palade, 1975 and by Rothman & Lenard, 1977) is that the polypeptide chain to be exported is formed on the ribosomes of the rough endoplasmic reticulum (RER) and directed toward the lumen during the elongation process. If carbohydrate is to be attached, N-acetylglucosaminyl and mannosyl residues are generally attached at this stage. More sugars may later be added by transferases as the proteins move from the rough to the smooth endoplasmic reticulum, then often through a Golgi apparatus and eventually to storage granules from which they are released following the appropriate physiological stimulus.

Excretion in yeasts and fungi should be rather similar to that in the more complex eukaryotes, but in the small, prokaryotic bacterial cell the total process must be considerably simpler. There is no evidence for the export of proteins by way of vesicles in bacteria. We have, therefore, considered the process as analogous to the initial stage in higher organisms, that is, the formation of the polypeptide chain on the RER and its passage into the lumen (Lampen, 1974). In the bacterium, export occurs through the plasma membrane out into the medium, so that one can often examine the newly formed molecules as they appear at the outer membrane surface and determine the mechanism of release.

Palade (1975) has suggested that the nascent chain passes through a hydrophilic pore with final folding of the molecule occurring in the lumen of the RER, thus trapping the new protein on the 'outer' side of the membrane. However, other studies show that the nascent chain is firmly held by the RER and may play a major role in anchoring the ribosome to the membrane (Adelman, Sabatini & Blobel, 1973). It seems more probable that the amino acid side chains of the entering NH_2-terminal portion of the polypeptide would interact with hydrophobic groups on the membrane proteins. As part of the extended chain nears the outer aqueous surface, its ionic groups become hydrated and would interact with other polar groups. This should reduce interactions with the lipophilic membrane components and lead to the development of a compact conformation with a hydrophobic interior and a hydrophilic surface and would, in a sense, pull the growing chain through the membrane. The fundamental concept is that export occurs through non-polar pores within clusters of proteins. The process might be energetically most favourable if the membrane proteins were already close enough that existing hydrophobic protein–protein interactions would simply be replaced by interactions between the proteins and the entering peptide chain. Such pores could well be present in bacillus membranes which have a protein content of 40% to 60%.

EXCRETION OF PENICILLINASE
BY *BACILLUS LICHENIFORMIS* 749

We chose the penicillinase of *Bacillus licheniformis* 749 for our investigation of enzyme excretion by a prokaryote. There is only a single structural gene (Sherratt & Collins, 1973) and the exoenzyme is a monomeric protein of known primary sequence and devoid of carbohydrate and S–S bridges (Ambler & Meadway, 1969). This highly soluble molecule is resistant to proteases and renatures readily after denaturation in 8 M urea, 6 M guanidine or hot 1% dodecyl sulphate. A substantial portion of the penicillinase activity remains attached to the outer surface of the cell membrane, especially during growth at slightly acid pH. Inducible (749) and penicillinase-constitutive (749/C) strains are available which produce large amounts of enzyme. Also there is a rather close coupling between synthesis and secretion with no evidence for an inactive macromolecular precursor (Sargent, Ghosh & Lampen, 1969). Treatment of protoplasts or of membrane fragments with trypsin releases essentially all of the bound penicillinase in a form identical with authentic exoenzyme except that it lacks the usual NH_2-terminal lysine residue.

During growth of the constitutive mutant 749/C, the level of membrane-bound enzyme is high in early exponential phase but later declines concomitantly with the acclerated appearance of exoenzyme. A pre-

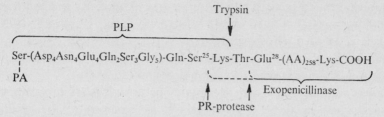

Fig. 1. Schematic structure of the membrane penicillinase of *B. licheniformis* 749/C. The relation to exopenicillinase and the phospholipopeptide (PLP) portion is illustrated. PA, phosphatidic acid residue.

cursor–product relationship for the two forms at pH values above neutrality was indicated by the ready conversion of membrane-bound enzyme to exoenzyme after de-novo synthesis had been halted by starvation for amino acids or treatment with chloramphenicol (Sargent, Ghosh & Lampen, 1968a, 1969). This relationship did not hold, however, in an early exponential-phase culture growing at pH 6.0 to 6.5; here most of the newly formed exoenzyme arose by de-novo synthesis and not by conversion of pre-existing (radioactively labelled) membrane penicillinase (Crane, Bettinger & Lampen, 1973).

NATURE OF THE MEMBRANE PENICILLINASE

Initial investigations of the membrane pencillinase from *B. licheniformis* 749/C revealed that it was a hydrophobic molecule of about 33000 daltons which retained radioactivity after purification from cultures containing $H_3^{33}PO_4$ or [2-^{3}H]glycerol (Sawai, Crane & Lampen, 1973). At the same time it appeared to contain all of the antigenic determinants characteristic of the exoenzyme (Crane & Lampen, 1974) and could be considered as consisting of two segments: one essentially equivalent to the hydrophilic exoenzyme (29500 daltons) and the other highly hydrophobic, perhaps phospholipid in nature. The purified membrane enzyme had the same molar specific activity as exoenzyme, as would be anticipated from the observation that release of exoenzyme by trypsin from protoplasts or from isolated membrane fragments occurs without a change in total penicillinase activity (Sawai & Lampen, 1974; Yamamoto & Lampen, 1976c).

A schematic structure of the membrane enzyme is presented in Fig. 1. Material purified from cells grown in the presence of [2-^{3}H]glycerol and ^{14}C-labelled amino acids was treated with trypsin to produce the exo-like enzyme and a fragment which apparently constituted the hydrophobic segment (Yamamoto & Lampen, 1976c). The major portion exhibited penicillinase activity and ^{14}C-activity but not tritium. The fragment contained both ^{3}H-activity and ^{14}C-labelled amino acids and yielded phosphatidylserine upon exhaustive digestion with pronase. We therefore termed the membrane enzyme a phospholipo-protein and the fragment a phospholipopeptide.

It was possible that the extra peptide portion of the membrane enzyme was attached to the COOH-terminus of exoenzyme since Kelly & Brammar (1973) had presented evidence that the first possible chain-terminating triplets in the mRNA for penicillinase are four and six codons respectively beyond the usual COOH-terminal lysine of the exoenzyme, and hence that this terminus must have arisen by specific hydrolysis of the initial penicillinase molecule. A COOH-terminal attachment of the phospholipopeptide became extremely unlikely, however, when the four COOH-terminal amino acids of both exo-enzyme and membrane enzyme were shown to be -Met-Asn-Gly-Lys-COOH (Yamamoto & Lampen, 1976b). Acid hydrolysis of the phos-pholipeptide produced only five types of amino acid residues (Ser, Gly, Asp, Glu, and Lys). The single lysine residue was the COOH terminus and proved to be identical with the NH$_2$-terminal lysine of the usual exoenzyme, thus establishing the NH$_2$-terminal location of the fragment.

The amino acid sequence of the phospholipopeptide (Yamamoto & Lampen, 1976a) is illustrated in Fig. 2. The peptide chain is relatively hydrophilic as at least 19 of the 25 amino acid residues are usually con-sidered to be polar. The hydrophobicity of the total phospholipopep-tide (and of the membrane enzyme) must therefore be primarily a func-tion of the NH$_2$-terminal phosphatidylserine residue. The existence of only 7 types of amino acid residues (Gly, Ser, Asp, Asn, Glu, Gln and Lys) indicated that the phospholipopeptide might have arisen from an oligopeptide structure, and the molecule can be presented (Fig. 2) as a repeating unit -Asp(Asn)-Glu(Gln)-Ser-Gly- which occurs seven times but with a number of modifications and deletions. The proposed modifi-cations seem reasonable in that they require only acceptable single-base changes in the corresponding mRNA. In addition one must postulate a duplication of the Glu^{19} residue and six deletions.

The polypeptide chain presumably produced by this repetitive process is not uniform in character. One can detect at least a highly polar region (residues 19 to 25) and a less polar portion (residues 1 to 18) which in-cludes all of the Gly residues and few dicarboxylic acids. One would not anticipate any significant selective pressure toward maintaining the precise sequence of the phospholipopeptide since the phosphatidylserine

$$\text{Ser}^1 \text{ --- Asn --- Asp --- Glu --- Gly}^5 \text{ --- Gly}$$

PA   Asp --- * --- Ser --- Gly.

Asn^{10}--- Gln --- Ser --- Gly

Asp --- * --- * --- Gly^{15}

Asn --- Gln --- Ser --- Glu --- Glu^{20}

Asn --- Glu --- * --- *

Asp --- Gln --- Ser^{25} --- * --- Lys-COO$^-$

Exoenzyme

Fig. 2. The amino acid sequence of the phospholipopeptide produced by trypsin cleavage of membrane penicillinase. The residues are arranged to indicate the possible repeated tetrapeptide structure. PA, phosphatidic acid residue. Asterisk indicates residue assumed to have been deleted.

residue is probably adequate to ensure attachment of the penicillinase to the membrane. Nevertheless the apparent changes in the repeating unit were not random, but rather led to a systematic divergence in polarity. There appears to have been an evolutionary selective pressure favouring this differentiation with its attendant effects on the affinity of these segments for the membrane and hence on the orientation of the total enzyme molecule.

CONVERSION OF MEMBRANE PENICILLINASE TO EXOENZYME

Role of penicillinase-releasing protease

A single enzyme currently given the trivial name penicillinase-releasing protease (PR-protease) appears to be almost exclusively responsible for the production of exopenicillinase.

During the conversion of strain 749/C cells to protoplasts nearly half of the membrane-bound penicillinase is released along with the vesicles and tubules (the so-called 'mesosome' fraction). This vesicle penicillinase was rapidly converted to a hydrophilic form during incubation at pH 8 to 9 and was at first thought to differ chemically from the material that remained with the plasma membrane (Sargent & Lampen, 1970). Further investigation showed the rapid formation of exoenzyme to be due to a soluble protease present in these preparations and the vesicle enzyme to be essentially identical with the plasma membrane enzyme (Traficante & Lampen, 1977a, b). This protease cleaves purified membrane penicillinase to yield the exoenzyme and releases exoenzyme from protoplasts or from cell membrane fragments. The enzyme does not appear in the medium in quantity until the early stationary phase of growth; in exponential cultures it is periplasmic in that it is liberated in a soluble form during protoplasting. PR-protease is active over the pH range 6 to 9 or 10 and is inhibited by diisopropylfluorophosphate, deoxycholate, quinacrine, or o-phenanthroline but not by ethylenediaminetetraacetic acid or SH-blocking reagents (Aiyappa, Traficante & Lampen, 1977; Traficante & Lampen, 1977a). It may be the only enzyme in these cultures capable of carrying out the conversion since the other proteases separated from PR-protease during its purification showed little, if any, activity in releasing penicillinase from membrane vesicles (the standard assay for monitoring purification).

A comparison of the process of exoenzyme release at acid and alkaline pH and by protoplasts or growing cultures strongly supports the concept that a single protease is functioning throughout. The exopenicillinases released had the same NH_2-terminal residues (lysine and some glutamic acid) as authentic exopenicillinase (Fig. 1). The sensitivity of the various release systems to inhibition by deoxycholate, quinacrine, and o-phenanthroline paralleled that of the purified PR-protease. Finally, the specificity of the PR-protease was determined by characterising the products of digestion of insulin and of ribonuclease B (Aiyappa & Lampen, 1977). Only peptide bonds involving the carboxyl groups of serine or threonine were hydrolysed, but there was no apparent specificity for the residue donating the amino group. The tetrapeptide segment containing the eventual NH_2-terminal lysine or glutamic acid residue of exopenicillinase is -Ser^{25}-Lys-Thr-Glu^{28}-, and cleavage of Ser^{25}-Lys^{26} or Thr^{27}-Glu^{28} must have occurred. There are other potentially sensitive bonds in the phospholipopeptide region of the membrane penicillinase, but as already noted the segment (residues 19 to 28) containing the two linkages known to be cleaved is polar and highly charged and may well project out of the membrane into the aqueous phase, whereas residues 1 to 18 are less polar and this segment should have a greater affinity for the lipid bilayer. Furthermore, exopenicillinase and the corresponding portion of the membrane enzyme are characteristically resistant to proteases, hence, the Ser^{25}-Lys^{26} and Thr^{27}-Glu^{28} bonds may be the only susceptible bonds in the membrane-bound enzyme that are readily accessible for cleavage by the PR-protease. These results offer a ready explanation of exopenicillinase release and of the role of PR-protease in penicillinase secretion.

Location of membrane penicillinase and PR-protease

The phospholipoprotein membrane penicillinase appears to be concentrated in or around those areas that are expelled as vesicles during the conversion of cells to protoplasts since these fragments contain about half of the total membrane enzyme and only about 15% of the total protein (Sargent, Ghosh & Lampen, 1968b). This localization is not only characteristic of the magnoconstitutive mutant 749/C but also exists from the beginning of the induction process in strain 749, in that cell samples taken throughout induction and then converted to protoplasts all released about half of their membrane-bound enzyme along with the vesicle fraction (G. E. Bettinger & J. O. Lampen, unpublished results). One might infer from this that the mesosome, whatever its structure, has a direct function in the excretion process, yet this seems unlikely since protoplasts synthesize and release penicillinase at about the same rate as did the cells from which they were derived.

There is indirect evidence that the cell-bound PR-protease may also be concentrated in or near the mesosomal area. Specifically, the exo-penicillinase released at pH 9 from a washed-cell suspension was derived mostly from the membrane fraction that would be expelled as vesicles during conversion of such organisms to protoplasts and not from the fraction that would remain with the protoplasts (Sargent et al., 1969). The situation is complex, however, in that the small amount of PR-protease activity which is retained by the protoplast can eventually convert about half of the protoplast-bound pencillinase to exoenzyme at pH 8.5 to 9.5 but not at pH 6.0 to 6.5, even though purified PR-protease is active at pH 6 and can release pencillinase from protoplasts at this pH (Aiyappa et al., 1977). The basis for this differential accessibility is as yet unclear.

BIOSYNTHESIS OF MEMBRANE PENICILLINASE

Nature of initial gene product

An essential step in determining the total pathway of penicillinase biosynthesis was to clarify the relationship between the phospholipoprotein membrane-bound enzyme and exopenicillinase. Dancer & Lampen (1975) investigated in-vitro penicillinase synthesis in extracts of B. licheniformis 749/C. This system (Davies, 1969) utilizes the 30000g supernatant (S-30) of cell lysates as the source of mRNA, tRNA, elongation factors, etc., and is considered capable only of completing polypeptide chains. The production of active penicillinase was inhibited by chloramphenicol. Purification of the product by affinity chromatography on Sepharose 4B-cephalosporin C and subsequent electrophoresis on polyacrylamide gels showed it to be the hydrophobic penicillinase containing tightly-bound phospholipid; the hydrophilic exoenzyme was barely detectable. These results indicate that nascent penicillinase,

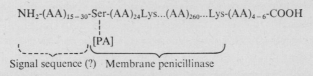

Fig. 3. Possible nature of the initial polypeptide product from the structural gene for penicillinase. PA, phosphatidic acid residue.

while still attached to the polysomes, already has a phospholipid group on the NH_2-terminal serine residue. Furthermore, they supplement the earlier evidence that at pH values above neutrality, much of the exoenzyme is produced by cleavage of the membrane penicillinase, but they do not prove that all exopenicillinase is produced *in vivo* by this pathway.

It is important to point out that the product isolated from the in-vitro system has been equated with membrane penicillinase largely on the basis of its strongly hydrophobic character which is presumably due to an attached phospholipid residue. It is, however, highly likely that the original product of translation (Fig. 3) contained four to six amino acid residues beyond the usual COOH-terminal lysine of penicillinase (Kelly & Brammar, 1973) and our procedures would not discriminate between this and the usual membrane enzyme. One must also consider the possibility that penicillinase, like the outer membrane lipoprotein (Wang *et al.*, 1977) and the alkaline phosphatase (Inouye & Beckwith, 1977) of *Escherichia coli*, carries a signal sequence at its NH_2 terminus that serves to orient the nascent chain to the proteins of the membrane pore and is subsequently removed (Blobel & Sabatini, 1971; Blobel & Dobberstein, 1975; Rothman & Lenard, 1977). Such sequences, usually about 15 to 30 residues long, are relatively hydrophobic though probably not enough to give the in-vitro product its overall hydrophobicity. These possibilities require further investigation.

Source of phosphatidylserine

We examined the phospholipid metabolism of *B. licheniformis* 749/C membranes in order to gain some information concerning the possible modes of attachment of the phospholipid residue in membrane penicillinase. Chloroform:methanol extracts of cells suspended in pH 7.0 buffer did not yield a detectable phosphatidylserine spot on thin-layer chromatography; however, phosphatidylserine was readily detected in extracts of cells suspended in pH 9 buffer (Dancer & Lampen, 1977). This indicated an alkali-labile linkage similar to that between tRNA and amino acids. When the tRNA fraction was prepared from washed 749/C membranes it also released phosphatidylserine at pH 9.0 but not at acid pH. It therefore seemed likely that tRNA with attached phosphatidylserine is a normal constituent of these cells, and that the phospholipid is bound in the same way as the usual amino acid. To obtain more material for characterization, the in-vitro production of phosphatidylseryl-tRNA from purified *B. licheniformis* tRNA was attempted. Cytidine triphosphate, phosphatidic acid, and ^{3}H-labelled seryl-tRNA were incubated with S-30 supernatant and the tRNA was then reisolated. This fraction contained no free phosphatidylserine but after incubation at pH 9 free phosphatidylserine was detected. A tentative reaction scheme parallels that for synthesis of phosphatidylserine:

$$CTP + phosphatidic\ acid \rightleftharpoons CDP\text{-diglyceride} + PP_i$$

$$CDP\text{-diglyceride} + seryl\text{-tRNA} \rightleftharpoons CMP + phosphatidylseryl\text{-tRNA}$$

Phosphatidylseryl-tRNA is an activated form of phosphatidylserine and protected from decarboxylation. Thus it is attractive as a phospholipid donor in the event that a phosphatidylseryl residue either serves to start the polypeptide chain or is added to an initially shorter chain (Fig. 4). It must be remembered, however, that the S-30 supernatant will also catalyse the transfer of a phosphatidic acid residue from CDP-diglyceride to free serine. Such a transfer may be involved if the initial penicillinase translation product carries a signal sequence and the critical serine residue is an internal residue in the initial polypeptide chain. Transfer of a phosphatidic acid residue to the serine would be analogous to the proposed transfer of a diglyceride residue to a cysteine during synthesis of the NH_2-terminal residue of the *E. coli* lipoprotein (Lin & Wu, 1976).

Site of penicillinase synthesis

Evidence is gradually accumulating that in bacterial protein excretion the polypeptide chain is formed on membrane-bound polysomes and extruded during the elongation process in a partially folded state with the final conformation attained at or on the outer membrane surface. For example, Bettinger and Lampen (1971, 1975) showed that the penicillinase of *B. licheniformis* 749/C is extruded in an incompletely folded form, and there are comparable data for the protease P produced by a *Sarcina* (Bissell, Tosi & Gorini, 1971) and for several enzymes produced by *B. amyloliquefaciens* (Sanders & May, 1975). According to a recent report by Smith, Tai, Thompson & Davis (1977) a number of proteins being synthesized by *E. coli* penetrate the plasma membrane

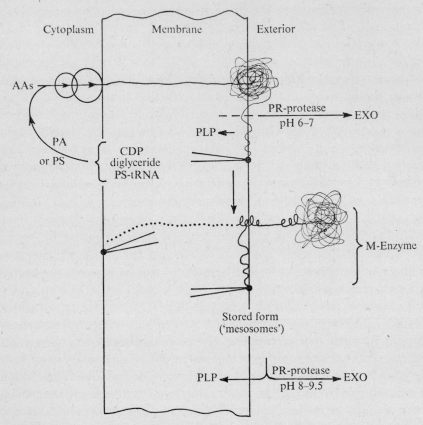

Fig. 4. Schematic representation of the various potential stages in the formation of exopeni-cillinase by *B. licheniformis* 749/C at acid and alkaline pH. See text for discussion. PA, phosphatidic acid residue; PS, phosphatidylseryl residue; PLP, phospholipopeptide formed by cleavage of membrane (*M*) penicillinase; EXO, exoenzyme.

and can be labelled by non-penetrating reagents while still attached to membrane-bound polysomes. One of the major proteins labelled in this manner was identified as the periplasmic enzyme alkaline phosphatase. Synthesis of alkaline phosphatase at the plasma membrane in *E. coli* is also strongly indicated by the work of MacAlister, Irwin & Costerton (1977*a*, *b*) who used ferritin-labelled antibodies to detect the enzyme in frozen sections of cells undergoing derepression of alkaline phosphatase synthesis. Finally Ghosh & Ghosh (1977) have employed similar im-muno-electron microscopic techniques and found that much of the penicillinase of strain 749/C is associated with the plasma membrane. Intriguingly, material reactive with the antiserum to the purified enzyme is also detectable in the cytoplasm as small clusters. Further studies on this system should be of great interest.

Overall process of exopenicillinase formation

Our present concept of the potential stages of exopenicillinase synthesis in the plasma membrane of *B. licheniformis* 749/C is illustrated in Fig. 4. The polypeptide chain is formed on a membrane-bound ribosome and extruded into the membrane with folding occurring at the outer surface. The phosphatidylserine residue at the NH_2-terminus of the membrane enzyme may be produced either by the transfer of a phosphatidic acid group from CDP-diglyceride or of a phosphatidylserine residue from phosphatidyseryl-tRNA. The work of Dancer & Lampen (1975) demonstrates that the phospholipid residue is added while the chain is still attached to the ribosome, but one can visualize variations of the transport process which would result in the phosphatidylseryl residue lying at either the outer or the inner face of the membrane and the adjacent, less polar segment of the phospholipopeptide portion either lying in the outer layer of the plasma membrane or spanning the membrane bilayer.

If penicillinase synthesis is going on at pH 7 or higher, most of the newly-formed enzyme remains bound to the membrane in a form containing the phospholipid residue and with any needed processing at the NH_2- or COOH-termini completed. The membrane enzyme can be cleaved by the PR-protease to form exopenicillinase (which passes through the cell wall and into the external medium) while the phospholipopeptide portion presumably is retained by the membrane, although its fate has not specifically been studied. Under these conditions, the great bulk of the exopenicillinase must arise by cleavage of the membrane enzyme, but there is evidence that during induction exoenzyme formation can begin without detectable lag (Chesbro & Lampen, 1968). Thus at pH 7 to 8, some cleavage of the newly formed molecules may occur before they have become stabilized as the typical membrane-bound form.

During penicillinase synthesis at slightly acid pH, most of the newly formed enzyme ends up in the plasma membrane as the typical membrane form (Sargent *et al.*, 1969) but this material (and pre-existing membrane enzyme) is cleaved very slowly by PR-protease, for reasons as yet unclear. At the same time exopenicillinase is being produced and this process (discussed above) is considered to involve the PR-protease. No specific information is available concerning the nature of the penicillinase polypeptide actually cleaved at acid pH, but a suggestion can be advanced based on the many similarities between the excretion processes in eukaryotic and prokaryotic cells. In eukaryotic cells, the

NH_2-terminal signal sequence is almost always removed *in vivo* before translation has reached the COOH-terminus of the total polypeptide (Blobel & Dobberstein, 1975; Campbell & Blobel, 1976). Pro-enzyme molecules, analogous to membrane penicillinase from the standpoint of their place in the overall synthetic process, are detectable in some but not all *in-vivo* systems.

I propose, as a working hypothesis, that at acid pH PR-protease removes the phospholipopeptide portion of 749/C penicillinase before translation of the total chain has been completed (or perhaps without the phospholipid residue having become attached to the polypeptide chain). This route to exoenzyme would be readily detectable only at acid pH since the formation of exoenzyme by cleavage of the membrane form is slow under these conditions, but it may also exist at higher pH even though masked by the rapid cleavage of the membrane penicillinase.

GENERAL SIGNIFICANCE OF PHOSPHOLIPOPROTEINS IN ENZYME EXCRETION

One enzyme, the membrane penicillinase of *B. licheniformis* 749, has now been shown to be a phospholipoprotein and its importance in the formation of the exoenzyme examined. The phospholipoprotein form is probably not the initial product of translation and may already have been processed at both termini. It is the direct precursor of exoenzyme in cultures growing at neutral or slightly alkaline pH, and it is likely that a very similar precursor is involved at acid pH. In addition, as recently discussed elsewhere (Lampen & Yamamoto, 1977), the putative mRNA corresponding to the phospholipopeptide portion of the membrane enzyme (75 nucleotides; almost 80% purines) should have a structure similar to that of poly(rA) and, presumably, a parallel affinity for membranes. This purine-rich sequence near the 5′-end of the mRNA for penicillinase may aid in orienting the nascent chain and ribosome to the plasma membrane. Finally, the phospholipopeptide portion can be considered as an anchor which holds the main catalytically active portion close to the cell membrane yet readily available to exogenous substrate, an arrangement of considerable potential advantage to the bacterium. Do other proteins of this type exist, perhaps as intermediates in the formation of other extracellular enzymes? Does the specific PR-protease which cleaves this phospholipoprotein have a more general function in enzyme excretion? Only very preliminary answers can be given to these questions.

We reported recently (Aiyappa & Lampen, 1976) that the plasma membrane of uninduced strain 749 contained a number of proteins which resembled membrane penicillinase and appeared to contain covalently bound phospholipids (contained label derived from $[2\text{-}^3\text{H}]$-glycerol). We are now convinced that the purification procedure used did not adequately remove the polyglycerol phosphate produced by strain 749 and as a result, the amount of these proteins was grossly overestimated.

In more recent studies (K. Izui and J. O. Lampen, unpublished results) the chloroform:methanol precipitates were repeatedly washed at pH 9.0 to remove contaminating polyglycerol phosphate, or *B. subtilis* strain B71 which forms mostly polyribitol phosphate rather than polyglycerol phosphate was tested. These experiments have shown that only small amounts of presumed phospholipoproteins are present.

In work still in progress we have substantially purified the membrane penicillinase from *B. licheniformis* strain 6346/C. This is the second type of penicillinase produced by various strains of *B. licheniformis*, and it shows major sequence homology with the 749/C enzyme (Ambler & Meadway, 1969). As might have been anticipated, the hydrophobic membrane-bound form appears to be larger by several thousand daltons than the exoenzyme and on treatment with trypsin yields an active hydrophilic protein of the same apparent size as the exoenzyme (P. S. Aiyappa & J. O. Lampen, unpublished results). We have not yet determined whether or not this penicillinase contains covalently bound phospholipid.

Another possible phospholipoprotein is the levansucrase of *B. subtilis* B (Caulfield, Chopra, Melling & Berkeley, 1976). The synthesis of this enzyme is highly sensitive to the antibiotic cerulenin which specifically inhibits fatty acid synthesis, and on this basis Caulfield *et al.* (1976) have suggested that levansucrase contains covalently linked lipid. Our recent observation (A. Thomas & J. O. Lampen, unpublished results) that the synthesis of 749/C membrane penicillinase is also quite sensitive to cerulenin is at least consistent with their suggestion. Characterization of the membrane-bound levansucrase should be of considerable interest.

Hydrophobic membrane-bound amylases have been detected in both *B. amyloliquefaciens* (Fernandez-Rivera Rio & Arroyo-Begovich, 1975) and *B. subtilis* (Nagata, Yamaguchi & Maruo, 1974). These forms are larger than the corresponding exoenzymes, but the published information concerning their properties does not specifically suggest that they are phospholipoproteins.

The production of PR-protease is not dependent on the concomitant production of penicillinase, since strain 749 uninduced for penicillinase synthesis forms as much PR-protease as does the penicillinase constitutive strain 749/C (Aiyappa *et al.*, 1977). Furthermore, *B. subtilis* B71 produces high levels of PR-protease but little (if any) penicillinase even following attempted induction while *B. amyloliquefaciens* 23844 produces a moderate amount of penicillinase and a relatively low level of PR-protease (A. Thomas and J. O. Lampen, unpublished results). In all of these bacilli the great bulk of the PR-protease is produced after penicillinase formation has essentially ceased. One must, therefore, conclude that only the small amount of the enzyme detected in the periplasm of the exponential phase cell is significant for the production of exopenicillinase.

The identification of any additional physiological roles for PR-protease may have to await the development of mutants lacking this enzyme. Several attempts to produce them have been unsuccessful. It should finally be noted that the rapid formation of PR-protease occurs only after cultures have reached early stationary phase and sporulation has begun. It is possible that PR-protease, and by inference phospholipoproteins, play some role in the complex process of sporulation.

REFERENCES

ADELMAN, M. R., SABATINI, D. D. & BLOBEL, G. (1973). Ribosome-membrane interaction. Nondestructive disassembly of rat liver rough microsomes into ribosomal and membranous components. *Journal of Cellular Biology*, **56**, 206–29.

AIYAPPA, P. S. & LAMPEN, J. O. (1976). Membrane-associated phopholipoproteins of *Bacillus licheniformis* 749. *Biochimica et Biophysica Acta*, **448**, 401–10.

AIYAPPA, P. S. & LAMPEN, J. O. (1977). Penicillinase-releasing protease of *Bacillus licheniformis* 749. Specificity for hydroxyamino acids. *Journal of Biological Chemistry*, **252**, 1745–7.

AIYAPPA, P. S., TRAFICANTE, L. J. & LAMPEN, J. O. (1977). Penicillinase-releasing protease of *Bacillus licheniformis*: Purification and general properties. *Journal of Bacteriology*, **129**, 191–7.

AMBLER, R. P. & MEADWAY, R. J. (1969). Chemical structure of bacterial penicillinases. *Nature, London*, **222**, 24–6.

BETTINGER, G. E. & LAMPEN, J. O. (1971). Evidence for the extrusion of an incompletely folded form of penicillinase during secretion by protoplasts of *Bacillus licheniformis* 749/C. *Biochemical and Biophysical Research Communications*, **43**, 200–6.

BETTINGER, G. E. & LAMPEN, J. O. (1975). Further evidence for a partially folded intermediate in penicillinase secretion by *Bacillus licheniformis*. *Journal of Bacteriology*, **121**, 83–90.

BISSELL, M. J., TOSI, R. & GORINI, L. (1971). Mechanism of excretion of a bacterial proteinase: factors controlling accumulation of the extracellular proteinase of a *Sarcina* strain (Coccus P). *Journal of Bacteriology*, **105**, 1099–1109.

BLOBEL, G. & DOBBERSTEIN, B. (1975). Transfer of protein across membranes. I. Presence of proteolytically processed and unprocessed nascent immunoglobulin light chains on membrane-bound ribosomes of murine myeloma. *Journal of Cell Biology*, **67**, 835–51.

BLOBEL, G. & SABATINI, D. D. (1971). Ribosome–membrane interactions in eukary-otic cells. In *Biomembranes*, **2**, ed. L. A. Manson, pp. 193–5. London: Plenum.

CAMPBELL, P. N. & BLOBEL, G. (1976). The role of organelles in the chemical modification of the primary translation products of secretory proteins. *FEBS Letters*, **72**, 215–26.

CAULFIELD, M., CHOPRA, I., MELLING, J. & BERKELEY, R. C. W. (1976). The effect of cerulenin on the liberation of extracellular inducible levansucrase by *Bacillus subtilis* B. *Proceedings of the Society for General Microbiology*, **3**, 91–2.

CHESBRO, W. R. & LAMPEN, J. O. (1968). Characteristics of secretion of penicillinase, alkaline phosphotase and nuclease by *Bacillus* species. *Journal of Bacteriology*, **96**, 428–37.

CRANE, L. J., BETTINGER, G. E. & LAMPEN, J. O. (1973). Affinity chromatography purification of penicillinase of *Bacillus licheniformis* 749/C and its use to meas-ure turnover of the cell-bound enzyme. *Biochemical and Biophysical Research Communications*, **50**, 220–7.

CRANE, L. J. & LAMPEN, J. O. (1974). *Bacillus licheniformis* 749/C plasma membrane penicillinase, a hydrophobic polar protein. *Archives of Biochemistry and Bio-physics*, **160**, 655–66.

DANCER, B. N. & LAMPEN, J. O. (1975). *In vitro* synthesis of hydrophobic penicil-linase in extracts of *Bacillus licheniformis* 749/C. *Biochemical and Biophysical Research Communications*, **66**, 1357–64.

DANCER, B. N. & LAMPEN, J. O. (1977). Membrane-bound penicillinase and phospholipid production in *Bacillus licheniformis* 749/C. In *Microbiology-1977*, ed. D. Schlessinger, pp. 100–3. Washington, DC: American Society for Microbiology.

DAVIES, J. W. (1969). Protein synthesis by cell free extracts of *B. licheniformis* 749/C. *Biochimica et Biophysica Acta*, **174**, 686–95.

FERNANDEZ-RIVERA RIO, L. & ARROYO-BEGOVICH, A. (1975). Evidence for the presence of α-amylase in the cell membrane of *Bacillus amyloliquefaciens*. *Biochemical and Biophysical Research Communications*, **65**, 161–9.

GHOSH, A. & GHOSH, B. K. (1977). Immuno-electron microscopic localization of penicillinase in *Bacillus licheniformis* 749/C. *Abstracts of the Annual Meeting of the American Society for Microbiology*, *K96*, p. 202. Washington, DC: American Society for Microbiology.

INOUYE, H. & BECKWITH, J. (1977). Synthesis and processing of an *Escherichia coli* alkaline phosphatase precursor *in vitro*. *Proceedings of the National Academy of Sciences, USA*, **74**, 1440–4.

KELLY, L. E. & BRAMMAR, W. J. (1973). A frameshift mutation that elongates the penicillinase protein of *Bacillus licheniformis*. *Journal of Molecular Biology*, **80**, 135–47.

LAMPEN, J. O. (1974). Movement of extracellular enzymes across membranes. In *Transport at the Cellular Level; Symposia of the Society for Experimental Bio-logy*, **28**, ed. M. A. Sleigh & D. H. Jennings, pp. 351–74. Cambridge University Press.

LAMPEN, J. O. & YAMAMOTO, S. (1977). Phospholipoprotein membrane penicil-linase of *Bacillus licheniformis* and its role in pencillinase secretion. In *Micro-biology-1977*, ed. D. Schlessinger, pp. 104–11. Washington, DC: American Society for Microbiology.

LIN, J. J.-C. & WU, H. C. P. (1976). Biosynthesis and assembly of envelope lipo-protein in a glycerol-requiring mutant of *Salmonella typhimurium*. *Journal of Bacteriology*, **125**, 892–904.

MACALISTER, T. R., IRWIN, R. T. & COSTERTON, J. W. (1977*a*). Cell-surface localized alkaline phosphatase of *Escherichia coli* as visualized by reaction product deposition and ferritin-labelled antibodies. *Journal of Bacteriology*, **130**, 318–28.

MACALISTER, T. J., IRWIN, R. T. & COSTERTON, J. W. (1977b). Immunological investigations of protein synthesis in *Escherichia coli*. *Journal of Bacteriology*, **130**, 329–38.

NAGATA, Y., YAMAGUCHI, K. & MARUO, B. (1974). Genetic and biochemical studies on cell-bound α-amylase in *Bacillus subtilis* Marburg. *Journal of Bacteriology*, **119**, 425–30.

PALADE, G. (1975). Intracellular aspects of protein synthesis. *Science*, **189**, 347–58.

ROTHMAN, J. E. & LENARD, J. (1977). Membrane asymmetry. The nature of membrane asymmetry provides clues to the puzzle of how membranes are assembled. *Science*, **195**, 743–53.

SANDERS, R. L. & MAY, B. K. (1975). Evidence for extrusion of unfolded extracellular enzyme polypeptide chains through membranes of *Bacillus amyloliquefaciens*. *Journal of Bacteriology*, **123**, 806–14.

SARGENT, M. G., GHOSH, B. K. & LAMPEN, J. O. (1968a). Characteristics of penicillinase release by washed cells of *Bacillus licheniformis*. *Journal of Bacteriology*, **96**, 1231–9.

SARGENT, M. G., GHOSH, B. K. & LAMPEN, J. O. (1968b). Localization of cell-bound pencillinase of *Bacillus licheniformis*. *Journal of Bacteriology*, **96**, 1329–38.

SARGENT, M. G., GHOSH, B. K. & LAMPEN, J. O. (1969). Characteristics of penicillinase secretion by growing cells and protoplasts of *Bacillus licheniformis*. *Journal of Bacteriology*, **97**, 820–6.

SARGENT, M. G. & LAMPEN, J. O. (1970). Organization of the membrane-bound penicillinases of *Bacillus licheniformis*. *Archives of Biochemistry and Biophysics*, **136**, 167–77.

SAWAI, T., CRANE, L. J. & LAMPEN, J. O. (1973). Evidence for phospholipid in plasma membrane penicillinase of *Bacillus licheniformis* 749/C. *Biochemical and Biophysical Research Communications*, **53**, 523–30.

SAWAI, T. & LAMPEN, J. O. (1974). Purification and characteristics of plasma membrane penicillinase from *Bacillus licheniformis* 749/C. *Journal of Biological Chemistry*, **249**, 2688–94.

SHERRATT, D. J. & COLLINS, J. F. (1973). Analysis by transformation of the penicillinase system in *Bacillus licheniformis*. *Journal of General Microbiology*, **76**, 217–30.

SMITH, W. P., TAI, P.-C., THOMPSON, R. C. & DAVIS, B. D. (1977). Extracellular labelling demonstrates nascent polypeptides traversing the membrane of *E. coli*. *Federation Proceedings*, **36**, 897.

TRAFICANTE, L. J. & LAMPEN, J. O. (1977a). Vesicle pencillinase of *Bacillus licheniformis*: Existence of periplasmic releasing factor(s). *Journal of Bacteriology*, **129**, 184–90.

TRAFICANTE, L. J. & LAMPEN, J. O. (1977b). Vesicle pencillinase of *Bacillus licheniformis* 749/C. Apparent identity with the plasma membrane enzyme. *Biochimica et Biophysica Acta*, **467**, 44–50.

WANG, S., INOUYE, S., SEKIZAWA, J., HALEGOUA, S. & INOUYE, M. (1977). Primary structure of the extended sequence of the phospholipoprotein of the *Escherichia coli* outer membrane. *Federation Proceedings*, **36**, 897.

YAMAMOTO, S. & LAMPEN, J. O. (1976a). Membrane penicillinase of *Bacillus licheniformis* 749/C: Sequence and possible repeated tetrapeptide structure of the phospholipopeptide region. *Proceedings of the National Academy of Sciences, USA*, **73**, 1457–61.

YAMAMOTO, S. & LAMPEN, J. O. (1976*b*). Purification of plasma membrane penicillinase from *Bacillus licheniformis* 749/C and comparison with exoenzyme. *Journal of Biological Chemistry*, **251**, 4095–101.

YAMAMOTO, S. & LAMPEN, J. O. (1976*c*). The hydrophobic membrane penicillinase of *Bacillus licheniformis* 749/C. Characterization of the hydrophilic enzyme and phospholipopeptide produced by trypsin cleavage. *Journal of Biological Chemistry*, **251**, 4102–10.

THE REVERSION OF BACTERIAL PROTOPLASTS AND L-FORMS

J. BARRIE WARD

Division of Microbiology, National Institute for Medical Research, Mill Hill, London NW7 1AA, UK

INTRODUCTION

It is now twenty-five years since the original observations (Tomcsik & Guex-Holzer, 1952; Weibull, 1953) that degradation by lysozyme of the wall of *Bacillus megaterium* incubated in a suitable protective medium, resulted in the formation of osmotically sensitive protoplasts. Subsequent investigations showed that such protoplasts were essentially free of wall material (McQuillen, 1960; Martin, 1963). In similar experiments with Gram-negative bacteria spherical osmotically sensitive organisms were again produced although in this case considerable amounts of wall material remained. To distinguish such organisms from the naked protoplasts obtained from *B. megaterium* and other Gram-positive organisms the term spheroplast was introduced (McQuillen, 1960). More recently Landman and his colleagues (Miller, Zsigray & Landman, 1967) following the action of lysozyme on chloramphenicol-treated *B. subtilis* have described a quasi-spheroplast. This is an osmotically-fragile form of the organisms with associated wall material occurring transiently during conversion of the bacillus to the wall-less protoplast.

One of the earlier observations on the properties of protoplasts was their inability to divide when incubated in liquid media (Weibull, 1953) although a few exceptions have been reported (McQuillen, 1960; Kusaka, 1967, 1971). The protoplasts do, however, retain for some time the ability to catalyse the balanced synthesis of macromolecules including peptidoglycan (Roth, Shockman & Daneo-Moore, 1971; Rosenthal & Shockman, 1975a, b; Rosenthal, Jungkind, Daneo-Moore & Shockman, 1975, Landman & Fox, 1977) and to complete the perhaps more complex biological activities of phage synthesis and induced enzyme formation (McQuillen, 1960; Martin, 1963). The reproducible division of bacterial protoplasts was first described by Landman & Halle (1963) when they incubated lysozyme-induced protoplasts of *B. subtilis* 168 on hypertonic soft-agar medium. Under such conditions a mass-conversion of the protoplasts to L-forms occurred. The L-forms could then be maintained in this wall-less form by repeated sub-culture under the same conditions at intervals of approximately 48 hours. In contrast, a similar plating of quasi-spheroplasts onto the hypertonic soft-agar gave rise only to bacilli (Miller *et al.*, 1967) whereas spheroplasts of gram-negative organisms grew both as bacterial and L-form colonies (Landman, 1968).

However, reversion of the *B. subtilis* L-forms did occur on prolonged (i.e. 3 days or more) incubation on soft-agar. That is the L-form colonies gradually became overgrown by bacillary revertants. Thus in this respect they resembled the unstable L-forms derived from other bacilli (Fodor & Rogers, 1966, Wyrick & Rogers, 1973) and *Staphylococcus aureus* (Chatterjee, Ward & Perkins, 1967). As described above, Landman & Halle (1963) were able to maintain their cultures for several months as L-forms by repeated sub-culture prior to the onset of reversion. These organisms did not lose their ability to revert. In subsequent publications (Landman, 1968, Landman & De Castro-Costa, 1976; De Castro-Costa & Landman, 1977) such cultures have been referred to as 'mass conversion stable L-forms' although 'stability' is clearly dependent upon the cultural conditions. On the other hand reversion could be prevented by incorporating any one of several D-amino acids into the soft-agar medium (Landman & Halle, 1963). Other investigators (Chatterjee *et al.*, 1967; Young, Haywood & Pollock, 1970; Wyrick & Rogers, 1973) have used penicillins, as antibiotics known to inhibit wall biosynthesis, to prevent the reversion of their unstable L-forms derived from *S. aureus*, *B. licheniformis* and *B. subtilis*. In contrast L-forms induced from protoplasts of both *Streptococcus pyogenes* (Gooder & Maxted, 1961) and *S. faecium* (King & Gooder, 1970 do not appear to revert even when plated on hypertonic agar in the absence of wall-inhibiting substances.

Unstable L-forms can become stabilised. This occurs when, after a variable number of sub-cultures in the presence of the inducing agent, they lose the ability to resynthesise their wall and thus become unable to revert to the bacterial form. Recent studies with several stable L-phase variants (stable L-forms) of *B. licheniformis* and *B. subtilis* have revealed all to have lesions in peptidoglycan biosynthesis (Ward, 1975). The majority of the L-phase variants were unable to synthesise diaminopimelic acid as a consequence of a defect in aspartate-β-semialdehyde-dehydrogenase activity. Others were defective, either in the formation of the nucleotide-linked precursors of peptidoglycan, or in phospho-*N*-acetylmuramyl-pentapeptide translocase, the first membrane-bound enzyme of peptidoglycan biosynthesis. Thus defined biochemical lesions have been described, which in the case of the stable L-form of *B. subtilis*

168, complement the genetic evidence (Wyrick, McConnell & Rogers, 1973) obtained earlier that truly stable L-forms arise as a result of a mutational event. More recently stable L-forms of both *S. pyogenes* (Reusch & Panos, 1976) and *S. faecium* (Gregory & Gooder, 1976) were shown to be defective in phospho-*N*-acetylmuramyl-pentapeptide translocase and hence were unable to synthesise peptidoglycan.

As mentioned briefly in this survey of the consequences of wall removal from a bacterial population, both protoplasts and unstable L-forms have the ability to revert to the bacterial form. It is a consideration of this reversion process from both the biochemical and ultrastructural aspects, that provides the main subject of this article.

THE EFFECT OF THE ENVIRONMENT ON REVERSION

Reversion of unstable L-forms, spheroplasts and protoplasts to the bacterial state has been described in general terms for a number of different organisms. Thus reversion of spheroplasts and unstable L-forms of *Proteus mirabilis* was followed by increase in resistance to osmotic shock (Alternbern, 1963). In addition there was also a rapid increase in acid-insoluble diaminopimelic acid during reversion, indicating the synthesis of peptidoglycan. Detailed investigations of the reversion process are however relatively few and appear to be restricted to protoplasts and unstable L-forms of *B. subtilis* (Landman, Ryter & Frehel, 1968; Landman & Forman, 1969; De Castro-Costa & Landman, 1977), *B. licheniformis* (Elliott, Ward & Rogers, 1975a, Elliott, Wyrick, Ward & Rogers, 1957b; Rogers, Ward & Elliott, 1976), *B. megaterium* (Hadlaczky, Fodor & Alfoldi, 1976), *S. faecium* (King & Gooder, 1970; Wyrick & Gooder, 1976) and *S. aureus* (Schönfeld & de Bruijn, 1973).

In each of these organisms protoplasts or unstable L-forms plated onto a suitable medium will resynthesise their walls. This process is accompanied by a morphological change from the spherical shape of the wall-less state, either protoplast or L-form, to, in the case of the bacilli, the rod-shape of the parent bacterium. Ultimately the reverting organisms also regain the ability to undergo a normal division. However, in those cases where it has been investigated, division only appears to occur some time after the wall of the reverting organisms has regained a normal ultrastructural appearance.

The detailed studies of Landman & Halle (1963) on the conversion of protoplasts of *B. subtilis* to unstable L-forms were accompanied by an equally detailed investigation of the factors controlling the reversion of these L-forms and protoplasts to bacilli. Thus the physical constitution of the medium was shown markedly to affect both the rate and extent of reversion. The appearance of bacilliary colonies was stimulated as the concentration of agar in the medium was increased from the customary 0.9% (w/v) to 2.5% and even further enhanced by the use of 25–35% gelatin as the solidifying agent. Gelatin had previously been used by Necas (1961, 1962) in his studies on the reversion of yeast protoplasts. In fact gelatin or hard-agar have been used to induce reversion in almost all the studies referred to earlier in this section.

A more detailed investigation of the effects of the physical state of the medium on reversion of both protoplasts and unstable L-forms of *B. subtilis* was made by Clive & Landman (1970). Growth of the wallless forms on the surface of certain membrane filters induced reversion to the bacillary state even though the supporting medium contained only 0.8% agar. Surprisingly reversion only occurred on cellulose filters, an observation which led the authors to investigate the effects of adding bacterial walls. A stimulation of reversion was obtained when walls of *B. subtilis* were included in the medium and this stimulation was even more marked when the walls were dried onto the surface of the filter prior to inoculation of the protoplasts. However, similar results were obtained when isolated walls were replaced by heat-killed cells of *B. subtilis*, various Gram-negative organisms and yeast. The stimulation of reversion appeared to be non-specific and clearly distinguished from the specific interaction of autolysins with exogenous wall, described below. Moreover, the increased reversion obtained when walls were present on the filter, compared with that obtained when they were incorporated into the medium, suggests a requirement for close contact between the protoplast and the exogenous material.

Thus the reversion of *B. subtilis* protoplasts and unstable L-forms is stimulated by five different environmental conditions; gelatin, hardagar, cellulose membrane filters, walls and heat-killed micro-organisms. The wide disparity in composition of these various substances effectively eliminates some chemical component as the reversion-inducing substance but rather implies that the physical properties of the various media promote reversion. In this context it is interesting to speculate that it is the fibrous nature of these various substances that makes them particularly effective. On the basis of their initial studies Landman & Halle (1963) had suggested that changes in medium consistency were responsible for the physical immobilisation of certain extracellular products necessary for wall synthesis. These would be retained in close proximity to the surface of the protoplasts allowing formation of peptidoglycan to proceed. Other mechanisms that were subsequently suggested (Clive &

Landman, 1970) included the inactivation either of lytic enzymes (i.e. preventing the degradation of wall as it was being synthesised) or of some repressor of wall biosynthesis. Underlying this last potential mechanism was the assumption that some stage of peptidoglycan biosynthesis is subject to repression and derepression. It was thought that repression occurred on removal of the wall during protoplast formation. Evidence to support this last hypothesis came from the finding that the synthesis of diaminopimelic acid is repressed in both protoplasts and unstable L-forms of *B. subtilis* (Bond, 1969). However, the wide variation in the conditions under which reversion is induced militates against such specific explanations. Probably of greatest importance is the observation (Clive & Landman, 1970) that reversion only ensues when the wall-less organisms are left undisturbed in close physical contact with a solid environment for a period, early in the reversion process.

MACROMOLECULAR BIOSYNTHESIS BY REVERTING PROTOPLASTS

Detailed investigation of macromolecular biosynthesis have been confined to studies utilising reverting protoplasts of *B. subtilis* (Landman & Forman, 1969) and *B. licheniformis* (Elliott *et al.*, 1975a). Although the two organisms are closely related, different methods have been used, making direct comparison difficult. Hence the results of both investigations are described.

Conditioning of protoplasts

Landman & Halle (1963) measured reversion as the number of bacilliary colonies appearing 3–14 days after the inoculation of either protoplasts or unstable L-forms onto the surface of gelatin or hard-agar medium. In these experiments the protoplasts were prepared by lysozyme-treatment of bacilli suspended in essentially minimal medium containing sucrose as the osmotic stabilising agent. Subsequently Landman & Forman (1969) found that the ability of *B. subtilis* protoplasts to undergo reversion was markedly enhanced if the protoplasts were first pre-incubated in a medium enriched with casein hydrolysate. This conditioning could be achieved either by the inclusion of casein hydrolysate in the protoplasting medium and prolonged incubation in the presence of lysozyme or alternatively by incubation of washed protoplasts under the appropriate conditions. However, in these experiments reversion was not measured by a direct plating of the conditioned protoplasts and following the appearance of bacilliary colonies. Instead, the conditioned protoplasts, prepared as described above, were first incubated for 60–90 min at 26 °C in a medium containing 25% gelatin. After this second incubation the gelatin cultures were melted and appropriate samples plated onto media containing either gelatin, 0.8% agar and D-methionine as an inhibitor of reversion (Landman & Halle, 1963) or 0.8% agar. After a further 48 hr incubation at 30 °C the plates were scored for the total L-form count (in the presence of D-methionine) and the bacilliary count (these colonies were found together with L-form colonies in the absence of D-methionine). From the number of bacilliary colonies obtained relative to the total L-form count the percentage of reversion was calculated. In this way three steps in the reversion process were distinguished: the conditioning of the protoplasts in the presence of casein hydrolysate (step 1); the initial stages of incubation in 25% gelatin (step 2); and the acquisition of osmotic resistance occurring on prolonged incubation in 25% gelatin (step 3). Under normal experimental conditions, i.e. 60–90 min incubation in 25% gelatin, step 3 would presumably not occur until some time after the reverting protoplasts were finally plated onto the solid medium.

However it will be recalled that protoplasts plated directly onto medium containing 0.8% agar will revert (Landman & Halle, 1963) although the process takes considerably less time on gelatin or hard-agar. Hence, it is important to define, in experimental terms, exactly what the reversion system of Landman and Forman demonstrates, and how it is separable into the three steps outlined above. Conditioned protoplasts plated directly onto soft-agar give rise only to L-form colonies within a 48 hr incubation period, although direct reversion to bacilli presumably occurs if incubation is prolonged. If on the other hand, the conditioned protoplasts are first incubated in 25% gelatin medium for 60–90 min, then an identical plating on soft-agar will result in the development of bacilliary colonies from 80 to 100% of the protoplasts. In these, and subsequent experiments by Landman and his colleagues, it is this commitment to form a bacilliary rather than an L-form colony that is described as reversion. Clearly the bulk of the macromolecular synthesis associated with wall formation must, under these conditions, occur on the surface of the agar plates. It is important however, to remember that incubation for 4–5 hr in the 25% gelatin medium leads to the acquisition of osmotic stability (step 3); that is 5.5–6.5 hr after the initial formation of the protoplasts. In their earlier investigation Landman et al. (1968) had reported reversion of protoplasts to bacilli when they were incubated in gelatin for 18–24 hr.

In contrast to the above results protoplasts prepared in minimal medium or prepared from organisms that had been pre-incubated with casein hydrolysate *prior* to protoplasting, were not committed to reversion until they had been incubated in the gelatin medium for 3–4 hr. Moreover, the conditioning process was blocked by chloramphenicol, puromycin and actinomycin D, whereas it was unaffected by inhibition of DNA synthesis or the addition of benzylpenicillin or lysozyme to the conditioning medium. Thus, the conditioning of protoplasts appears to involve a period of active protein and RNA synthesis. The absence of any demonstrable effect following the inclusion of either benzylpenicillin or lysozyme in the conditioning medium, suggests that peptidoglycan biosynthesis at this stage is not necessary for the subsequent reversion of the protoplasts.

Perhaps related to the conditioning process, but clearly distinguished from it particularly by a lack of inhibition by chloramphenicol, is the 'pre-conditioning' of protoplast membranes described by Reynolds (1971). In this case the capacity of *B. megaterium* protoplast membranes to catalyse in-vitro biosynthesis of peptidoglycan was markedly enhanced if the protoplasts were first incubated in a rich growth medium. In *B. licheniformis* (Elliott *et al.*, 1975*a*, *b*) the protoplasts for reversion were prepared in a rich medium, conditions almost certainly fulfilling the requirements for protein synthesis. There is no evidence that conditioning affects the rate of reversion when protoplasts are plated directly onto hard-agar. If there is a requirement for conditioning, or more probably for the protoplasts to be in an active metabolic state to ensure relatively rapid reversion, then it seems likely that the *B. licheniformis* protoplasts studied by Elliot *et al.* (1975*a*, *b*) would meet such criteria.

The biosynthesis of wall during reversion

During incubation on 25% gelatin, reversion of appropriately conditioned *B. subtilis* protoplasts remained unaffected by the presence of chloramphenicol or by the inhibition of DNA synthesis. Reversion was however delayed by both puromycin and actinomycin D and inhibited by benzylpenicillin and to a lesser extent by D-cycloserine (Landman & Forman, 1969). Clearly peptidoglycan synthesis occurs during this period of incubation and if left undisturbed then the protoplasts will reform their walls and revert to bacilli (Landman *et al.*, 1968). However, under the experimental conditions normally followed, the protoplasts are, at the time of plating onto soft-agar, still osmotically sensitive although as subsequent growth shows they have become committed to form bacilli.

More recently, De Castro-Costa & Landman (1977) have described evidence for an inhibitor of reversion (reversion inhibitory factor – RIF) which they conclude is one or both of the autolysins of *B. subtilis*. A more detailed consideration of these results is given below in the section dealing with the role of autolysins in reversion. The inhibition of reversion was overcome by incubation of conditioned protoplasts with trypsin. If the protoplasts were then washed and incubated under conditions where protein synthesis was inhibited, either in the presence of chloramphenicol or in the absence of an essential amino acid, then reversion (the commitment to form a bacilliary colony) occurred in liquid media (Landman & Fox, 1977). The synthesis of peptidoglycan under these conditions was followed by the incorporation of N-acetyl-[^{3}H]-glucosamine into polymeric material. After incubation for 30 min, 70% of the polymeric material was associated with the protoplast membranes, the remaining 30% being found in the supernatant. The average biosynthetic glycan chain length of the newly-synthesised peptidoglycan, based on the ratio of free reducing groups of muramic acid to total muramic acid incorporated (Ward, 1973) was 17 disaccharide units whereas the average in-vivo glycan chain length was 2.6 disaccharide units (O. E. Landman & S. M. Fox, personal communication). This latter value is based on the ratio of total free reducing groups of both muramic acid and glucosamine to total muramic acid incorporated. The large difference found between the two measurements implies that on average each glycan chain has been cleaved 6.5 times by an endo-N-acetylglucosaminidase. Autolysis of *B. subtilis* peptidoglycan has been shown to result from the combined action of such an N-acetylglucosaminidase together with an N-acetylmuramic acid-L-alanine amidase (Brown & Young, 1970; Hughes, 1970).

On continued incubation additional peptidoglycan was synthesised, although surprisingly the average glycan chain lengths did not alter. Cross-linking of the peptide side-chains had occurred, 21% and 29% cross-linkage being found after 30 min and 120 min incubation. These values contrast with that of 55–65% found in peptidoglycan isolated from *B. subtilis* itself (Hughes, 1970). Thus the peptidoglycan synthesised by the reverting protoplasts was significantly less cross-linked than that of the bacillus and had been subject to considerable degradation by at least one of the known autolysins. Concomitant loss of peptide side-chains through amidase activity would not be detected from material radioactively labelled only in the glycan.

The biosynthesis of wall polymers and protein has been studied during reversion of *B. licheniformis* protoplasts incubated on 2.5% agar (Elliott *et al.*, 1975*a*). When incubated at 35 °C for 18 hr almost all the protoplasts reverted to form normally dividing bacilli. In these experiments protoplasts, prepared by lysozyme-treatment in rich medium osmotically stabilised with 0.5 M sucrose, were spread onto the hard-agar containing one or more of the following radioactive precursors, N-acetyl-[^{14}C]glucosamine, [2-^{3}H]glycerol or [^{3}H] trytophan, to measure the synthesis of peptidoglycan, teichoic acid and protein respectively. Reverting protoplasts were scraped from the surface of the agar and newly synthesised wall was isolated as material insoluble in sodium dodecyl sulphate (4% w/v). Peptidoglycan was obtained after treatment of such wall preparations with trichloroacetic acid and trypsin to remove associated secondary wall polymers (teichoic and teichuronic acids) and protein respectively. Teichoic acids were isolated from wall preparations by extraction with alkali (Hughes & Tanner, 1968). After certain periods of incubation soluble wall material was isolated by freezing and thawing the agar plates on which the reverting protoplasts had been incubated. Polymeric material was isolated from the soluble fraction by chromatography on columns of Sephadex G25.

Incorporation of the appropriate radioactive precursors into peptidoglycan and teichoic acid occurred within 40 min of the start of incubation. In both wall polymers and protein an exponential incorporation of radioactivity was obtained with doubling times of about 60 min. Chemical analysis of peptidoglycan isolated at various times of incubation from 2–18 hr showed it to be similar in composition to that isolated from *B. licheniformis*. Moreover, both teichoic and teichuronic acids were present in the wall preparations in approximately normal proportions. In this respect the control of wall formation present in the parent bacillus appeared also to be operating in the reverting protoplasts.

These observations are somewhat surprising when considered in the light of the results obtained for the average glycan chain lengths and the extent of cross-linking of the newly synthesised peptidoglycan. As reversion proceeded both the length of the glycan chains and the number of cross-links present increased, until after 12 hr incubation the values found in each case were close to those obtained with peptidoglycan isolated from the parent bacillus (Table 1). After 3 hr incubation however, the average biosynthetic glycan chain length (10 disaccharide units) was very small and the extent of cross-linkage (28%) low when compared to the values ultimately obtained. Clearly the situation existing with respect to peptidoglycan synthesis in these early stages of reversion differs markedly from that present both in the later stages and in the parent bacillus. The most obvious difference is the absence of pre-formed wall and thus the acceptor for newly synthesised peptidogly-

Table 1. *The extent of cross-linking and the average glycan chain lengths of protoplast-associated and soluble peptidoglycan from* B. licheniformis *protoplasts reverting on 2.5% agar*

Time of reversion (hr)		Extent of cross-linking		Average glycan chain lengths (disaccharide units) deduced from	
		HNO₂[a]	Muramidase[b]	Muramitol: Muramic acid ratio	Total amino sugar alcohol: Muramic acid ratio
3	P[c]	28	24	10	5
	S	—	—	5	2.4
6	P	40	47	44	24
9	P	57	—	133	71
12	P	57	66	200	103
	S	—	—	71	4.7
18	P	58	68	272	128

Data taken from Elliott *et al.* (1975*a*).

[a] % of diaminopimelic acid amino groups blocked by cross-linking.

[b] % of the peptidoglycan present as dimer after chromatography of Streptomyces muramidase digests of walls isolated from reverting protoplasts.

[c] P, isolated as insoluble material from protoplasts; S, isolated as soluble material from the medium.

can. In addition there is the possibility that removal of the wall results in some unknown change in the spatial relationship of the various peptidoglycan synthesising enzymes. These changes would be expected to exert their maximal effect in the protoplasts and the early stages of reversion.

The peptidoglycan isolated from *B. licheniformis* protoplasts after 3 hr incubation on hard-agar was very similar in structure to peptidoglycan isolated from protoplasts of *B. subtilis* incubated in liquid medium (Landman & Fox, 1977). However, the two preparations differed markedly in the extent to which they had been subject to autolytic degradation. The glycan chains synthesised by the *B. licheniformis* protoplasts had on average been subject to the hydrolysis of a single bond by an *N*-acetylglucosaminidase in contrast to the 6.5 breaks found in *B. subtilis*. This in *B. licheniformis* the in-vivo glycan chain lengths found were approximately half the lengths to which the glycan chains had been biosynthesised. This relationship between biosynthetic and in-vivo chain lengths was true for all preparations of peptidoglycan isolated from the reverting protoplasts and walls of the vegetative organisms. Again, the mechanism which determines the fine structure of the peptidoglycan in the parent bacillus appears to function in the reverting protoplast.

The synthesis of peptidoglycan during the early stages of reversion in *B. licheniformis* and by the *B. subtilis* protoplasts incubated in liquid medium is also similar in that 30% of the newly synthesised material is found in a soluble form. In *B. licheniformis* the average glycan chain length of the soluble peptidoglycan was approximately half that of the insoluble polymer (Table 1). Evidence for the involvement of the major autolysin, the *N*-acetylmuramyl-L-alanine amidase, in release of soluble peptidoglycan came from the finding that peptide side-chains, measured as diaminopimelic acid were missing from 33% of the muramic acid residues. Soluble peptidoglycan was not present in the agar on which protoplasts had been incubated for 6 and 9 hr but reappeared after 12 hr incubation. This material which represented only 8% of the incorporated radioactivity, lacked approximately 19% of the peptide side-chains again measured as diaminopimelic acid. More marked was the decrease in the in-vivo glycan chain length resulting from extensive hydrolysis by *N*-acetylglucosaminidase. Thus the soluble peptidoglycan preparations differ in the extent to which they have been modified by the known autolysins of *B. licheniformis* although the significance of the differences observed remains unclear.

ULTRASTRUCTURAL STUDIES ON REVERSION

In contrast to the limited information available concerning the biochemical events underlying reversion, considerably more attention has been paid to the morphological aspects of this complex process. However, as would be expected from the limitations of the techniques involved, many of these studies are concerned primarily with the latter stages of reversion when the wall being biosynthesised already has the characteristics of that isolated from the vegetative organism. In an attempt to overcome certain of these problems, recent studies (Elliott *et al.*, 1975*b*; Wyrick & Gooder, 1976) have used ferritin-conjugated antibody directed against peptidoglycan as a specific label of wall material (Plate 1 *b*, *c*).

Earlier, Kusaka (1971) following the growth and division of *B. megaterium* by electron microscopy observed the presence of a 'fibrous coat' surrounding the protoplasts after 2 hr incubation. This coat did not form in the presence of benzylpenicillin suggesting that it may have been newly synthesised wall. After 12 hr growth in the absence of antibiotic the protoplasts had a thick compact wall. Division of these walled forms occasionally occurred although reversion to a normal rod morphology was not seen.

These studies had been preceded by an ultrastuctural investigation of the reversion of *B. subtilis* protoplasts on 25% gelatin plates (Landman *et al.*, 1968). This had revealed, after 9 hr incubation, the presence of a thin layer of wall surrounding some of the protoplasts. Reversion was however extremely variable perhaps as a result of the concentrated inocula required for direct examination. As a result, reverted protoplasts with clearly recognisable walls were present, together with wall-less protoplasts, in the same field.

In *B. licheniformis* the use of ferritin-conjugated antibody allowed the demonstration of wall material at much earlier stages of reversion. Although it should be remembered that the actual times of incubation in this system are not comparable with those of reversion in *B. subtilis*, in view of the different media and temperatures employed. After 3 hr incubation on the surface of hard-agar the *B. licheniformis* protoplasts were characterised by their pleomorphic shapes (Plate 1*a*). Occasionally fibrillar material was seen in association with membranes of the reverting protoplasts. Some of this material reacted with the ferritin-labelled antibody and appeared to be degraded by lysozyme. Since at this time 30% of the newly synthesised peptidoglycan was present as a soluble form it seems likely that the fibrils are related to this. The possibility that some of the soluble peptidoglycan arises by shearing of fibrils from the protoplasts during harvesting cannot be excluded. After 6 hr incubation the surface of the protoplasts was covered with loosely organised wall material seen in freeze-etched preparations (Plate 1*d*) as a fringe of variable thickness extending up to 400 nm from the surface. The peptidoglycan present did not have shape maintaining properties since the protoplasts rounded-up on being washed from the surface of the agar with osmotically stabilised medium (Plate 1*b*, *c*). The associated biochemical studies showed that although the extent of cross-linkage of the peptidoglycan was approaching that found in the bacillary wall, the average glycan chain lengths were still quite small. Clearly recognisable wall, present as a thin compact layer surrounding the protoplasts was seen in preparations sectioned *in situ* at 9 hr. As reversion proceeded the wall thickened, often unevenly, and again the organisms became extremely pleomorphic, where septation was observed it was often irregular (Plate 2*a*, *b*). At this time and probably somewhat earlier although this was not tested, the protoplasts regained their resistance to osmotic lysis. The morphological changes occurring during this and subsequent stages were very similar to those reported earlier with protoplasts of *B. subtilis* (Landman *et al.*, 1968). One major difference was the finding, particularly in preparations freeze-etched *in situ*, of fibrous material both covering and extending from the wall (Plate 2*c*, *d*). Fibres which reacted with ferritin-labelled antibody along their length (Plate 3*a*) were recovered from the supernants after washing organisms scraped from the surface of the agar. They were also found in the supernant solutions from SDS-treated organisms (the initial step in isolation of peptidoglycan). Chemical examination showed these fibres to be identical in composition to wall isolated at the same time from the bacilli (reverted protoplasts). In negatively stained preparations (Plate 3b) the fibres appeared to have a complex structure probably made up of many smaller units. It is at this time that soluble peptidoglycan reappears in the agar on which the protoplasts have been incubated. Whether there is any correlation between the presence of fibrous material associated with the wall and the soluble peptidoglycan remains unclear. Clearly one possibility is that the fibres can either be incorporated into the wall or subject to autolytic degradation resulting in the formation of soluble peptidoglycan. Some of the organisms at later stages of reversion (14–18 hr) when freeze-etched, also exhibited fibres which were present along the longitudinal axis of the bacilli. Other studies on the reversion of both yeast and plant protoplasts have demonstrated the presence of fibrous wall components at some stage of the process (for recent studies see Peberdy, Rose, Rogers & Cocking, 1976). Despite the major differences that exist in the wall structure of these various organisms, these observations suggest that the formation of fibres of wall material in reversion is probably a general phenomenon.

After 12–14 hr incubation, the peptidoglycan of the reverting organisms had essentially the same structure with respect to average glycan chain length and extent of cross-linkage as that isolated from the parent bacillus. At this time however, both wall thickening and septation were still irregular although the bacilli were beginning to stretch out into the normal rod-shapes (Plate 3c). Only after a further 4–6 hr incubation did a regular pattern of division and the rod morphology become re-established (Plate 3d).

The sequence of events in wall formation observed in reverting protoplasts of B. licheniformis and B. subtilis is in marked contrast to that found in S. faecium (Plate 4) (Wyrick & Gooder, 1976). Whereas the bacilli revert rapidly and directly from protoplast to bacillus, the streptococci appear to undergo a transition to L-form growth prior to the onset of reversion (Plate 4a). Newly synthesised wall material was first detected through the use of ferritin-labelled antibody as loosely organised material associated with the membrane of the reverting protoplast (L-form). As reversion proceeded the walls appeared as short fragments having the triple-layered structure and thickness of normal wall (Plate 4b), in contrast to the thin continuous layer observed in

bacilli. These fragments then increased in length although in reverting
organisms sectioned *in situ* on the surface of 35% gelatin, they often did
not maintain contact with the membrane surface (Plate 4*c*, *d*, *e*).
Whether this appearance results from manipulations involved in preparation of the material for electron microscopy or reflects the situation
existing in the reverting organism remains unclear. The formation of
new wall as fragments rather than as a continuous layer such as that
found in the bacilli, may arise from localised sites of synthesis. Evidence
from *S. faecalis* for the existence of specific sites concerned with peripheral wall formation has been described by Shockman and his
colleagues (Higgins & Shockman, 1970; Shockman, Daneo-Moore &
Higgins, 1975 see also pp. 154–9). These sites are located in close
proximity to the position of the nascent cross-walls.

However streptococci also contain other potential sites of wall synthesis involved for instance in wall thickening which are located over
the entire membrane surface, although in the growing organism only
the specific sites for peripheral wall synthesis appear to be active (see
pp. 158–9). With our present limited knowledge of the mechanisms
underlying wall synthesis both during reversion and in normal vegetative growth any attempt to interpret the events observed in the reversion
of *S. faecium* protoplasts would be purely speculative.

In all three systems normal division and morphology were only re-
established some time after the organisms appeared to have a complete
wall. Similar observations have been made during reversion of spheroplasts from *E. coli* (Schwarz & Leutgeb, 1971). Confirming earlier observations (Bauman & Davis, 1957), Schwarz and Leutgeb converted a
mutant of *E. coli* auxotrophic for diaminopimelic acid, into spheroplasts. This was done by suspending the organism in suitably protective
medium in the absence of the amino acid. Peptidoglycan synthesis was
then restarted by the addition of diaminopimelate to the medium.
Chemical analysis of the newly synthesised peptidoglycan revealed it to
be of normal composition although both the bacteria and the murein
sacculus (i.e. peptidoglycan and covalently linked lipoprotein) isolated
from them retained the spherical shape of the spheroplasts. The osmotic
stability of the organisms was regained after only 20 min from the reinitiation of peptidoglycan synthesis whereas the normal rod morphology was only restored after at least 2 hr. Moreover this shape
change was said to occur in the absence of cell division. Similarly,
spherical *E. coli*, induced by incubation of exponentially growing
cultures in the presence of the β-lactam antibiotic mecillinam (FL. 1060),
have been shown to revert to the normal rod-morphology in the presence

of chloramphenicol (James, Haga & Pardee, 1975). Thus it appears that neither protein synthesis nor cell division are necessary for the shape-transition to occur. Goodell & Schwarz (1975) have also investigated the conversion of mecillinam-induced spheres of *E. coli* to rod-morphology. In contrast to the findings described above (James *et al.*, 1975) and to earlier work from their own laboratory (Schwarz & Leutgeb, 1971) they found the shape-change to be accompanied by cell division and to occur over a period of several generations. At this time the reasons underlying this apparent discrepancy remain unclear.

In terms of reversion to the normal shape, these observations suggest that peptidoglycan synthesis has a role in the morphogenesis of bacteria. In each example wall synthesis has been disturbed by complete removal of pre-existing wall in the protoplasts of the bacilli and *S. faecium* and in more subtle ways in the induction of spheroplasts and spherical forms of *E. coli*. As reversion proceeds new wall is synthesised and eventually the protoplasts and spheroplasts regain their resistance to osmotic lysis. At this time the chemical characteristics of the newly synthesised peptidoglycan, with respect to average glycan chain length and extent of cross-linkage, appear to be close to, or identical with, peptidoglycan isolated from vegetative bacteria. However, the organisms have not regained a normal morphology. As described above shape changes apparently occur in the absence of protein synthesis and cell division. This would suggest that restoration of a normal shape results from continued peptidoglycan synthesis which as reversion proceeds gives a wall more closely resembling, in fine structure, that found in the vegetative organism.

THE ROLE OF AUTOLYSINS IN REVERSION

In the preceding sections the involvement of autolysins during reversion, particularly in the release of soluble peptidoglycan, has been mentioned briefly. The action of the major autolysin of *B. licheniformis* and *B. subtilis*, the *N*-acetylmuramyl-L-alanine amidase, has been implicated in this release by the finding that soluble peptidoglycan isolated from the medium after 3 hr and 12 hr incubation was deficient in peptide side-chains. Material released after 12 hr had in addition been subject to extensive degradation by the second autolysin of bacilli, an endo-*N*-acetylglucosaminidase. These enzymes are membrane-bound in proto-plasts and L-phase variants of *B. licheniformis* and *B. subtilis* and can be transferred from the membrane to exogenous wall by incubation of the wall-less organisms in the presence of wall (Forsberg & Ward, 1972;

M. V. Hayes & J. B. Ward, unpublished observations). This affinity for
wall was used to investigate further the role of amidase in the release
of soluble peptidoglycan (Elliott *et al.*, 1975*a*). In fact, the inclusion of
isolated walls from *B. licheniformis* into the reversion medium abolished
and reduced by 85%, the release of material after 3 hr and 12 hr incubation, respectively. Similarly, the reversion of an amidase-deficient
mutant of *B. licheniformis* proceeded without the release of soluble
peptidoglycan. In contrast, when medium containing 0.8% agar was
used, conditions under which reversion is slower and considerably more
variable, soluble peptidoglycan was isolated in much greater quantities
and was present throughout reversion. Thus after 3 hr incubation,
53–67% of the total polymeric material synthesised by *B. licheniformis*
protoplasts was in a soluble form and this value fell only to 21% after
12 hr incubation. Clearly, these results point to enhanced autolytic
activity at the lower agar concentration but the reasons underlying this
increase remain unclear.

More recently De Castro-Costa & Landman (1977) have implicated
one or both of the autolysins as the reversion inhibitory factor (RIF)
of *B. subtilis*. Earlier experiments had shown that the protoplast concentration employed during the incubation in 25% gelatin (step 2 of the
reversion sequence) was critical (Landman *et al.*, 1968). Thus reversion,
the commitment to form a bacilliary colony upon subsequent plating,
occurred with protoplast densities of less than 10^6 ml^{-1} but not at 10^7
ml^{-1}. Inhibition of reversion at low protoplast densities was also obtained on the addition of protoplast lysates to the gelatin medium. The
inhibitory factor present in such lysates was sensitive to heat, to ionic
but not to non-ionic detergents and could be inactivated by trypsin.
Moreover, trypsin-treatment of intact conditioned protoplasts caused
a marked stimulation of reversion. As described earlier, the synthesis
of peptidoglycan, implied from the inhibitory effects of lysozyme, and
benzylpenicillin, appeared to be the dominant feature of step 2. Therefore it seemed probable that the inhibitory factor in some way interfered
with this process. Further experiments revealed that the inhibitory factor
had several properties in common with the autolysins of *B. subtilis*.
These included a high affinity for exogenous wall; reversion was stimulated by the inclusion of walls in the gelatin medium; and the finding
that both the autolysin of *B. subtilis* and the activity of the inhibitory
factor were inhibited by gelatin. In addition the inclusion of trypsin in
0.7% agar plates resulted in a marked stimulation of the direct reversion
of 'mass-conversion' L-forms of *B. subtilis*. Similarly mutants of *B.
subtilis* defective in the glucosylation of the wall teichoic acid, were found

to revert more rapidly than *B. subtilis* itself. Such mutants have previously been shown to be deficient in autolytic activity (Young, 1967). On the basis of these results De Castro-Costa & Landman (1977) proposed that reversion inhibitory factor (autolysin) acts by preventing the accumulation of peptidoglycan at the membrane surface. They concluded that reversion occurred when the balance between peptidoglycan synthesis and the autolytic degradation of newly synthesised material moved in favour of synthesis. Such a movement could result either from increased wall synthesis or, alternatively, by inhibition or lowering of the autolytic activity. In this context high concentrations of gelatin were shown to inhibit autolysin whereas the presence of exogenous walls in the medium stimulated reversion. However, the modification of autolytic activity by such means cannot be the sole explanation for the onset of reversion. Protoplasts of both *B. subtilis* and *B. licheniformis* are known to revert on hard-agar in the absence of either gelatin or exogenous walls (Elliott *et al.*, 1975*a*). The isolation and characterisation of soluble peptidoglycan from various stages of reversion shows that both autolysins are active under these conditions. The stimulation of reversion by added wall can also be obtained by using instead whole micro-organisms including Gram-negative bacteria and yeast (Clive & Landman, 1970). Thus a lowering of the autolysin concentration, such as occurs by a specific binding with exogenous wall, does not appear to be involved. In fact the wide variety of substances and conditions known to stimulate reversion argues against any direct interaction with the autolysins themselves. It seems more probable that they act by causing a physical separation of the autolysins and their substrate. One might expect this separation to be least effective at the earlier stages of reversion when membrane-bound autolysins and newly synthesised peptidoglycan are in close contact. This was the time when the greatest proportion of soluble peptidoglycan was found. Similarly the lowering of the agar or gelatin concentration or the incubation of protoplasts in liquid medium would also be expected to affect separation. Where investigated these conditions have led to an increased formation of soluble peptidoglycan.

I thank Dr I. D. J. Burdett and Dr T. S. J. Elliott for help with preparation of the illustrations and Dr H. J. Rogers for his continued interest.

Plate 1 (*a, b*) and Plate 3(*a, b, c*) are taken from Elliott *et al.* (1975*b*) and are published with permission of the American Society for Microbiology.

REFERENCES

AINSWORTH, S. K. & KARNOVSKY, M. J. (1972). An ultrastructural staining method for enhancing the size and electron opacity of ferritin in thin sections. *Journal of Histochemistry and Cytochemistry*, **20**, 225–9.

ALTERNBERN, R. A. (1963). Reversion of L-forms and spheroplasts of *Proteus mirabilis*. *Journal of Bacteriology*, **85**, 269–72.

BAUMAN, N. & DAVIS, B. D. (1957). Selection of auxotrophic bacterial mutants through diaminopimelic acid or thymine deprival. *Science*, **126**, 170.

BOND, E. C. (1969). Gene activity following cell wall removal in the wild-type and in a mutant of *Bacillus subtilis* capable of growth in liquid media in the L-form. Ph.D. Thesis, Georgetown University, Washington, D.C., USA.

BROWN, W. C. & YOUNG, F. E. (1970). Dynamic interaction between cell wall polymers, extracellular proteases and autolytic enzymes. *Biochemical and Biophysical Research Communications*, **38**, 564–8.

CHATTERJEE, A. N., WARD, J. B. & PERKINS, H. R. (1967). Synthesis of mucopeptide by L-form membranes. *Nature, London*, **214**, 1311–14.

CLIVE, D. & LANDMAN, O. E. (1970). Reversion of *Bacillus subtilis* protoplasts to the bacilliary form induced by exogenous cell wall, bacteria and by growth in membrane filters. *Journal of General Microbiology*, **61**, 233–41.

DE CASTRO-COSTA, M. R. & LANDMAN, O. E. (1977). Inhibitory protein controls the reversion of protoplasts and L-forms of *Bacillus subtilis* to the walled state. *Journal of Bacteriology*, **129**, 678–89.

ELLIOTT, T. S. J., WARD, J. B. & ROGERS, H. J. (1975*a*). Formation of wall polymers by reverting protoplasts of *Bacillus licheniformis*. *Journal of Bacteriology*, **124**, 623–32.

ELLIOTT, T. S. J., WYRICK, P. B., WARD, J. B. & ROGERS, H. J. (1975*b*). Ultrastructural study of the reversion of protoplasts of *Bacillus licheniformis* to bacilli. *Journal of Bacteriology*, **124**, 905–17.

FODOR, M. & ROGERS, H. J. (1966). Antagonism between vegetative cells and L-forms of *Bacillus licheniformis* strain 6346. *Nature, London*, **211**, 658–9.

FORSBERG, C. W. & WARD, J. B. (1972). *N*-acetylmuramyl-L-alanine amidase of *Bacillus licheniformis* and its L-form. *Journal of Bacteriology*, **110**, 878–88.

GOODELL, E. W. & SCHWARTZ, U. (1975). Sphere-rod morphogenesis of *Escherichia coli*. *Journal of General Microbiology*, **86**, 201–9.

GOODER, H. & MAXTED, W. R. (1961). External factors influencing structure and activities of *Streptococcus pyrogenes*. In *Microbial reaction to Environment*; *Symposia of the Society for General Microbiology*, **11**, ed. G. G. Meynell & H. Gooder, pp. 151–73. Cambridge University Press.

GREGORY, W. W. & GOODER, H. (1976). Peptidoglycan and rhamnose synthesis in a stable L-phase variant of *Streptococcus faecium* strain F24. *Abstracts of the Annual Meeting of the American Society for Microbiology*, Abstract D36.

HADLACZKY, G., FODOR, K. & ALFOLDI, L. (1976). Morphological study of the reversion to the bacilliary form of *Bacillus megaterium* protoplasts. *Journal of Bacteriology*, **125**, 1172–9.

HIGGINS, M. L. & SHOCKMAN, G. D. (1970). Model for cell wall growth of *Streptococcus faecalis*. *Journal of Bacteriology*, **101**, 643–8.

HUGHES, R. C. (1970). Autolysis of isolated cell walls of *Bacillus licheniformis* NCTC 6346 and *Bacillus subtilis* Marburg strain 168. Separation of the products and characterisation of the mucopeptide fragments. *Biochemical Journal*, **119**, 849–60.

HUGHES, R. C. & TANNER, P. J. (1968). The action of dilute alkali on some bacterial cell walls. *Biochemical and Biophysical Research Communications*, **33**, 22–8.

JAMES, R., HAGA, J. Y. & PARDEE, A. B. (1975). Inhibition of an early event in the cell division cycle of *Escherichia coli* by FL 1060, an amidino-penicillanic acid. *Journal of Bacteriology*, **122**, 1283–92.

KING, J. R. & GOODER, H. (1970). Reversion to the streptococcal state of enterococcal protoplasts, spheroplasts and L-forms. *Journal of Bacteriology*, **103**, 692–6.

KUSAKA, I. (1967). Growth and division of protoplasts of *Bacillus megaterium* and inhibition of division by penicillin. *Journal of Bacteriology*, **94**, 884–8.

KUSAKA, I. (1971). Electron-microscopic observations on growing and dividing protoplasts of *Bacillus megaterium*. *Journal of General Microbiology*, **63**, 199–202.

LANDMAN, O. E. (1968). Protoplasts, spheroplasts and L-forms viewed as a genetic system. In *Microbial protoplasts, spheroplasts and L-forms*, ed. L. B. Guze, pp. 319–32. Baltimore: Williams & Wilkins Co.

LANDMAN, O. E. & DE CASTRO-COSTA, M. R. (1976). Reversion of protoplasts and L-forms of bacilli. In *Microbial and Plant Protoplasts*, ed. J. F. Peberdy, A. H. Rose, H. J. Rogers, & E. C. Cocking, pp. 201–17. London and New York: Academic Press.

LANDMAN, O. E. & FORMAN, A. (1969). Gelatin-induced reversion of protoplasts of *Bacillus subtilis* to the bacilliary form: Biosynthesis of macromolecules and wall during successive steps. *Journal of Bacteriology*, **99**, 576–89.

LANDMAN, O. E. & FOX, S. M. (1977). Peptidoglycan synthesis during reversion of *Bacillus subtilis* protoplasts in liquid media. *Abstracts of the Annual Meeting of the American Society for Microbiology*. Abstract K174.

LANDMAN, O. E. & HALLE, S. (1963). Enzymically and physically induced inheritance changes in *Bacillus subtilis*. *Journal of Molecular Biology*, **7**, 721–38.

LANDMAN, O. E., RYTER, A. & FREHEL, C. (1968). Gelatin-induced reversion of protoplasts of *Bacillus subtilis* to the bacilliary form: electronmicroscopic and physical study. *Journal of Bacteriology*, **96**, 2154–70.

MARTIN, H. H. (1963). Bacterial protoplasts – a review. *Journal of Theoretical Biology*, **5**, 1–34.

MCQUILLEN, K. (1960). Bacterial protoplasts. In *The Bacteria*, 1, ed. I. C. Gunsalus & R. Y. Stanier, pp. 249–359. London and New York: Academic Press.

MILLER, I. L., ZSIGRAY, R. M. & LANDMAN, O. E. (1967). The formation of protoplasts and quasi spheroplasts in normal and chloramphenicol-pretreated *Bacillus subtilis*. *Journal of General Microbiology*, **49**, 513–5.

NECAS, O. (1961). Physical conditions as important factors for the regeneration of naked yeast protoplasts. *Nature, London*, **192**, 580–1.

NECAS, O. (1962). The mechanism of regeneration of yeast protoplasts. 1. Physical conditions. *Folia Microbiologica, Praha*, **8**, 256–62.

PERBEDY, J. F., ROSE, A. H., ROGERS, H. J. & COCKING, E. C. (eds.) (1976). *Microbial and Plant Protoplasts*. London and New York: Academic Press.

REUSCH, V. M. & PANOS, C. (1976). Defective synthesis of lipid intermediates for peptidoglycan formation in a stabilised L-form of *Streptococcus pyogenes*. *Journal of Bacteriology*, **126**, 300–11.

REYNOLDS, P. E. (1971). Peptidoglycan synthesis in Bacilli. II. Characteristics of protoplast membrane preparations. *Biochimica et Biophysica Acta*, **237**, 255–72.

ROGERS, H. J., WARD, J. B. & ELLIOTT, T. S. J. (1976). Biosynthesis of wall polymers by bacterial protoplasts. In *Microbial and Plant Protoplasts*, ed. J. F. Peberdy, A. H. Rose, H. J. Rogers & E. C. Cocking, pp. 219–35. London and New York: Academic Press.

ROSENTHAL, R. S. & SHOCKMAN, G. D. (1975a). Characterisation of the presumed peptide cross-links in the soluble peptidoglycan fragments synthesised by protoplasts of *Streptococcus faecalis*. *Journal of Bacteriology*, **124**, 410–18.

ROSENTHAL, R. S. & SHOCKMAN, G. D. (1975*b*). Synthesis of peptidoglycan in the form of soluble glycan chains by growing protoplasts (autoplasts) of *Streptococcus faecalis*. *Journal of Bacteriology*, **124**, 419–23.

ROSENTHAL, R. S., JUNGKIND, D., DANEO-MOORE, L. & SHOCKMAN, G. D. (1975). Evidence for the synthesis of soluble peptidoglycan fragments by protoplasts of *Streptococcus faecalis*. *Journal of Bacteriology*, **124**, 398–409.

ROTH, G. S., SHOCKMAN, G. D. & DANEO-MOORE, L. (1971). Balanced macromolecular biosynthesis in 'protoplasts' of *Streptococcus faecalis*. *Journal of Bacteriology*, **105**, 710–17.

SCHÖNFELD, J. K. & DE BRUIJN, W. C. (1973). Ultrastructure of the intermediate stages in the reverting L-phase organisms of *Staphylococcus aureus* and *Streptococcus faecalis*. *Journal of General Microbiology*, **77**, 261–71.

SCHWARTZ, U. & LEUTGEB, W. (1971). Morphogenetic aspects of murein structure and biosynthesis. *Journal of Bacteriology*, **106**, 588–95.

SHOCKMAN, G. D., DANEO-MOORE, L. & HIGGINS, M. L. (1975). Problems of cell wall and membrane growth, enlargement and division. *Annals of the New York Academy of Science*, **235**, 161–96.

TOMCSIK, J. & GUEX-HOLZER, S. (1952). Änderung der Struktur der Bakterienzelle im Verlauf der Lyzozym-Einwirkung. *Schweizerische Zeitschrift für allgemeine Pathologie und Bacteriologie*, **15**, 517–25.

WARD, J. B. (1973). The chain length of the glycans in bacterial cell walls. *Biochemical Journal*, **133**, 395–8.

WARD, J. B. (1975). Peptidoglycan synthesis in L-phase variants of *Bacillus licheniformis* and *Bacillus subtilis*. *Journal of Bacteriology*, **124**, 668–78.

WEIBULL, C. (1953). The isolation of protoplasts from *Bacillus megaterium* by controlled treatment with lysozyme. *Journal of Bacteriology*, **66**, 688–95.

WYRICK, P. B. & GOODER, H. (1976). Reversion of *Streptococcus faecium* cell wall-defective variants to the intact bacterial state. *Les Colloques de l'Institut National de la Santé et de la Recherch Médicale*, 21–25 September, 1976, vol. 64, pp. 59–88.

WYRICK, P. B. & ROGERS, H. J. (1973). Isolation and characterisation of cell wall-defective variants of *Bacillus subtilis* and *Bacillus licheniformis*. *Journal of Bacteriology*, **116**, 456–65.

WYRICK, P. B., McCONNELL, M. & ROGERS, H. J. (1973). Genetic transfer of the stable L-form to intact bacterial cells. *Nature, London*, **244**, 505–7.

YOUNG, F. E. (1967). Requirement of glycosylated teichoic acid for absorption of phage in *Bacillus subtilis* 168. *Proceedings of the National Academy of Sciences, USA*, **58**, 2377–84.

YOUNG, F. E., HAYWOOD, P. & POLLOCK, M. (1970). Isolation of L-forms of *Bacillus subtilis* which grow in liquid medium. *Journal of Bacteriology*, **102**, 867–70.

EXPLANATION OF PLATES

PLATE 1

Reverting protoplasts of *B. licheniformis* after 3–6 hr incubation. (*a*) Sectioned *in situ* after 3 hr incubation on the agar surface showing deformation of shape. (*b, c*) Sections of protoplasts after 6 hr incubation, removed from the agar surface and resuspended in liquid medium: (*c*) treated directly with ferritin-conjugated antibody directed against peptidoglycan showing association of ferritin particles with the protoplast surface; (*b*) treated first with unconjugated antibody directed against peptidoglycan and then with ferritin-conjugated antibody showing the absence of ferritin particles. (*d*) protoplasts incubated for 6 hr, freeze-etched *in situ* showing the fringe of wall material and shape of the organisms. The arrow shows the direction of shadowing the replicas.

PLATE 1

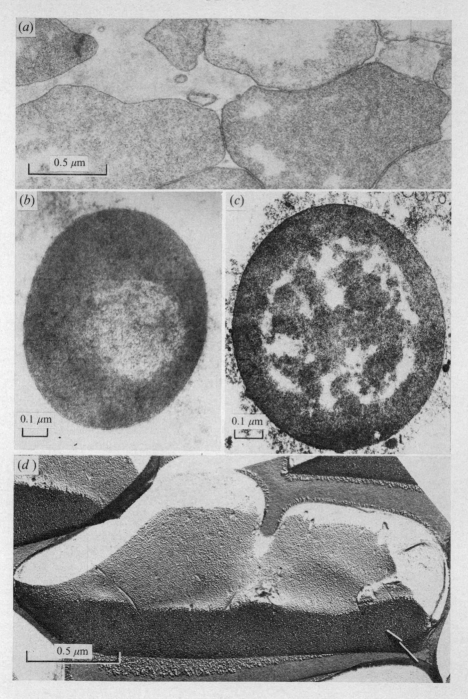

(a)

0.5 μm

(b)

0.1 μm

(c)

0.1 μm

(d)

0.5 μm

PLATE 2

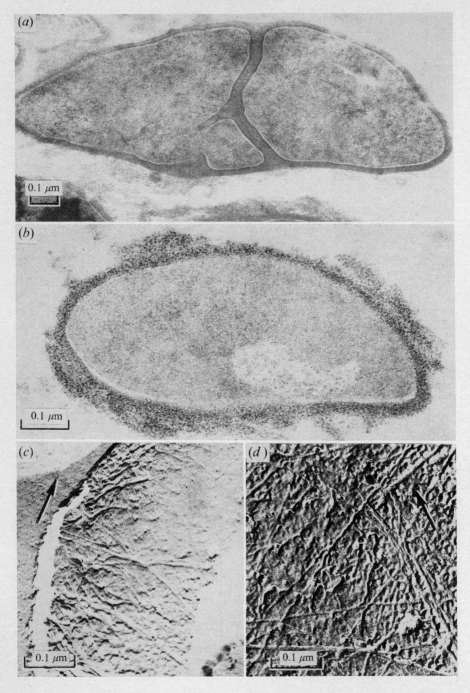

(a)

0.1 μm

(b)

0.1 μm

(c)

0.1 μm

(d)

0.1 μm

PLATE 3

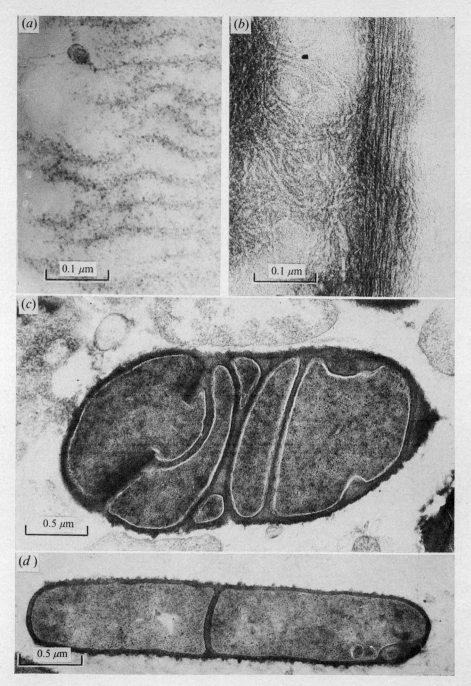

PLATE 4

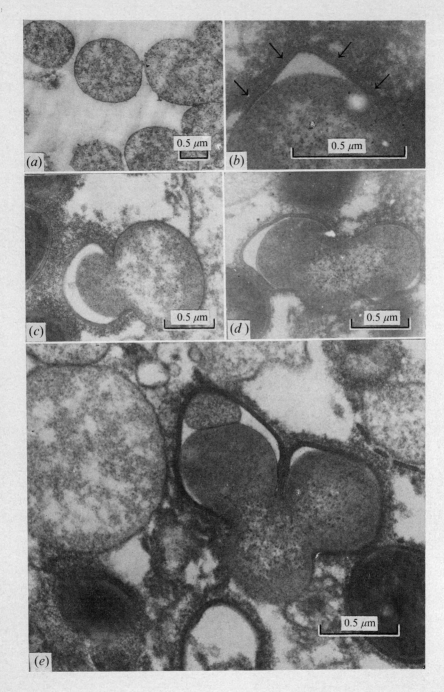

(a)

(b)

0.5 μm

0.5 μm

(c)

(d)

0.5 μm

0.5 μm

(e)

0.5 μm

PLATE 2

Reversion of *B. licheniformis* protoplasts after 9–12 hr of incubation. Sections of organisms incubated for 10–11 hr showing abnormal septation (*a*) and thickened walls which reacted with the ferritin-conjugated antibody throughout their depth (*b*). This section was stained with bismuth subnitrite to enhance the contrast (Ainsworth & Karnovsky, 1972). Protoplasts after 12 hr incubation freeze-etched while still on the agar surface, showing fibres extending from the wall (*c, d*). The arrows indicate the direction of shadowing the replicas.

PLATE 3

Fibres isolated from *B. licheniformis* protoplasts incubated for 12 hr by treatment with sodium dodecyl sulphate: (*a*) sectioned after treatment with ferritin-conjugated antibody showing association of ferritin particles along the length of the fibres; (*b*) negatively-stained with phosphotungstic acid – the fibres appeared to be in bundles with some unravelling having occurred; (*c*) sectioned *in situ* after 12 hr incubation – septation is still abnormal with thickened walls although the organisms are becoming more rod-shaped; (*d*) organisms after 16 hr incubation – with a relatively normal morphology.

PLATE 4

Reversion of *S. faecium* protoplasts on gelatin medium: (*a*) section through microcolony after 12 hr incubation; (*b, c, d*) sections through organisms in sequential stages of reversion. All microcolonies were treated with ferritin-conjugated antibody directed against the streptococcal wall. Arrows indicate short but triple-layered pieces of wall. As reversion proceeds these pieces increase in length but do not always appear to remain associated with the membrane. (*e*) Section through a reverting organism after 18 hr incubation. The colony was treated with ferritin-conjugated antibody directed against the wall and the thin-sections were post-stained with uranyl acetate and lead citrate. These electron micrographs were kindly provided by Drs Priscilla B. Wyrick and H. Gooder.

STRUCTURE AND FUNCTION OF BACTERIAL FLAGELLA

M. SIMON, M. SILVERMAN, P. MATSUMURA, H. RIDGWAY, Y. KOMEDA AND M. HILMEN

*Department of Biology, University of California,
San Diego, La Jolla, California 92093, USA*

INTRODUCTION

Bacterial flagella function to endow the cell with motility and chemotaxis. One of the goals of our work is to determine the mechanisms involved in these functions. Our approach thus far has been to study the structure of the organelle and its components and to try to deduce the mechanisms that operate in flagellar function from the nature of flagellar related structures. Electron microscopic observations have allowed the delineation of the structure of the organelle. Furthermore, it is possible using biochemical techniques, to isolate relatively intact organelles and to dissociate them into their sub-unit polypeptide components (DePamphilis & Adler, 1971*a*; Dimmitt & Simon, 1971; Hilmen, Silverman & Simon, 1974). The filament derived from *Escherichia coli* is a long helical structure composed of a single subunit, flagellin, with a molecular weight of approximately 54000 (Kondoh & Hotani, 1974). It ends in the hook structure which is also composed primarily of a single subunit with a molecular weight of 42000 (Silverman & Simon, 1972; Kagawa, Asakura & Iino, 1973). Dissociation of the rest of the basal structure results in nine other polypeptides (Hilmen & Simon, 1976). This basal complex is composed of four rings and a rodlike structure that have been shown to be associated with the bacterial membrane (DePamphilis & Adler, 1971*b*, *c*). The M- and S-rings are associated with the inner membrane. The P-ring may be associated with the peptidoglycan layer and the L-ring is associated with the lipopolysaccharide outer membrane of the cell.

On a gross level, it is clear that the flagellar filament functions by rotating (Silverman & Simon, 1974*a*; Berg, 1974). The helical structure of the filaments allows them to form bundles with individually rotating flagella that propel the cell through the medium. Flagellar rotation can be made perceptible by tethering the bacterium via its flagellar organelle to a glass slide surface or by attaching small beads to the flagellar filament (Silverman & Simon, 1974*a*). Furthermore, it can be shown that chemotaxis is a function of the frequency of reversal of flagellar rotation (Larsen *et al.*, 1974). One way of observing the frequency of flagellar rotation reversal is to add attractants or repellants to tethered bacteria. An increase in the concentration of attractant causes a transient suppression of clockwise rotation and the bacteria rotate only in the counter-clockwise mode for a period of time that is proportional to the increase in attractant concentration (Berg & Tedesco, 1975). An increase in repellant concentration causes transient rotation in the clockwise direction.

The immediate behaviour of the organelle that results in both motility and chemotaxis can be observed and the structure of the organelle can be analysed. This information allows us to ask specific questions about the mechanisms that underlie these functions. What drives flagellar rotation? Where precisely is the motor located? How is it coupled to the flagellar structure and to its energy source? What controls the reversal of flagellar rotation and how do attractants and repellants regulate the frequency of reversal of rotation? The isolation of the organelle does not allow us to answer these questions directly since the structure that has been isolated does not contain all of the components necessary for motility and chemotaxis. If the isolated flagellar organelle contained the proteins that were responsible for motility, we would expect that mutants that were paralysed, i.e. Mot⁻ mutants, should result in the absence of one of the components of the basal structure. However, when Mot⁻ amber mutants were tested, they were found to contain all of the same components as the wild-type flagellar basal structure (Hilmen & Simon, 1976). This suggests that there are various of components responsible for flagellar function that have not been isolated along with the structure. In order to obtain a more complete catalogue of the components that are involved in the function and assembly of the flagellar structure and to understand the relationship between these components and the organelle, genetic approaches were employed. A large number of mutants that were defective in motility, chemotaxis and the ability to form a flagellar structure, were isolated. All of these mutants were classified into functional groups by using specific genetic complementation tests and each of these groups was mapped on the *E. coli* genome. Fig. 1 shows a map with thirty flagellar related loci. These include the following classes: *fla* genes, Fla⁻ mutants that are defective in the ability to assemble a flagellum; *mot* genes which are responsible for the rotation of the flagellar filament, i.e., mutants in these genes have intact filaments, but are paralysed; and *che* genes, mutants in these genes being defective in their chemotactic response to a

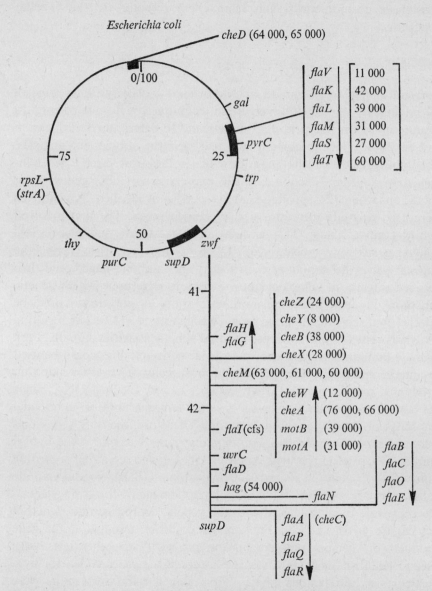

Fig. 1. The map of flagellar related genes in *E. coli*. The genes that map between *gal* and *trp* are in region I. The 42 000 dalton polypeptide corresponds to the hook gene product and has been assigned to the *flaK* gene. The other polypeptides are products of this region but have not been assigned to individual genes. The genes that map between *flaH* and *uvrC* are in region II. The apparent molecular weight of each gene product is listed alongside the gene. The region between *uvrC* and *supD* is referred to as region III.

large number of compounds. They appear to carry defects in the transmission of chemotactic signals from chemoreceptors to the flagellar
apparatus.

GENE PRODUCT IDENTIFICATION

Conventional genetic techniques could be used to classify and arrange all
of the flagellar genes. However, these techniques could not be used to
identify all of the polypeptide products and to determine their location
and function. Genetic cloning was used for the subsequent analysis.
Recombinant DNA techniques allow the isolation of specific genes involved in flagellar function and the expression of these genes under
conditions where their products can be specifically labelled. Two systems
were used to study the expression of cloned genes. The first involved
hybrid lambda phage. The lambda acted as vehicle and dispensable
lambda genes were replaced with DNA from *E. coli* that carried the
flagellar genes (Silverman *et al.*, 1976*a*). The lambda phage could then
be used directly in complementation tests to determine which genetic
functions it carried. Furthermore, a variety of deletions could be
selected that removed different parts of the inserted DNA. The resulting
virus was tested to determine its residual genetic activities. Finally, each
of these lambdas carrying known flagellar gene activities could be used
to infect ultraviolet (UV) irradiated host cells. The UV-irradiation
eliminated protein synthesis that was specified by the host DNA. Since
the host was a lambda lysogen the resident lambda repressor eliminated
transcription from the lambda genome. Therefore, the only genes that
were transcribed were the *E. coli* flagellar genes carried on the hybrid
lambda. These gene products could be labelled and separated according
to molecular weight on sodium dodecyl sulphate (SDS)–polyacrylamide
gel electrophoresis (PAGE). The polypeptide bands found in electrophoresis were then correlated with the genetic complementation activities carried by the phages. Thus, bands could be assigned as the gene
products of specific complementation groups. Two other tests were
used to establish the flagellar specific nature of the genes coded for by a
particular hybrid lambda phage. First, flagellar mutant strains were
UV-irradiated and then superinfected by lambda phages that were
shown to carry the genetic activity missing in the mutant. The behaviour
of the defective cell was then examined to see if the activity was restored
by the infecting phage. Most of the phages were able to restore functional
activity to the mutant cell within a very short time after infection (Silverman, Matsumura & Simon, 1976*b*). This indicated that the product

made from the genes introduced on the lambda vehicle was able to function in the UV-irradiated cell. A second test involved the use of mutants in the *flaI* gene. This gene is a master control switch which acts as a positive regulator of flagellar gene expression. The synthesis of most of the flagellar gene related products requires the presence of the intact *flaI* gene product (Silverman & Simon, 1974*b*). Therefore, in control experiments, lambda phages carrying flagellar specific genetic activities were used to infect the Fla⁻ mutant and the absence of a specific band in the FlaI⁻ background and its presence in the FlaI⁺ strain indicated that the band was probably flagellar specific.

It was not possible to obtain protein synthesis with all of the genes using the lambda vehicles. Therefore, a second system was used. It involved colicinogenic factor El as vehicle and fragments derived from the whole *E. coli* genome as passenger. A collection carrying all of the *E. coli* genes was prepared by Clarke & Carbon (1976). This collection of 2000 strains was screened and all of the strains carrying plasmids that had genetic complementation activity for flagellar genes were selected and purified. These plasmids were introduced into a minicell forming strain. The minicells could be purified, and the proteins synthesised in them specifically labeled. Since the only DNA present was the plasmid that managed to be segregated during minicell formation, minicell protein synthesis was coded for by plasmid genes. Comparison of plasmids that carried overlapping regions of the genome with a variety of flagellar genes allowed the assignment of specific polypeptides to specific genes.

Plate 1 shows an example of the results of the application of the first approach to the identification of genes in region II of the flagellar map. At the top is a list of the genetic complementation activities that were carried on the lambda phages. Below is an autoradiograph of the polyacrylamide gel electrophoresis pattern of the products made in UV-irradiated cells. Specific radioactive bands can be correlated with each of the genetic activities. These bands always appear when the hybrid λ carries the specific genetic activity and are always absent when that genetic activity is absent. Fig. 1 summarizes the molecular weights of the polypeptides which were assigned to each of the genes using this technique. A 31000 molecular weight protein correponds to *motA*, 39000 to *motB*, two polypeptides were found to correspond to *cheA*, one that was 76000 molecular weight and the other 66000 molecular weight (Silverman & Simon, 1976). When the peptide maps derived from these two proteins were examined, it was found that all of the peptide bands derived from the 66000 molecular weight protein appeared in the 76000 molecular weight protein. On the other hand, there were a few peptides that were part of the pattern obtained with the 76 000 molecular weight protein that were not found in the digestion of the 66 000 molecular weight protein. Furthermore, measurements of the coding capacity of the DNA that corresponded to the *cheA* gene indicated that it was sufficient to code for a single polypeptide of 76 000 daltons. Therefore, we propose that the smaller polypeptide was derived from the large one either by proteolysis or by the presence of a second initiation point within the *cheA* gene. The *cheW* gene product had a molecular weight of 12 000. The polypeptides corresponding to *motA*, *motB*, *cheA*, *cheW* were part of a co-transcribed unit (Silverman & Simon, 1976). The *cheM* gene is adjacent to the *cheW* gene and it codes for synthesis of a polypeptide with an approximate molecular weight of 61 000. However, this polypeptide appears on polyacrylamide gel electrophoresis as three bands, one with an apparent molecular weight of 60 000, one at 61 000 and another at 63 000 (Silverman, Matsumura, Hilmen & Simon, 1977). The peptide maps derived from these three bands coincide and estimates of the coding capacity of the *cheM* gene indicate that it is only able to program the synthesis of a polypeptide of about 60 000 daltons. Therefore, we conclude that the three bands are modified forms of the same polypeptide. Kort, Goy, Larsen & Adler (1975) showed that there was an integral membrane protein that was methylated in response to chemotactic signals. This protein was called methyl accepting chemotaxis protein (MCP). The product of the *cheM* gene corresponds to part of the MCP (Silverman & Simon, 1977*b*). The *cheX* gene product has a molecular weight of 28 000, the *cheB* gene product has a molecular weight of 38 000, the *cheY* gene product has an apparent molecular weight of 8 000 and the *cheZ* gene product has a molecular weight of 24 000 (Silverman & Simon, 1977*a*).

The use of the plasmid approach has allowed the identification of the gene products in region I (Matsumura, Silverman & Simon, 1977). Thus far, genetic tests have defined six complementation groups in this region (Komeda, Silverman & Simon, 1977). They all appear to be co-transcribed. In initial experiments, it was difficult to get a clear assignment of these gene products using the hybrid lambda system. The UV-irradiated cells have relatively low capacity for protein synthesis and these gene products are not ordinarily translated at a very high level. Therefore, these genes were obtained on a number of plasmids and the plasmids were used to code for protein synthesis in minicells. Plate 2(*b*) shows the two-dimensional gel electrophoresis map of the products formed in minicells containing a plasmid that carries the genes in region I. Plate 2(*a*) compares this two-dimensional map to the map of the labelled polypeptides obtained by the dissociation of purified hook-basal body complex. The polypeptides of 60000, 42000, 31000, 27000 and 11000 molecular weight all have exactly the same position on the two-dimensional gels. A 39000 molecular weight component also appears on SDS—polyacrylamide gels and its position is the same as that of the 39000 polypeptide found in the flagellar basal complex. Thus, while all of the gene products have not been specifically assigned to a given cistron, it is clear that this plasmid which carries six flagellar genes is responsible for the synthesis of at least six of the components of the hook-basal body complex, including the hook subunit gene product. Comparison of the hook protein made in specific mutants with the wild-type hook protein indicates that *flaK* is the structural gene for the hook subunit (Y. Komeda, *et al.*, personal communication).

Another gene product, the *cheD* product, was identified using a combination of the two approaches described here. A specific plasmid from the Clarke and Carbon collection was shown to carry the gene. This plasmid was used as a source of DNA to synthesize hybrid lambda carrying *cheD*. Deletions of the lambda phage were then prepared and their expression studied in UV-irradiated cells. The product of the *cheD* gene appears as a group of bands with molecular weights corresponding to 64000 and 65000. These polypeptides probably represent the products of a single gene with *cheD* activity. They are methylated (Silverman & Simon, 1977*b*) and have been referred to as MCP (1).

Thus far, many of the products of the genes that are involved in flagellar assembly and function have been identified. There still are a number of genes that have not been assigned products and there are a number of polypeptides that have not been given specific gene designations. However, the results up to this point suggest that it should be possible, using these techniques, to identify the products of all of the genes that have so far been studied.

LOCATION

The hybrid lambda programs the synthesis of specific flagellar polypeptides. The peptides are labelled and then are able to function. This suggests that they are located in the position in the cell where they are necessary for flagellar function. By studying the distribution of the radioactive proteins in hybrid-lambda infected cells, it should be possible to locate the polypeptides and learn more about the way in which they function. In order to do this, UV-irradiated cells were infected with hybrid lambda, and the cells were broken open and the distribution of the labelled polypeptides studied by fractionating the cell into periplasmic, cytoplasmic and membrane fractions (H. Ridgway, *et al.*, unpublished observations). Plate 3 shows the results of a fractionation experiment and the radioautogram of the distribution of the labelled specific polypeptides in the different subcellular fractions. In general, flagellin and most of the flagellar structural components are found associated with the outer membrane fraction (DePamphilis & Adler, 1971*c*). On the other hand, the two *mot* polypeptides, *motA* and *motB* and the *cheM* and *cheD* gene products are all found exclusively in the inner membrane. They are, therefore, integral membrane proteins. On the other hand, the *cheA*, *cheW*, *cheY* and *cheZ* gene products are found associated almost exclusively with the cytoplasm. The *cheX* and *cheB* gene products are found mostly in the cytoplasm. However, some of the material is also associated with the membrane.

The *mot* proteins may either be localized around the base of the flagellum or they may be distributed throughout the membrane structure. The *cheM* and *cheD* proteins also function in the membrane, and they represent a relatively significant fraction of the total membrane protein (H. Ridgway, *et al.*, unpublished observations). On the other hand, many of the other chemotaxis-specific polypeptides appear to be cytoplasmic products and they may function by either being transiently associated with the inner membrane or by forming cytoplasmic intermediates that are necessary for the transmission of the chemotactic signals received on the outer surface of the cell to the flagellar filament. In fact, it now becomes easier to direct questions about the function associated with the flagellar organelle. If *motA* and *motB* are integral membrane proteins and are not isolated along with the flagellar apparatus, how do they function? They may act as mediators of ion flow or they may act directly by interaction with the basal component of the flagellum. What are the phenotypes of the *cheM* and the *cheD* genes? Their products are integral membrane proteins which stand between the chemoreceptors generally found in the periplasmic space or on the outer surface in the membrane and a variety of chemotactic specific proteins which are found in the cytoplasm or transiently attached to the inner cell membrane. How do they transmit the signal from the surface receptors to the flagellar apparatus?

FUNCTIONS OF FLAGELLAR GENE PRODUCTS

The general function of some of the gene products can be deduced from the phenotypes of mutants carrying defects in the genes. There are a number of convergent pathways that are required for chemotaxis, motility, the assembly of the flagellar structure and the regulation of flagellar assembly. Chemotaxis essentially involves two pathways: one that functions to receive chemotactic signals and the other that is involved in processing signals from chemoreceptors and transmitting the signal to the flagellar apparatus. Work in the laboratories of Adler (1975) and Koshland (1977) has shown clearly that there are a large variety of specialized chemoreceptors that effectively bind attractants and that are required by the bacterium in order to respond to a given chemical. Furthermore, there is evidence that there are intermediate functions that collect information from these chemoreceptors. Defects in these functions result in mutants that do not respond to a given chemical or to a subclass of chemicals, for example, mutants that do not respond to either galactose or ribose (Ordal & Adler, 1974; Strange & Koshland, 1976). Information gathered by the chemoreceptors is then fed into a general chemotaxis pathway. Hence, mutants in these genes are unresponsive to all attractants and repellants. They have been called 'general chemotaxis mutants'. They constitute a pathway whose function it is to transmit signals from the chemoreceptor pathway to the effector organelle, i.e. the flagellar rotor. In the process, the signals from chemoreceptors must be translated into a form that will allow the flagellar rotor to respond, appropriate to chemoreceptor events (Parkinson, 1977).

It is interesting to look at mutations which apparently define the overlap between two pathways. These types of mutations appear to have pleiotropic effects. For example, the *cheM* and *cheD* genes lie between the chemoreceptor pathway and the general chemotaxis pathway (Silverman & Simon, 1977*b*). The products determined by these genes are integral membrane proteins; the *cheM* product appears on acrylamide gels as a series of polypeptide bands with molecular weights in the range 60000 to 63000, and the *cheD* gene product appears as a series of bands with molecular weights from 63000 to 65000. These correspond to MCP initially described by Kort *et al.* (1975). Addition of attractants and repellants to the cell result in changes in the pattern of methylation of these gene products (Springer, Goy, Reader & Adler, 1977). However, many mutations in either the *cheD* or the *cheM* do not result in the absence of chemotaxis. Mutants in the *cheM* gene have a phenotype referred to as TAR⁻. These mutants (Silverman & Simon, 1977*b*; Springer *et al.*, 1977) were unresponsive to aspartate and to a series of other attractants and repellants. On the other hand, they did respond to serine and a number of sugars as well as to other repellants. Furthermore, mutants in the *cheD* gene had the complementary phenotype, i.e. they were unresponsive to serine, but they were able to respond to the aspartate group of attractants and repellants. This phenotype was referred to as TSR⁻. Double mutants did not respond to any attractant or repellant. This suggests that the *cheM* (*tar*) and *cheD* (*tsr*) genes function as a transition between the chemoreceptor pathway and the general chemotaxis pathway. They are integral membrane proteins and they each collect signals from a subset of receptors and then transmit those signals to the general chemotactic system, most of which is found in the cytoplasm of the cell. During this process, the *cheM* and the *cheD* proteins are also methylated and presumably demethylated to restore them to their initial condition. The function of the chemoreceptors may involve the binding of specific ligands and changes in the conformation of the protein. These conformational changes could then be transmitted and eventually result in a change in the conformation of the *cheM* or *cheD* protein. At this point, the conformational change could become translated into a chemical change, which results in the methylation and demethylation of the *cheM* and *cheD* proteins and presumably in the generation of another chemical signal, perhaps specific ion flow that is further mediated by all of the other chemotaxis genes. This signal could directly affect the flagellar rotor (Silverman & Simon, 1977*b*). There is some evidence that it may also affect the motility gene products. Szmelcman & Adler (1976) presented evidence for *mot* dependent changes in ion distribution upon the addition of chemoattractants or repellants. There may, therefore, be an overlap between the *che* system and the *mot* system. These relationships are schematically outlined in Fig. 2. There are other examples of overlapping functions in the genes involved in flagellar structure. There is a clear overlap between the *che* system and the structure of the flagellar rotor. One of the genes, *flaA*, has been found to have multiple phenotypes. Mutants in this gene may either lack a flagellar structure or they may have an intact flagellar structure and be defective in chemotaxis (Silverman & Simon, 1973). This gene has two phenotypes and thus presumably represents the function connecting the flagellar rotor structure and the chemotaxis system. There are a variety of other genes that appear to be directly involved in the formation of the flagellar rotor structure. Many of the genes in region I code for polypeptides that are structural components of a flagellar basal region and the hook. The *hag* gene is a structural gene for the flagellar filament and presumably, some of the *fla* genes in region III are also involved in coding for structural elements that compose the flagellar rotor. Some of these appear to be involved in the overlap between the *mot* function and

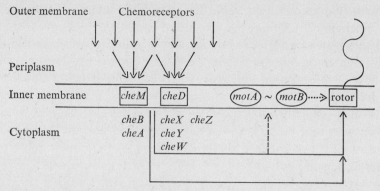

Fig. 2. A scheme showing the flow of information through the chemotaxis system. A large variety of chemoreceptors bind attractants and repellants. These signals are processed in the periplasmic space and on the cell membrane. They are eventually transmitted to either the *cheM* or *cheD* protein or to both. These proteins act as the terminal receptors and transfer the signal to the cytoplasm or inner membrane. Both the information and the receptor undergo further processing via the *cheB*, *cheX*, *cheZ* and *cheA*, *cheY* and *cheW* gene products. Some of these are transiently associated with the membrane and others are in the cytoplasm. The input signal is eventually transmitted to the flagella where it can result in an immediate increase or decrease in the frequency of changes in the direction of flagellar rotation and it can maintain the changed frequency level for a specific time period before the initial frequency is restored. The terminal signal may act on both the rotor and on the *mot* gene products.

the flagellar structural function. In *Salmonella abortesqui*, mutants in the *flaAIII* gene have two phenotypes, some of them are unable to form flagellar structures at all, while others form intact flagellar structures but are not motile (Yamaguchi, Iino, Horiguchi & Ohta, 1972). Thus, they show the paralysed phenotype that is generally characteristic of *mot* mutants. They may represent a function shared by the *mot* pathway and the flagellar structure. Finally, with respect to regulatory functions, there are some genes which appear to regulate the synthesis of all of the proteins in the *che-fla-mot* pathways, for example, the *flaI* gene. Mutants in this gene eliminate the synthesis of all of the flagellar structural components and of the *mot* and *che* gene products. Other regulatory genes have a more limited function; thus, mutants in the *flaE* gene are unable to regulate the length of the flagellar hook structure (Silverman & Simon, 1974). Mutations in other *fla* genes repress the synthesis of flagellin (Suzuki & Iino, 1975).

The examination of the flagellar organelle and its function leads us to a complex system which involves gene products that are located in the cytoplasm, the inner cell membrane, and the outer membrane. The function of the organelle depends not only on the structural components of the flagellum, but also on integral membrane proteins and cytoplasmic components. Genetic approaches allow us to define and identify the elements that are involved in the structure and function of the organelle, and biochemical techniques allow us to locate these components. It is necessary now to understand the molecular relationships between them, both in terms of the interactions between components, and in terms of the flow of information and energy necessary for flagellar function. At present we have a general view of gene functions. It will be necessary to establish specifically what the roles of the genes in these pathways are. The answers that we seek are in molecular terms and the questions that we must now ask concern the biochemistry of the components of the bacterial flagellar system.

This work was supported by a grant from the National Science Foundation, BMS 13-01606, and by a grant from the National Institutes of Health, AI 13008.

REFERENCES

ADLER, J. (1975). Chemotaxis in bacteria. *Annual Review of Biochemistry*, **44**, 341–56.

BERG, H. C. (1974). Dynamic properties of bacterial flagellar motors. *Nature, London*, **249**, 77–9.

BERG, H. C. & TEDESCO, P. M. (1975). Transient response to chemotactic stimuli in *E. coli*. *Proceedings of the National Academy of Sciences, USA*, **72**, 3235–9.

CLARKE, L. & CARBON, J. (1976). A colony bank containing synthetic *Col* E1 hybrid plasmids representative of the entire *E. coli* genome. *Cell*, **9**, 91–9.

DEPAMPHILIS, M. L. & ADLER, J. (1971a). Purification of intact flagella from *E. coli* and *B. subtilis*. *Journal of Bacteriology*, **105**, 376–83.

DEPAMPHILIS, M. L. & ADLER, J. (1971b). Hook-basal body complex of flagella. *Journal of Bacteriology*, **105**, 384–95.

DEPAMPHILIS, M. L. & ADLER, J. (1971c). Attachment of flagellar basal bodies to the cell envelope. *Journal of Bacteriology*, **105**, 396–407.

DIMMITT, K. & SIMON, M. (1971). Purification and thermal stability of intact *B. subtilis* flagella. *Journal of Bacteriology*, **105**, 369–75.

HILMEN, M., SILVERMAN, M. & SIMON, M. (1974). The regulation of flagellar formation. *Journal of Supramolecular Structure*, **2**, 360–74.

HILMEN, M. & SIMON, M. (1976). Motility and the structure of bacterial flagella. In *Cell Motility*, ed. R. Goldman, T. Pollard & J. Rosenbaum, **3**, pp. 35–45. Cold Spring Harbor Laboratory, USA.

KAGAWA, H., ASAKURA, S. & IINO, T. (1973). Serological study of bacterial flagellar hooks. *Journal of Bacteriology*, **113**, 1474–81.

KOMEDA, Y., SILVERMAN, M. & SIMON, M. (1977). Genetic analysis of region I flagellar mutants in *E. coli* K-12. *Journal of Bacteriology*, in press.

KONDOH, H. & HOTANI, H. (1974). Flagellin from *E. coli* K12. *Biochimica et Biophysica Acta*, **336**, 119–26.

KORT, E. N., GOY, M. F., LARSEN, S. H. & ADLER, J. (1975). Methylation of a membrane protein involved in bacterial chemotaxis. *Proceedings of the National Academy of Sciences, USA*, **72**, 3939–43.

KOSHLAND, D. E., JR. (1977). A response regulator model in a simple sensory system. *Science*, **196**, 1055–63.

LARSEN, S. H., READER, R. W., KORT, E. N., TSO, W. W., & ADLER, J. (1974).

Change in direction of flagellar rotation is the basis of the chemotactic response. *Nature, London,* **249**, 74–7.

MATSUMURA, P., SILVERMAN, M. & SIMON, M. (1977). Expression of the flagellar hook gene on hybrid plasmids in minicells. *Nature, London,* **265**, 758–60.

ORDAL, G. W. & ADLER, J. (1974). Properties of mutants in galactose taxis and transport. *Journal of Bacteriology,* **117**, 517–26.

PARKINSON, J. S. (1977). Behavioural genetics in bacteria. *Annual Review of Genetics,* in press.

SILVERMAN, M., MATSUMURA, P., DRAPER, R., EDWARDS, S. and SIMON, M. (1967a). Expression of flagellar genes carried by bacteriophage lambda. *Nature, London,* **261**, 248–50.

SILVERMAN, M., MATSUMURA, P. & SIMON, M. (1976b). The identification of the *mot* gene product with *Escherichia coli* – lambda hybrids. *Proceedings of the National Academy of Sciences, USA,* **73**, 3126–30.

SILVERMAN, M., MATSUMURA, P., HILMEN, M & SIMON, M. (1977). Characterization of lambda *Escherichia coli* hybrids carrying chemotaxis genes. *Journal of Bacteriology,* **130**, 877–87.

SILVERMAN, M. & SIMON, M. (1972). Flagellar assembly mutants in *E. coli. Journal of Bacteriology,* **112**, 986–93.

SILVERMAN, M. & SIMON, M. (1973). Genetic analyses of bacteriophage mu-induced flagellar mutants in *E. coli. Journal of Bacteriology,* **116**, 114–22.

SILVERMAN, M. & SIMON, M. (1974a). Flagellar rotation and the mechanism of bacterial motility. *Nature, London,* **249**, 73–4.

SILVERMAN, M. & SIMON, M. (1974b). Characterization of *Escherichia coli* flagellar mutants that are insensitive to catabolite repression. *Journal of Bacteriology,* **120**, 1196–203.

SILVERMAN, M. & SIMON, M. (1976). Genes controlling motility and chemotaxis in *E. coli. Nature, London,* **264**, 577–9.

SILVERMAN, M. & SIMON, M. (1977a). Identification of polypeptides necessary for chemotaxis in *E. coli. Journal of Bacteriology,* **130**, in press.

SILVERMAN, M. & SIMON, M. (1977b). Chemotaxis in *E. coli*: methylation of *che* gene products. *Proceedings of the National Academy of Sciences, USA,* in press.

SPRINGER, M., GOY, M., READER, R. & ADLER, J. (1977). Sensory transduction in *E. coli. Proceedings of the National Academy of Sciences, USA,* in press.

STRANGE, P. G. & KOSHLAND, D. E. (1976). Receptor interactions in a signalling system. *Proceedings of the National Academy of Sciences, USA,* **73**, 762–6.

SUZUKI, T. & IINO, T. (1975). Absence of messenger RNA specific for flagellin in non-flagellate mutants of *Salmonella. Journal of Molecular Biology,* **95**, 549–56.

SZMELCMAN, S. & ADLER, J. (1976). Change in membrane potential during bacterial chemotaxis. *Proceedings of the National Academy of Sciences, USA,* **73**, 4387–91.

YAMAGUCHI, S., IINO, T., HORIGUCHI, T. & OHTA, K. (1972). Genetic analyses of *fla* and *mot* cistrons closely linked to H1 in *Salmonella abortesqui* and its derivatives. *Journal of General Microbiology,* **70**, 59–75.

EXPLANATION OF PLATES

PLATE 1

Specific protein synthesis by hybrid lambda carrying region II flagellar genes in UV-irradiated bacteria. The preparation of the deleted hybrid λ phage and the conditions of infection of *E. coli* K12 159 carrying lambda have been described (Silverman & Simon, 1977a). The genetic complementation activity carried by a specific λ phage is listed above the radioautograph. The gene specific proteins are designated by the letters and they correspond to the following molecular weights: a *cheA* 76000; b *cheA* 66000; c *cheM* 63000, 61000, 60000; d *hag* 54000; f *motB* 39000; g *cheB* 38000; h *motA* 31000; i *cheX* 28000; k *cheW* 12000; l *cheY* 8000.

PLATE 2

Two-dimensional acrylamide gel pattern of hook-basal structure components compared to polypeptides synthesized in minicells carrying a plasmid with region I genes. (*a*) The isolation of ^{35}Sulphur labelled hook-basal body complex and the dissociation and electrophoresis in two-dimensions of the basal components has been described (Hilmen & Simon, 1976). The horizontal direction corresponds to isoelectric focusing in the pH range from 5.0 (left) to 7.0 (right). The vertical direction is SDS-PAGE.

(*b*) The extract of minicells derived from a parent strain carrying a plasmid with the region I genes (Matsumura, *et al.*, 1977). Some of the major products have the same mobilities as the hook-basal body components. These are indicated with arrows and they correspond to molecular weights of, (from the top), 60000, 42000, 31000, 27000 and 11000. There is also a spot at 35000 that may be specified by flagellar related genes carried on this plasmid.

PLATE 3

The distribution of *mot* and *che* proteins during cell fractionation. The results of SDS-PAGE of cells and cell fractions after infection with specific hybrid λ phage and labelling with [^{35}S]methionine. Each well contains 100 μg of protein. The top figure shows the gel after staining with Coomassie blue. The lower figure shows the corresponding autoradiograph. Column 1 shows the whole cells; column 2, the total cell extract after treatment with lysozyme EDTA; column 3, the periplasmic proteins; column 4, the total cytoplasm after removal of the membrane fraction; column 5, total membrane fraction; column 6, purified inner membrane fraction; column 7, mixture of inner and outer membrane; column 8, purified outer membrane fraction; column 9, molecular weight markers. The details of the cell fractionation are to be published elsewhere (H. Ridgway, M. Silverman & M. Simon, submitted for publication).

PLATE 1

λ fla 52 (motA, motB, cheA, cheW, cheX)

λ fla 52 Δ1(motA, motB, cheA, cheW, cheX)

λ fla 52Δ 22 (motA, motB, cheA, cheX)

λ fla 57 Δ 27 (motA, motB, cheA)

λ fla 57 Δ 5 (motA, motB, cheX)

λ fla 57 Δ 11 (motA, cheX)

λ fla 52 Δ 2 (cheX)

λ fla 57 Δ21 (cheW, cheX)

λ fla 42 (hag, cheW, cheX, cheB, cheY)

a
b
c
d
e
f
g
h
i

k

l

PLATE 2

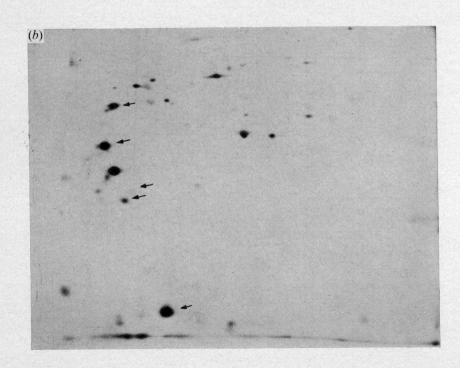

PLATE 3

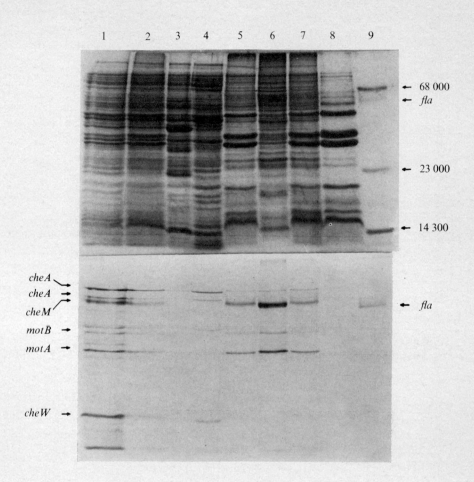

LEPTOSPIRAL MOTILITY

HOWARD C. BERG*, DAVID B. BROMLEY†
AND NYLES W. CHARON†

* Department of Molecular, Cellular and Developmental Biology,
University of Colorado, Boulder, Colorado 80309, USA

† Department of Microbiology,
West Virginia University Medical Center,
Morgantown, West Virginia 26506, USA

INTRODUCTION

The mechanism of motility and the hydrodynamics of motion of flagellated bacteria are now well understood. The helical flagella are driven at their base by a rotary motor (Berg & Anderson, 1973; Silverman & Simon, 1974; Berg, 1975a). The thrust generated by rotation of the flagellar filaments is balanced by viscous drag due to translation of the body of the cell, while the torque generated by rotation of the flagellar filaments is balanced by viscous drag due to rotation of the body of the cell (Chwang & Wu, 1971). When a microorganism swims at a constant velocity, the net force on it due to viscous shear in the surrounding medium is zero. Only viscous forces matter; inertial forces are entirely negligible (Taylor, 1952).

In spirochetes, some other thrust-generating mechanism must be at work. Although the axial filaments are structurally and chemically similar to bacterial flagella, they do not extend out into the external medium; they are held in juxtaposition to the cell body by an external sheath (for reviews and symposia, see Breznak, 1973; Smibert, 1973; Johnson, 1976; Canale-Parola, 1977).

In theory, a spirochete could swim by propagating a helical wave if it were able to spin about its local body axis (Wang & Jahn, 1972; Chwang, Winet & Wu, 1974). The torque generated by the spin would balance that due to the propagation of the wave, just as the torque generated by the rotation of the body of the flagellated bacterium balances that due to the motion of its flagella. This model would require that the protoplasmic cylinder be equipped with some sort of contractile apparatus. No such apparatus has been found, as yet, in any spirochete.

It is more reasonable to suggest that the motion is produced by axial filaments rotating in a manner similar to the flagella of other bacteria.

Models of this kind have been proposed (Jarosch, 1967; Berg, 1976). If the external sheath is loosely fitting, rotation of the axial filaments may cause the sheath to slip relative to the protoplasmic cylinder, like the tread on a tank (Berg, 1976). If so, the cell will rotate about its helical axis in the opposite direction and, in effect, screw its way through the medium. The torque required to rotate the cell about its helical axis is provided by circumferential viscous shear due to the roll of the sheath. This model runs into serious difficulty if the sheath has specific attachments to the protoplasmic cylinder (e.g. Hollande & Gharagozlou, 1967). It also fails to explain many of the movements of *Leptospira*.

In recent studies of non-motile mutants of *Leptospira interrogans* serotype *illini*, axial filaments from cells that had lost their hooked ends were found to be approximately straight, rather than coiled, as in the wild type (Bromley & Charon, 1977). This not only provides genetic evidence that the axial filaments are directly involved in the motility of *Leptospira*, it also suggests that they are relatively stiff in comparison to the protoplasmic cylinder, and that their shape is crucial for motility. Starting from these observations, we show how movements typical of *Leptospira* may be generated by rotation of the axial filaments. Elements of the model are contained in a theory proposed by Jarosch (1967); however, he was not aware of the existence of the external sheath in *Leptospira* or of the requirement that the axial filaments be coiled and stiff.

MORPHOLOGY

Shapes commonly assumed by *Leptospira* in aqueous suspension are shown in Plate 1(*a*) The double hook configuration is common at rest, the single hook configuration during translational movement (Noguchi, 1918; Jarosch, 1967; Cox & Twigg, 1974). The two axial filaments arise subpolarly at either end of the cell, as in other spirochetes, but extend only a short distance toward the middle (Plate 1*b*). The filaments neither meet nor overlap (Birch-Andersen, Hovind Hougen & Borg-Petersen, 1973). If a cell is treated extensively with deoxycholate, the axial filaments are freed from the confinements of the sheath and coil up, like the balance spring of a clock (Nauman, Holt & Cox, 1969; Birch-Andersen *et al.*, 1973). Apparently, such extensive coiling is a unique attribute of leptospiral axial filaments; it is not as marked in bacterial flagella or in axial filaments from treponemes (Birch-Andersen *et al.*, 1973).

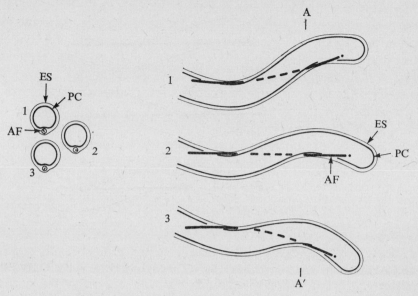

Fig. 1. Gyratory motion of one end of a cell generated by rotation of a curved axial filament (AF) when the external sheath (ES) and protoplasmic cylinder (PC) are held fixed near the middle of the cell. The cell is viewed in transverse section (left) and from the side (right). Three stages of one cycle are shown, 120° apart (1, 2, 3). The axial filament is rotating clockwise, as indicated by the tick marks. The plane of the section (AA′) is indicated in the side view. Side views 1 and 3 are displaced vertically for ease of viewing; in 1, the free end of the cell bends upwards and into the page; in 2, out of the page; in 3, downwards and into the page. The external sheath and protoplasmic cylinder flex; they do not rotate. The external sheath may adhere to the protoplasmic cylinder, but not in the vicinity of the axial filament. (Compare Jarosch, 1967, Fig. 92b.)

BASIC MOVEMENTS

We assume that the axial filaments are free to rotate within the confines of the external sheath, that is, that they do not adhere to the external sheath or the protoplasmic cylinder. Two basic movements are possible, depending upon the external constraints. If the external sheath and protoplasmic cylinder are held fixed near the middle of the cell, then each end of the cell will gyrate, as shown in Fig. 1. If, on the other hand, the axial filament is held fixed – we will see how this may occur later on – then the protoplasmic cylinder will roll about the filament, as shown in Fig. 2. In either case, provided that the axial filament is curved and relatively stiff, the protoplasmic cylinder must flex. In Fig. 1, it simply bends laterally (like the rubber tube in Taylor's model of the swimming spermatozoon; see Taylor, 1952, Fig. 3; or Berg, 1975b, p. 39). If the axial filament is helical, with a wavelength longer than that of the protoplasmic cylinder, then the protoplasmic cylinder will propagate a

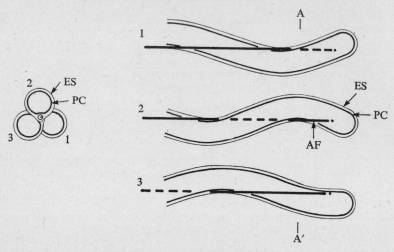

Fig. 2. Roll of the external sheath (ES) and protoplasmic cylinder (PC) about the axial filament (AF) when the axial filament is held fixed. The cell is viewed as in Fig. 1. The roll is
counterclockwise. (Compare Jarosch, 1967, Fig. 92a.)

long-wavelength helical wave. This is a situation in which a spiralled
filament produces secondary spiral waves.

We show in the sections that follow how these basic movements may
account for the manoeuvres typical of *Leptospira*. We assume that the
motors driving the axial filaments may turn either clockwise (CW) or
counter-clockwise (CCW), as is the case for flagellar motors in other
bacteria. In defining CW and CCW, we adopt the frame of reference of
an observer riding on the outer surface of the cell looking at the shaft
of the motor as it emerges from the protoplasmic cylinder. We refer to
a helix as right (or left) handed if an object moving along the helix
away from an observer spirals clockwise (or counter-clockwise).

NON-TRANSLATIONAL MOVEMENT

The motors at either end of the cell are opposed. If the axial filaments
both turn in the same direction (both CW or CCW), the ends of the
cell will gyrate in opposite directions. The basic movement is that of
Fig. 1. The torques due to viscous drag at the ends of the cell balance
when the centre of the cell is stationary. This motion has been described
in detail by a number of workers (Noguchi, 1918; Jarosch, 1967;
Cox & Twigg, 1974; Kaiser & Doetsch, 1975). It is shown schematically
in Fig. 3(*a*).

The shape of each axial filament will depend, in part, on its dynamic
load; this will depend, in turn, on its direction of rotation. Therefore,

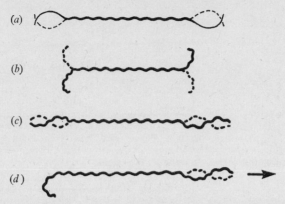

Fig. 3. Movement of *Leptospira* in free solution. (*a*) Rapid non-translational movement, according to Cox and Twigg (1974); the ends blur into the familiar 'button hole' figure. (*b*, *c*) Variations in non-translational movement expected if the shapes of the axial filaments depend on their direction of rotation. (*d*) Translational movement, according to Cox and Twigg (1974); the cell swims to the right (arrow), the anterior end propagating a helical wave toward the posterior end.

the shapes of cells in non-translational movement should vary, depending upon whether the axial filaments are both turning CW or CCW. Given the shapes observed for cells in translational movement (below) we predict hook–hook and spiral–spiral configurations (Figs. 3(*b*) and (*c*)). We have observed both of these configurations in serotype *illini*; both have been seen by others (e.g., Jarosch, 1967, Figs. 83 g, m).

Variations on this theme will occur whenever cells adhere to solid surfaces. If a cell sticks in the middle, both ends are free to gyrate, and they may do so in either direction. We have seen this with serotype *illini*; the ends of a cell may flex even when it adheres to the surface throughout most of its length; the centre remains stationary. In theory, if a cell sticks at one end, the entire body should gyrate. Rotation of the proximal axial filament will carry the cell through a wide arc; rotation of the distal filament may wave the distal end. The sheath and the protoplasmic cylinder simply flex.

TRANSLATIONAL MOVEMENT

If the axial filaments turn in opposite directions (one CW, the other CCW), both ends of the cell will gyrate in the same direction. This can occur in free solution only if the external sheath and protoplasmic cylinder roll the *other* way. The basic movements are those of Fig. 1 and Fig. 2. The torques due to the gyrations of the ends of the cell are balanced by the torque due to the roll of the cylinder. This is shown schematically in Fig. 4.

Fig. 4. Schematic representation of one end of a cell undergoing translational movement. The torque due to the gyration of the body about its long axis (a) is balanced by the torque due to the roll of the sheath and protoplasmic cylinder about the filament (b). The gyration is generated by the rotation of the filament, as shown in Fig. 1. The roll is illustrated in Fig. 2. The gyration and roll are in opposite directions.

The leading end of the cell is spiral-shaped, the trailing end hooked-shaped, Fig. 3(d) (Noguchi, 1918; Jarosch, 1967; Cox & Twigg, 1974). Thrust may be generated in two ways: by backwards propagation of the anterior long-wavelength helical wave and by roll of the short-wavelength helical cylinder. The locomotion will be most efficient if the two mechanisms are synergetic, that is, if the spiral filament is right handed, its motor turns CW, and the body helix is left handed *or* if the spiral filament is left handed, its motor turns CCW, and the body helix is right handed. If both motors change their direction of rotation, the cell swims in the same manner but in the opposite direction (as if reflected in a mirror normal to its long axis).

The trailing end of a swimming leptospire can move in a variety of ways. The simplest case occurs when the motor driving the hooked filament jams. Then, as the cylinder rolls, it will carry the filament with it. The hook acts as a rotational anchor; it moves with the protoplasmic cylinder, i.e. in a direction opposite to that of the spiral filament. This behaviour has been observed by Cox & Twigg (1974). Alternatively, if the motor driving the hooked filament does not jam, the hook will rotate in the same direction as the spiral filament. In this case, the sheath and protoplasmic cylinder roll the other way, as shown in Fig. 4. Finally, if the motor driving the hooked filament and the protoplasmic cylinder turn at the same rate, the hook will remain stationary; the trailing end of the cell will not gyrate. In this case, the motor driving the hooked filament will not generate any torque. (Note, however, that torque will be generated when the cell moves along a solid surface, i.e. when lateral movement of the hook is blocked.)

In a medium of low viscosity, most of the thrust is generated by backward propagation of the anterior spiral wave. Cox & Twigg (1974) described a leptospire moving with a large piece of colloidal graphite fixed to the tip of the trailing end. The graphite did not rotate. This showed, as they argued, that the thrust was generated at the anterior end. But they concluded, more generally, that the hook plays an essential role of preventing rotation of the whole body. This is not the case. Leptospires swim perfectly well without a trailing hook. We have seen cells of serotype *illini* moving in a straight-spiral configuration, spiral end first. These cells presumably have only one axial filament (or only one curved axial filament). The spiral wave motion and the roll of the protoplasmic cylinder are both quite vigorous. Such a cell may switch abruptly to a straight–hook configuration. The hook gyrates and the cylinder rolls, but now the cell remains stationary. This shows that the motor driving the axial filament continues to run, even when the filament is hooked shaped, but that the gyration of the hook and the counter-roll of the protoplasmic cylinder generate little thrust. When the motor reverses and the hook reverts to a spiral, the cell swims away. The cell advances and pauses but does not retreat. It advances only when propagating the anterior spiral wave.

These schemes fail if the filaments are straight or if they flex more readily than the protoplasmic cylinder. At least one end of the cell must gyrate. Note, however, that this is not the case if the external sheath slips about the protoplasmic cylinder; viscous shear due to the slip will cause the protoplasmic cylinder to roll rigidly, without any gyrations or moving secondary waves (Berg, 1976). A number of spirochetes appear to swim in this fashion, e.g., *Cristispira* (Chwang *et al.*, 1974, as cited by Berg, 1976).

MOVEMENT IN SEMI-SOLID MEDIUM

Spirochetes have the remarkable ability of boring their way through a semi-solid medium in a corkscrew-like manner. In *Leptospira*, the body may be bent in a serpentine fashion, while the short-wavelength helix moves with little or no slip (Noguchi, 1918; Jarosch, 1967; Cox & Twigg, 1974). Serotype *illini* swims in this manner in 1% methyl cellulose (viscosity 1 P). The spiral appears as before, but the hook is more nearly aligned with the cell axis (as in Fig. 3(*a*) rather than 3(*b*)). The rates of gyration and roll of the cylinder are reduced, whereas the translational velocity is markedly enhanced (see also Kaiser & Doetsch, 1975).

In a semi-solid medium, most of the thrust is generated by the roll of the protoplasmic cylinder. Cells without a trailing hook swim well in the straight–spiral configuration, as before, spiral end first. But now, when they switch to the straight–hook configuration, they swim equally well straight-end first. Either way, one end of the cell gyrates while the protoplasmic cylinder screws its way through the medium. Since the cell swims about as well in either direction, i.e. whether the gyrating end is in the hook or spiral configuration, most of the thrust must be generated by the roll of the cylinder. Since the cell swims spiral end first, as it does in a medium of low viscosity, the wave-propagation and the roll are, indeed, synergetic.

A helical-shaped object cannot develop thrust in a true liquid without circumferential slip (Chwang *et al.*, 1974). Thus, methyl cellulose must be gel-like (semi-solid). Given an applied torque, a microscopic helix immersed in such a medium evidently threads its way through the interstices in the gel. This kind of motion is not unique to spirochetes. *Spirillum volutans* moves this way in 5% gelatin (Metzner, 1920). Only the torque generating mechanisms differ. *Leptospira* uses the mechanism of Fig. 2. *Spirillum* rotates its external flagellar filaments. The mechanism of Fig. 2 should be particularly efficient if, as one might expect in a semi-solid medium, the lateral motion of the end of the cell is hindered more than the roll of the protoplasmic cylinder. (Jarosch's interpretation of motion in a semi-solid medium is rather different; his spirochetes are propelled by fluid pumped along the axes of the filaments; see Jarosch, 1967, Fig. 91*e*.)

Non-translational movement also occurs in a semi-solid medium; it is essentially the same as that in a non-viscous medium. The ends of the cell gyrate in the hook–hook or spiral–spiral configuration. The centre of the cell remains stationary.

CONCLUSIONS

We have shown how movements typical of *Leptospira* may be generated by rotation of the axial filaments. Differences between the mechanism of locomotion of *Leptospira* and that of other spirochetes appear as a natural consequence of differences in morphology. In *Leptospira*, the axial filaments are relatively stiff compared to the protoplasmic cylinder, while the external sheath is more adherent. The protoplasmic cylinder of *Leptospira* is comparatively thin; its axial filaments are comparatively short. A striking variety of movements can be engendered by subtle changes in internal or external constraints.

It should be noted that most of our observations have been made with serotype *illini*, a leptospire in a markedly different genetic group than most other known leptospires (Brendle, Rogul & Alexander, 1974). However, we believe that the model will apply to the hooked motile leptospires of the various genetic groups.

Other motility mutants would be helpful in testing the model. Mutants of rod-shaped bacteria have been found in which the flagellar motors only run CW or CCW (Larsen *et al.*, 1974). Similar mutants should exist in *Leptospira*. We should be able to find cells locked into the hook–hook, the spiral–spiral, and possibly even the hook–spiral configurations. Mutants should exist with only one axial filament or with axial filaments of various shapes. In the wild type, the handedness of the anterior spiral filament should be opposite to that of the protoplasmic cylinder. Mutants might be found in which this is not the case.

Antibody-coated latex beads also would be helpful in testing the model. During translational movement, beads attached anywhere on the surface of the cell should rotate about the axis of the protoplasmic cylinder in a direction opposite to that of the anterior spiral. During non-translational movement, they should not rotate. They should rotate when the cell moves along a solid surface, even when the hook appears stationary, but not if the tip of the cell is anchored, as in the experiment of Cox and Twigg (1974) with colloidal graphite.

It may be possible to remove the external sheath and to tether an axial filament. If so, the cell body should spin, now CW, now CCW, as in other bacteria (Silverman & Simon, 1974).

The mechanism of motility and the hydrodynamics of motion of *Leptospira* do not appear to differ fundamentally from those in other bacteria. The flagella of *Leptospira* rotate intracellularly rather than extracellularly. We wonder how this variation might have come about.

This work was supported by grants from the US National Science Foundation (BMS75-05848 to H. C. B.) and the West Virginia University Medical Corporation (2-210-1615-103 to N. W. C.). We thank Oleg Carleton for stimulating discussions and Dr Herbert Voelz for assistance with the electron microscope.
Received 29 June 1977.

REFERENCES

Berg, H. C. (1975a). Bacterial behaviour. *Nature, London,* **254**, 389–92.

Berg, H. C. (1975b). How bacteria swim. *Scientific American,* **233** (2), 36–44.

Berg, H. C. (1976). How spirochetes may swim. *Journal of Theoretical Biology,* **56**, 269–73.

Berg, H. C. & Anderson, R. A. (1973). Bacteria swim by rotating their flagellar filaments. *Nature, London,* **245**, 380–2.

Birch-Andersen, A., Hovind Hougen, K. & Borg-Petersen, C. (1973). Electron microscopy of *Leptospira* 1. Leptospira *Strain Pomona. Acta Pathologica et Microbiologica Scandinavica Section B,* **81**, 665–76.

Brendle, J. J., Rogul, M. & Alexander, A. D. (1974). Deoxyribonucleic acid hybridization among selected leptospiral serotypes. *International Journal of Systematic Bacteriology,* **24**, 205–14.

BREZNAK, J. A. (1973). Biology of nonpathogenic, host-associated spirochetes. *CRC Critical Reviews of Microbiology*, **2**, 457–89.

BROMLEY, D. B. & CHARON, N. W. (1977). The role of the axial filament in the motility and morphology of *Leptospira*. *Abstracts of the Annual Meeting of the American Society of Microbiology*. J4.

CANALE-PAROLA, E. (1977). Physiology and evolution of spirochetes. *Bacteriological Reviews*, **41**, 181–204.

CHWANG, A. T., WINET, H. & WU, T. Y. (1974). A theoretical mechanism of spirochete locomotion. *Journal of Mechanochemistry and Cell Motility*, **3**, 69–76.

CHWANG, A. T. & WU, T. Y. (1971). A note on the helical movement of microorganisms. *Proceedings of the Royal Society of London B*, **178**, 327–46.

COX, P. J. & TWIGG, G. I. (1974). Leptospiral motility. *Nature, London*, **250**, 260–1.

HOLLANDE, M. A. & GHARAGOZLOU, I. (1967). Morphologie infrastructurale de *Pillotina calotermitides* nov. gen., nov. sp., Spirochaetale de l'intestin de *Calotermes praecox*. *Comptes rendus de l'Académie des Sciences, Paris*, **265**, Ser. D, 1309–12.

JAROSCH, R. (1967). Studien zur Bewegungsmechanik der Bakterien und Spirochäten des Hochmoores. *Österreichische botanische Zeitschrift*, **114**, 255–306.

JOHNSON, R. C., ed. (1976). *The Biology of Parasitic Spirochetes*. New York: Academic Press.

KAISER, G. E. & DOETSCH, R. N. (1975). Enhanced translational motion of *Leptospira* in viscous environments. *Nature, London*, **255**, 656–7.

LARSEN, S. H., READER, R. W., KORT, E. N., TSO, W.-W. & ADLER, J. (1974). Change in direction of flagellar rotation is the basis of the chemotactic response in *Escherichia coli*. *Nature, London*, **249**, 74–7.

METZNER, P. (1920). Die Bewegung und Reizbeantwortung der bipolar begeisselten Spirillen. *Jahrbuch für wissenschaftliche Botanik*, **59**, 325–412.

NAUMAN, R. K., HOLT, S. C. & COX, C. D. (1969). Purification, ultrastructure, and composition of axial filaments from *Leptospira*. *Journal of Bacteriology*, **98**, 264–79.

NOGUCHI, H. (1918). Morphological characteristics and nomenclature of *Leptospira* (*Spirochaeta*) *icterohaemorrhagiae* (Inada and Ido). *Journal of Experimental Medicine*, **27**, 575–92.

SILVERMAN, M. & SIMON, M. (1974). Flagellar rotation and the mechanism of bacterial motility. *Nature, London*, **249**, 73–4.

SMIBERT, R. M. (1973). Spirochaetales, a review. *CRC Critical Reviews of Microbiology*, **2**, 491–552.

TAYLOR, G. (1952). The action of waving cylindrical tails in propelling microscopic organisms. *Proceedings of the Royal Society of London A*, **211**, 225–39.

WANG, C.-Y. & JAHN, T. L. (1972). A theory for the locomotion of spirochetes. *Journal of Theoretical Biology*, **36**, 53–60.

EXPLANATION OF PLATE

PLATE 1

(a) Serotype *illini* cells fixed in 1% glutaraldehyde, suspended in 1% methyl cellulose, and photographed under the oil immersion lens of a Leitz darkfield microscope. (b) Serotype *illini* cells treated for 30 min with 2% deoxycholate, stained with 1% phosphotungstic acid, and photographed in the electron microscope. The structures seen are the protoplasmic cylinder (PC), the axial filament (AF), and the external sheath (ES).

PLATE 1

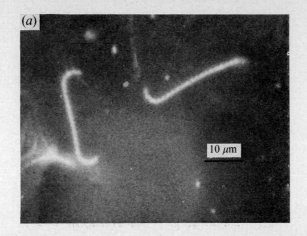

(a)

10 μm

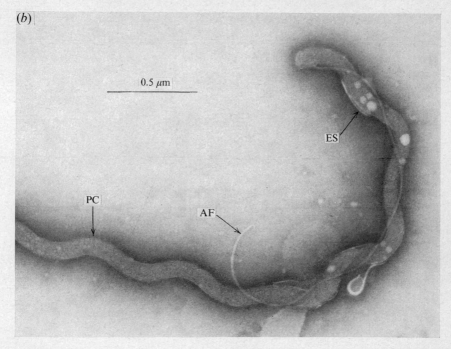

(b)

0.5 μm

ES

PC

AF

SPORE SPECIFIC STRUCTURES AND THEIR FUNCTION

D. J. ELLAR

*Department of Biochemistry, University of Cambridge,
Tennis Court Road, Cambridge CB2 1QW, UK*

INTRODUCTION

'Successful architecture is more than just a proper relationship between structure and function (Woods, 1965).' This was an aphorism coined thirteen years ago in the fifteenth symposium in this series, by the late D. D. Woods as he lamented the austere tone which he felt was then creeping into scientific publications. In encouraging scientists to allow more of the intrinsic beauty of microbes to show through in their writing, he enquired how many microbiologists derived aesthetic enjoyment from observing their organisms in the phase contrast microscope. Most microbiologists undoubtedly remain as fascinated as ever today by the successful architecture of prokaryotes, although the instruments for observing this architecture are likely to be less immediately visually rewarding than the microscope. Today the aesthetic enjoyment is frequently to be found in probing microbial architecture at the sub-microscopic level and discerning functional relationships between the different cellular structures. This survey will review aspects of this approach as it has developed in the study of bacterial endospores.

GENERAL FEATURES OF SPORULATION

Outside the requirements of a brief description of the overall sporulation sequence, no attempt will be made to deal comprehensively with all known spore specific structures. Unfortunately, for a number of these structures such as spore coat proteins, we do not yet have a sufficiently detailed chemical picture to be able to relate structure to spore function. The emphasis in this survey will therefore be on those spore specific structures such as cortex and membranes where our knowledge of macromolecular architecture enables us at least to predict and test likely functions. One structure which will be included in this approach is the developing spore (forespore) itself. With the recent discovery of methods for isolating the forespore from the sporulating cell (sporangium) at all stages of morphogenesis (Andreoli *et al.*, 1973; Ellar &

Posgate, 1974), it has become possible to ask direct questions about the division of labour between the two compartments of the sporangium. In attempting to ascribe characteristic spore properties such as octanol resistance, germinability, heat resistance, etc. to one or more spore structures, it is invaluable to be able to look for these properties in forespores, in which these structures are missing or incomplete. As will be seen later, this approach has enabled us to detect the presence of a typical spore germination response in forespores isolated before coat synthesis, calcium accumulation or dipicolinic acid (DPA) synthesis are complete and before the spore acquires heat resistance. Additional spore specific components such as DPA, low molecular weight acetic acid soluble proteins and spore water will be discussed, because of suggestions that these may be involved in creating unique spore structures or alternatively interacting with spore structures, to establish the typical dormancy and resistance of bacterial spores.

While this survey will concentrate largely on spore formation in bacilli, it would be inappropriate, in a symposium of this society, to overlook the fact that endospores with similar characteristics are formed by a variety of other micro-organisms, including clostridia, sarcinae and thermophilic actinomycetes. Fortunately Slepecky (1972) has provided an excellent review of the ecology of bacterial spore formers which includes a survey of their distribution among different genera.

When an essential metabolite in the growth environment of spore forming bacteria is depleted below a certain level, these organisms respond by beginning the ordered sequence of biochemical and morphological events which characterise sporulation. Vegetative growth and division cease and an asymmetric spore specific cell division occurs to produce a discrete new cell (forespore) within the cytoplasm of the original organism (mother cell). This forespore is unique among micro-organisms in being surrounded by two membranes which show opposite surface polarity relative to each other (Wilkinson, Deans & Ellar, 1975). It will be seen later that this spore specific arrangement of membranes probably has an important influence on the mechanisms and control of spore formation.

The mature spore is undoubtedly well equipped for its cryptobiotic role. Not only are bacterial endospores among the most resistant life forms, with remarkable tolerance of heat, radiation and bactericidal agents, but they also exhibit no detectable metabolism. They are therefore capable of extreme dormancy and can remain in this state for considerable periods. In this dormant state what might be termed the sensory apparatus of the spore remains alert however. This apparatus enables the spore to respond within minutes to germinants such as L-alanine and reinitiate the machinery for vegetative growth. Since all these typical spore properties of resistance, extreme dormancy and germinability must have their origin in the biochemistry and molecular architecture of the spore, there is ample justification for examining spore specific structures in detail. Moreover the bacterial spore may be a valuable model from which to learn more of the creation and maintenance of dormancy in the wider range of dormant forms existing in nature. These include the spores of yeasts and fungi, protozoal and nematode cysts, plant seeds, dormant buds and tubers, insects in diapause and hibernating mammals. The mature unfertilised eggs of animals and amphibia also share features in common with dormant cells (Keynan, 1968). Sub-populations of dormant cells exist among some animal cell types such as lymphocytes and bone marrow stem cells. Small but significant fractions of leukaemic and other neoplastic cells have been found to remain dormant in the host for long periods. This has important clinical implications since dormant non-proliferating cells are resistant to permissible doses of cytotoxic drugs (Clarkson, 1974). Sporulation and germination are, however, of even more general interest, since they constitute well-defined systems of differentiation and dedifferentiation respectively. The time ordered sequence of biochemical and morphological changes which characterises these systems, coupled with the relative simplicity and genetic manipulability of micro-organisms, offer excellent opportunities to investigate the regulation of selective gene expression. The following reviews and reports will be found to deal in detail with the various aspects of sporulation and germination referred to above and others which are relevant to the later discussion: cytology of spores and sporulation, Young & Fitz-James (1959a, b), Holt, Gauthier & Tipper (1975), Ellar & Lundgren (1966), Ellar, Lundgren & Slepecky (1967), Aronson & Fitz-James (1976), Walker, (1970), Holt & Leadbetter (1969), Cross, Davies & Walker (1971); spore specific structures and their chemistry, Tipper & Gauthier (1972); initiation and control of sporulation, Szulmajster (1973), Doi & Sanchez-Anzaldo (1976), Mandelstam (1976), Freese (1977), Hanson (1975); genetics of sporulation, Piggot & Coote (1976); spore dormancy and germination, Gould (1969, 1970), Lewis (1969); spore water and physical properties, Marshall & Murrell (1970).

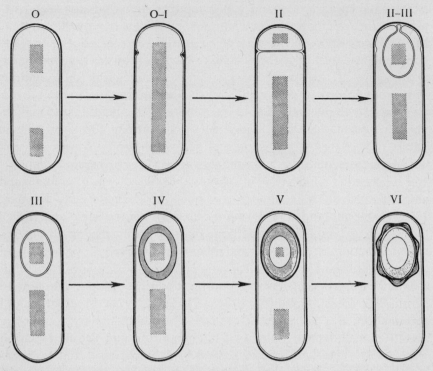

Fig. 1. Generalised diagram of the major morphological events occurring during sporulation. Individual stages are indicated by Roman numerals.

OUTLINE OF MORPHOLOGICAL CHANGES DURING SPORULATION

In general terms, the major morphological changes during sporulation are similar in all the endospore forming bacteria. On the basis of electron microscopic observations it is possible to divide this morphogenesis into a number of stages (Ryter, Ionesco & Schaeffer, 1961; Schaeffer, Ionesco, Ryter & Balassa, 1963; Ryter, 1965; Fitz-James, 1965). These are illustrated diagrammatically in Figs. 1 and 2. When the growth of vegetative cells (Stage O) becomes limited by the depletion of an essential metabolite, sporulation is initiated. The next stage of sporulation (Stage I) has been designated as the period during which the bacterial chromosome changes conformation from the compact vegetative form to an axially disposed filamentous structure. If this change is sporulation specific, the altered DNA conformation would probably represent the earliest spore specific structure. However, on the basis of work with mutants (Mandelstam, Kay & Hranueli, 1975) and microscopic analysis (Greene, Holt, Leadbetter & Slepecky, 1971) there

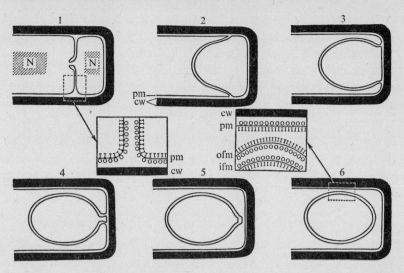

Fig. 2. Formation of the bacterial forespore during States II–III. Diagram designed to show the polarity or 'sidedness' of the membranes involved. N, nuclear material; pm, plasma membrane; cw, cell wall; ofm, outer forespore membrane; ifm, inner forespore membrane. In the two insets in the middle of the diagram the symbols O and I represent the 'sidedness' of the membranes; O represents the outer surface of the plasma membrane, I the inner surface.

is some doubt as to whether this structure is the result of the expression of spore specific genes. This doubt illustrates the common problem of distinguishing between events during sporulation which reflect the general response of vegetative bacteria to limiting growth conditions and those events which are unique to spore formation. This difficulty is most acute in consideration of the biochemical events in sporulation, such as variation in solute and ion pools, enzyme levels and nucleic acid synthesis. Mandelstam (1976) has recently examined this problem and defined three categories of events associated with sporulation. The first category comprises all events which are essential to the development process. The second category contains events which are triggered by first category events, but are not part of the primary sequence, and the third category contains those general events which occur when vegetative bacteria are exposed to the limiting growth conditions necessary to induce sporulation. Since this survey is primarily concerned with the unique structures which develop during sporulation, these problems of classification are minimised. However as mentioned above, the Stage I axial DNA structure may be an exception to this, since it could result from the changes in intracellular solute and ion levels that might be expected during the shift down conditions at the end of Stage O. In Stage II of sporulation the cell is partitioned by a double membrane septum asymmetrically disposed towards one cell pole (Figs. 1 and 2). At the same time an accompanying DNA division ensures that each compartment contains at least one complete chromosome. Although this septum probably represents a modified cell division (Hitchins & Slepecky, 1969) it is clearly a spore specific event. In most organisms the sporulation septum shows no sign of the peptidoglycan layer between the invaginating membranes which is characteristic of a vegetative transverse septum. In *Bacillus sphaericus* however, material resembling peptidoglycan can be seen at this time (Holt *et al.*, 1975). An association of mesosomes with this asymmetric division has been well documented, but any conclusions from these observations must be evaluated in the light of the recent report that the appearance of the structures is an artefact generated by fixation or any other manipulation which interrupts normal growth of the organism (Higgins, Tsien & Daneo-Moore, 1976).

The smaller compartment formed by the Stage II division is gradually engulfed within the larger compartment (mother cell) by what appears to be a continued proliferation of the septal membranes and a translocation of the base of the septum toward the cell pole (Fig. 2). After engulfment is complete, the new intracellular compartment (forespore) exists within the mother cell as a discrete cell bounded by two membranes (Stage III). These are designated as the inner and outer forespore membranes (Fig. 2). The former will later emerge as the cytoplasmic membrane of the germinating spore.

During Stage IV, peptidoglycan is laid down between the inner and outer forespore membranes. There is evidence from some organisms, notably *B. sphaericus* (Holt *et al.*, 1975), that this occurs in two phases. The first phase involves deposition of vegetative cell-type polymer (primordial cell wall) which will become the cell wall of the germinating spore. In the second phase the spore specific peptidoglycan (cortex) is deposited as an electron transparent layer external to the primordial cell wall. Also during Stage IV the forespore begins to accumulate calcium and DPA and acquires a phase grey appearance.

Although Stage V is generally associated with synthesis of the bulk of the proteinaceous spore coats external to the outer forespore membrane, there is now biochemical and cytological evidence that deposition of at least part of the coat layers occurs during Stages III and IV. During Stage V calcium and DPA accumulation continue, the forespore bebecomes phase white and acquires octanol resistance and the capacity for germination (see later).

In Stage VI, the forespore acquires typical heat resistance and the

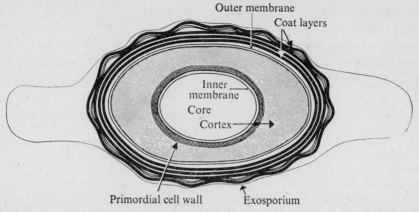

Fig. 3. Composite diagram drawn from a variety of *Bacillus* spp. to show the characteristic morphological features found in dormant spores.

synthesis of all visible spore specific structures appears complete. The mother cell loses its phase dark appearance and the forespore is maximally phase white. Stage VII denotes the period during which the mother cell lyses releasing the mature spore.

In those organisms which possess a loosely fitting exosporium (Fig. 3) or a parasporal crystal, these structures begin to appear during Stage III or even during Stage II in case of the latter (Aronson & Fitz-James, 1976). It is important to note that this division of sporulation into stages is often one of convenience since the synthesis and/or modification of most spore specific structures probably occurs over several stages. It is valuable however in locating the earliest time of appearance of a particular structure or a spore property, e.g. germinability. Examination of a number of sporulation events such as spore dehydration (Leman, 1973), variation in protein synthesis and turnover rates (Eaton & Ellar, 1974) and changes in oxidative capacity (Wilkinson & Ellar, 1975) reveal that these events are occurring gradually and continuously throughout morphogenesis. Other events however such as cytochrome *c* synthesis (Wilkinson & Ellar, 1975) and the appearance of germinability and octanol and heat resistance in sporulating *B. megaterium*, can be assigned to a distinct stage (Ellar *et al.*, 1975).

SPORULATION SPECIFIC MEMBRANE STRUCTURE AND FUNCTION

Reversal of membrane polarity

The first sporulation specific structures to be considered in detail are the two membranes which surround the forespore at Stage III. Although the presence of the double membrane around the developing spore is in itself unique, the orientation of the membrane surfaces relative to each other is even more important in considering the biosyntheses and transport processes that are essential to spore morphogenesis. From the nature of the membrane invagination in Stage II which gives rise to the forespore membranes, it appears that the outer forespore membrane has reversed surface polarity relative to both the inner forespore membrane and the mother-cell plasma membrane (Fig. 2). Thus, the inner forespore membrane surface that originated as the cytoplasmic surface of the mother-cell plasma membrane remains opposed to the spore cytoplasm. Figure 2 shows that by the nature of the engulfment process the outer surface of the forespore outer membrane also arises from what was previously the cytoplasmic surface of the mother-cell membrane. It follows that in the double membrane of the completed forespore, what were previously the outer surfaces of the mother-cell membrane now face each other. If each of the forespore membranes retains the normal mechanisms for active transport, they would be expected to operate in opposite directions (Freese, 1972). If this is indeed the situation *in vivo*, it is probable that transport from mother-cell cytoplasm into the spore could only occur by passive or facilitated diffusion (Ellar, 1978*a*). In the case of the sporulation specific calcium accumulation, it will be seen later that this prediction has now been confirmed (Hogarth, Deans & Ellar, 1977*a*).

Biochemical confirmation of the reversed polarity of forespore membranes has been obtained by exploiting the impermeability of the intact bacterial membrane to such hydrophilic substrates as ATP and NADH (Ellar, *et al.*, 1975; Wilkinson, *et al.*, 1975). With the development of a method for isolating *B. megaterium* forespores (Ellar & Posgate, 1974) it was possible to use this experimental approach to compare polarity in intact forespores and sporangia. The results showed that adenosine triphosphatase, NADH dehydrogenase and L-malate dehydrogenase activities were undetectable in assays with intact sporangia, as would be expected from the location of these enzymes on the cytoplasmic surface of the mother-cell plasma membrane. In contrast, these enzymes were readily assayed in intact forespores, consistent with a reversed polarity of the outer forespore membrane.

With this knowledge of membrane polarity, we can now take up the question of transport between mother cell and forespore. Information on the permeability properties of forespores is limited but Eaton & Ellar (1974) found that though spore specific protein synthesis continues from State III to VI in the forespore, protein turnover is confined to the mother-cell compartment. This suggests that amino acids from the mother cell can be accumulated by the forespore. In the case of certain amino acids such as methionine and arginine, the observed lack of the appropriate biosynthetic enzyme in the mature spore (Setlow & Primus, 1975) and forespore (Singh, Setlow & Setlow, 1977) indicates that such transport may be essential. Exogenous uracil and DPA have also been reported to be taken up by sporulating cells and incorporated into the forespore (Ryter, Bloom & Aubert, 1966; Halvorson & Swanson, 1969). Sporulation specific Ca^{2+} uptake is the most obvious instance where forespore accumulation must occur. Throughout vegetative growth and prior to Stage III of sporulation, the intracellular Ca^{2+} levels are extremely low. At the end of this Stage there is a rapid increase in Ca^{2+} accumulation, which results in the uptake of almost all the Ca^{2+} contained in the sporulation medium during the remaining sporulation stages (Hogarth et al., 1977a). The accumulated Ca^{2+} continually moves into the forespore compartment (La Nauze, Ellar, Denton & Posgate, 1974) so that in the mature spore it accounts for 2% on average of spore dry weight and exceeds the mole sum of all the other inorganic spore cations (Murrell, 1969). Like other non sporing micro-organisms that have been studied, vegetative bacilli are characterised by an energy-dependent Ca^{2+} efflux mechanism (Silver, Toth & Scribner, 1975; Bronner, Nash & Golub, 1975). Consequently the shift to Ca^{2+} influx during sporulation may require either the synthesis during Stage III of new membrane proteins to catalyse this uptake, or modification of the vegetative efflux mechanism to allow it to reverse direction. Alternatively the spore specific uptake may proceed by facilitated diffusion rather than active transport. Whichever mechanism is operating, it is important in probing the division of labour between forespore and mother cell to determine its location. We have already seen that the reversed polarity of forespore membranes appears to preclude any active transport and since calcium is known to enter the forespore, the mechanism of this uptake is of added interest.

Sporulation specific calcium transport

At approximately the same stage in sporulation that Ca^{2+} uptake begins, the cell begins to synthesise DPA (pyridine-2,6-dicarboxylic acid). This compound is unique to sporulation and its synthesis continues in parallel with Ca^{2+} accumulation. In the spores of most species, the amount of DPA is such as to give an approximately 1:1 molar ratio with Ca^{2+} (Murrell, 1969). Several studies have indicated that DPA occurs as a chelate in the spore. It is almost invariably extracted from spores as the calcium–DPA chelate and occasionally chelated with other divalent metals. When a wide range of DPA analogues were tested, only 4H-pyran-2,6-dicarboxylic acid was able to replace DPA in restoring sporulation to a *B. megaterium* mutant incapable of DPA synthesis (Fukuda, Gilvarg & Lewis, 1969). This observed association of Ca^{2+} and DPA has stimulated speculation about a role for the latter in Ca^{2+} transport (Eisenstadt & Silver, 1972; La Nauze *et al.*, 1974).

With the availability of a method of forespore isolation (Ellar & Posgate, 1974) the question of Ca^{2+} uptake could be tackled directly. La Nauze *et al.* (1974) first showed that *B. megaterium* forespores isolated at Stages IV, V and VI could not continue Ca^{2+} accumulation after removal from the sporangia. In these in-vitro forespore incubations, the Ca^{2+} concentration (0.1 mM) was identical to that in the normal sporulation medium. This result suggested that some structural and/or functional component in the mother cell was required to support forespore Ca^{2+} uptake. Attempts to satisfy this requirement by the addition of oxidisable substrates, ATP or DPA, to isolated forespores were unsuccessful (Ellar *et al.*, 1975). When measurements were made of the Ca^{2+} concentration in mother-cell cytoplasm during sporulation (La Nauze *et al.*, 1974; Hogarth *et al.*, 1977a) it was found to be in the range 3–9 mM. It was therefore essential to re-examine the Ca^{2+} uptake capacity of isolated forespores at considerably higher concentrations than the 0.1 mM originally employed. In the presence of 7.5 mM external Ca^{2+}, isolated forespores were found to accumulate it at rates equivalent to those observed in the intact sporulating cell (Hogarth *et al.*, 1977a). The K_m for this uptake was found to be 2.1×10^{-3} M. The finding that the rate and extent of calcium accumulation by forespores *in vitro* was dependent on the external Ca^{2+} concentration, coupled with the fact that the concentration of this cation in mother-cell cytoplasm was between 3 and 9 mM, suggested that uptake of Ca^{2+} by forespores occurred by a carrier-mediated transfer down a concentration gradient. In contrast, as was indicated above, intact sporangia normally accummulated Ca^{2+} during sporulation from a medium containing 0.1 mM of the cation. Kinetic analysis of sporangial uptake gave a K_m of 3.1×10^{-5} M (Ellar *et al.*, 1975; Hogarth *et al.*, 1977a). From this comparatively low K_m value and the inhibitor sensitivity, saturability and calcium specificity of the sporangial uptake system, it was apparent that a carrier-mediated active transport of Ca^{2+} occurred at some site in the sporulating cell. The objective now was to identify on which cell membrane(s) this site was located.

This was achieved by comparing the effects of a range of inhibitors on Ca^{2+} uptake and O_2 consumption by intact sporangia and forespores

(Hogarth *et al.*, 1977*a*; C. Hogarth & D. J. Ellar, in preparation). KCN at 1 mM completely inhibited uptake of the cation by sporangia and severely reduced their O_2 uptake. Antimycin A (40 μM) was equally inhibitory for Ca^{2+} uptake. These inhibitors had a significantly different effect on forespores however. Both Ca^{2+} uptake and O_2 consumption of isolated forespores were unaffected by 1 mM KCN. Even the presence of 20 mM KCN failed to inhibit Ca^{2+} accumulation by forespores and 40 μM Antimycin A had no effect on either O_2 consumption or Ca^{2+} uptake. From these results it was clear that transport from the sporulation medium into the intact sporangium could be completely inhibited by concentrations of inhibitors which have no effect on Ca^{2+} uptake or respiration of isolated forespores. These findings identify the mother-cell plasma membrane as the site for the energy-dependent component of Ca^{2+} transport. Furthermore, the observation that uptake of the cation by isolated forespores is unchanged, when forespore respiration is severely depressed by the addition of 100 μM 2-heptyl-4-hydroxyquinoline *N*-oxide (HOQNO) plus 10 mM KCN, is further evidence that uptake into the forespore compartment is not an active transport process. These data reveal a co-ordination and division of labour between sporangial structures to achieve Ca^{2+} uptake. The ion is actively transported from the medium into the mother-cell cytoplasm and subseqently moves into the forespore by a facilitated diffusion mechanism. The measured K_m for this forespore uptake (2.1×10^{-3} M) is compatible with the figure of 3–9 mM for the calcium concentration in the surrounding mother-cell cytoplasm.

So far no involvement of DPA in Ca^{2+} transport has been observed. In considering a possible role for DPA in the facilitated diffusion of calcium across the double forespore membrane, two observations are important: firstly, addition of exogenous DPA had no significant effect on the rate or extent of calcium accumulation by either isolated forespores or intact sporangia (Ellar *et al.*, 1975; Hogarth *et al.*, 1977*a*), and secondly, in the sporulating cell, the total content of DPA is confined to the forespore compartment (Ellar & Posgate, 1974). The latter observation allows us to propose a possible mechanism for ensuring that a downhill concentration gradient is maintained in the forespore compartment throughout sporulation, despite the very high levels of calcium which occur in the mature spore. This model which is illustrated in Fig. 4, proposes that the active transport mechanism in the mother-cell plasma membrane is responsible for concentrating Ca^{2+} to a level of 3–9 mM in the cytoplasm surrounding the forespore. From this location the cation moves into the forespore compartment by facilitated

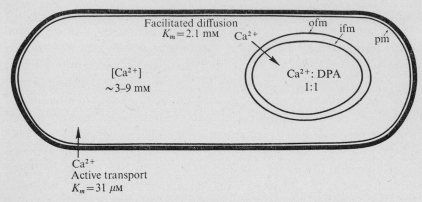

Fig. 4. Model proposed for Ca^{2+} accumulation during sporulation. pm, plasma membrane of mother cell; ofm, outer forespore membrane; ifm, inner forespore membrane; DPA, dipicolinic acid (pyridine-2, 6-dicarboxylic acid).

diffusion where it chelates with DPA. The synthesis of DPA in parallel with Ca^{2+} uptake throughout sporulation, would allow the forespore to continue withdrawing Ca^{2+} from the surrounding cytoplasm by chelation of the incoming cation, thereby maining a low *free* Ca^{2+} concentration in the forespore (La Nauze *et al.*, 1974; Hogarth *et al.*, 1977*a*). This model also provides an explanation both for the unique presence of DPA in spores and also for its occurrence in an approximately 1:1 mole ratio with calcium.

We have seen that the site of the sporulation specific transition from Ca^{2+} efflux to active transport of this cation resides in the mother-cell membrane. The question of whether *any* active transport is possible across the reversed polarity double forespore membranes remains unresolved and the transport kinetics of amino acids and other metabolites must now be studied. If the forespore is incapable of any active transport supported by either endogenous respiration or by exogenous substrate oxidation, this will be an important consideration in our attempts to learn more about the control of morphogenesis. In this context, Singh *et al.*, (1977) have reported that while the ATP:adenine nucleotide ratio in whole cells of *B. megaterium* QM B1551 remained fairly constant during sporulation, this ratio decreased dramatically in the forespore compartment around the time at which DPA began to accumulate. Prior to this stage it is possible that the forespore retains the capacity for ATP synthesis. Certainly by Stage V of sporulation, the specific activity of oxygen uptake by isolated forespores is much less than that of intact sporangia, when both are oxidising endogenous substrates (Wilkinson *et al.*, 1975; Hogarth *et al.*, 1977*a*).

Morphogenesis of the membrane-bound electron transport system during sporulation

Consideration of the respiratory capacity of forespores prompts questions about the electron transport systems operating in the two compartments during sporulation. The early report of Doi & Halvorson (1961) had indicated that *B. cereus* T spores contained no cytochromes and that spore electron transport was mediated by a soluble NADH-flavoprotein oxidase, in contrast to the membrane bound cytochrome system in the vegetative cell. These results implied that during sporulation a striking atrophy of the electron transport system occurs to produce a considerably modified dormant spore membrane. In *B. subtilis*, other workers had succeeded in demonstrating cytochromes in the spore (Keilin & Hartree, 1949; Tochikubo, 1971; Weber & Broadbent, 1975). Once again, the availability of isolated forespores stimulated a re-examination of this question (Wilkinson & Ellar, 1975; Hogarth, Wilkinson & Ellar, 1977*b*). If drastic changes were indeed occurring in membrane oxidative capacity during sporulation, it should be possible by dissecting the cell in this way, to determine whether these changes were confined to the forespore membranes. The results of these experiments with *B. megaterium* (Wilkinson & Ellar, 1975) showed that cytochromes were present in the dormant spore membranes, although at much lower concentrations than in either mother-cell membranes or membranes of vegetative cells. Comparisons of forespore and mother-cell membranes throughout sporulation indicated that cytochrome synthesis ceases in the forespore compartment after Stage III. The only oxidase activity detectable in dormant spore membranes was NADH oxidase. These findings support the view that an atrophy of membrane electron transport and oxidative capacity is occurring in the forespore compartment during sporulation. However, since sporulation occurs in response to growth limitation, it is important to establish that this change is genuinely sporulation specific and not merely an adaptation to the changing growth conditions. One striking change in the electron transport systems of *B. megaterium* for which a new protein is required, is the synthesis of cytochrome *c* during Stages I–II (Wilkinson & Ellar, 1975; Hogarth *et al.*, 1977*b*). This cytochrome is found in both forespore and mother-cell membranes from Stage III onwards but is completely absent from vegetative cell membranes. A proposed role for this cytochrome in sporulation will be discussed later.

Spore membranes and cyanide resistance

In their study of sporulation specific Ca^{2+} uptake, Hogarth *et al.*
(1977*a*) had detected a striking difference in the cyanide sensitivity of
intact sporangia and isolated forespore endogenous respiration. With
1 mM KCN, oxygen uptake by sporangia was inhibited 80%, while
forespore respiration was unaffected. Increasing the KCN concentration to 20 mM produced only 47% inhibition of oxygen uptake by the
isolated forespores. Andreoli *et al.* (1975) had earlier reported that respiration of isolated Stage III forespores was unaffected by 0.33 mM KCN
and Wilkinson & Ellar (1975) had found that dormant spore membrane
NADH oxidase was only 35% inhibited by 1 mM KCN. These results
suggested that the changes in cyanide sensitivity might be restricted to
the forespore membranes. However as subsequent experiments with *B.
megaterium* have shown (Hogarth *et al.*, 1977*b*), the membrane which
appears to be modified in its cyanide resistance is in fact the mother-
cell membrane. While these studies established that the NADH oxidase
activity of forespore inner membranes was significantly more resistant
to KCN than the same activity in the mother-cell plasma membrane, they
also revealed that during *vegetative* growth of this organism, the plasma
membrane displays the same high resistance to cyanide that is found in
the forespore membrane during sporulation. Thus the cyanide sensitivity
of the mother-cell membrane must require some membrane modification
to occur as an early sporulation event. This change in cyanide sensitivity
was found to occur in the period between Stages II and III, which is also
the time during which de-novo synthesis and incorporation of cytochrome *c* was observed (see earlier). The membrane modifications responsible for these differing cyanide sensitivities have been identified
and discussed (Hogarth *et al.*, 1977*b*). We can now ask if these modifications could be involved in any specific way with spore morphogenesis.
Jones, Brice, Downs & Drozd (1975) have shown that *B. megaterium,
B. subtilis* and *B. licheniformis* possess only two proton translocation
loops, when grown vegetatively, because their membranes lack cytochrome *c* and pyridine nucleotide transhydrogenase. Bacilli are therefore
potentially less efficient at conserving energy than other micro-organisms
which can use cytochrome *c* in a third proton translocation loop. Perhaps as Jones *et al.* (1975) have suggested, the natural habitat of bacilli
is sufficiently rich in organic material to allow them to use these substrates with low efficiency. In the restricted conditions of a sporulation
medium, such poor energy conservation could be a serious disadvantage.
Therefore the incorporation of cytochrome *c* into the membranes at

Stage II–III of sporulation in *B. megaterium* could markedly improve energy conservation at this time by providing the organism with a third coupling site.

It is particularly difficult to assess the amount of O_2 which is available to the inner forespore membrane even in a well-aerated culture, since both the mother-cell membrane and the outer forespore membrane possess NADH oxidase activity (Hogarth *et al.*, 1977*b*). If however the forespore compartment is relatively anaerobic compared to the surrounding cytoplasm, the presence of a cyanide-resistant oxidative pathway in this compartment could be an important asset, since Meyer & Jones (1973) have suggested that the high oxygen affinity of a cyanide resistant oxidase may permit oxidative phosphorylation to continue at low oxygen concentrations.

SPORE CORTEX STRUCTURE AND BIOSYNTHESIS

The spore cortex is another spore specific structure which presents us with interesting problems of cellular organisation and structure/function relationships. One such problem is the fact that spore peptidoglycan assumes a spherical or ellipsoidal shape compared to the cylindrical form of vegetative cell walls. Detailed analyses of cortex structure in *B. subtilis* (Warth & Strominger, 1969, 1972) and *B. sphaericus* (Linnett & Tipper, 1976) have exposed several important differences in cortex peptidoglycan compared to the vegetative polymer, which suggest that the spore polymer may possess unique properties (Rogers, 1977). Most vegetative peptidoglycans are based upon glycan chains of alternating residues of β-1,4-linked *N*-acetylglucosaminyl and *N*-acetylmuramyl residues, with each muramic acid residue substituted with a peptide. However in *B. subtilis* cortex, about half of the muramic acid residues are present as the spore specific muramic acid lactam (Warth & Strominger, 1969). The lactam is not randomly located in the glycan chain, but occurs predominantly at every alternate disaccharide (Tipper & Gauthier, 1972). Of the remaining muramic acid residues, about 30% are substituted with a tetrapeptide and the rest terminate in a single L-alanine substituent with its carboxyl group free (Warth & Strominger, 1969; Tipper & Gauthier, 1972). The cross-linking of spore peptidoglycan is considerably less than that of the vegetative polymer. For *B. sphaericus* cortex only 6% cross-links per disaccharide were found, compared to an average value of 60% for a vegetative cell wall (Tipper & Gauthier, 1972). Although various accessory polymers such as teichoic acids are generally found covalently linked to peptidoglycan in vegetative cell walls, these polymers are not found in spore cortex (Rogers, 1977).

In an earlier section it was noted that in some spores, there is evidence, recently discussed by Rogers (1977), to suggest that the layer of peptidoglycan closest to the forespore inner membrane differs from the bulk of the cortex and retains the chemical structure of vegetative peptidoglycan. This inner cortex layer has been termed primordial cell wall, because of the reports that it develops into the cell wall of the outgrowing cell after germination (Tipper & Linnett, 1976; Fitz-James & Young, 1969; Murrell, 1967). The vectorial trans-membrane nature of peptidoglycan synthesis forces us once again to consider the implications of the reversed polarity of forespore membranes for cortex assembly. Microscopic studies have established that this polymer is deposited between the two membranes and it was therefore reasonable to propose (Tipper & Gauthier, 1972) that cortex is formed from precursors in the mother-cell cytoplasm by the enzymes of the outer forespore membrane. Moreover, synthesis of the primordial cell wall of *B. sphaericus* would be the responsibility of the forespore inner membrane, utilising a separate precursor pool located in the forespore compartment. *B. sphaericus* was a particularly useful organism in which to test this hypothesis, since the vegetative peptidoglycan is cross-linked between L-lysine residues in the peptide by D-alanyl-D-isoasparaginyl residues and is devoid of diaminopimelic acid, whereas the spore cortex contains meso-diaminopimelic acid and lacks lysine and aspartic acid (Linnett & Tipper, 1976). In addition there is evidence that the primordial wall in this organism can be distinguished from the bulk of the cortex as a more lysozyme-resistant vegetative-type polymer containing lysine and D-aspartate (Linnett & Tipper, 1976). With this chemical background, Tipper & Linnett (1976) predicted that the spore specific enzyme which incorporates meso-diaminopimelic acid into soluble cortex precursors (meso-diaminopimelyl ligase), would be found only in the sporangial cytoplasm. As a prerequisite to this experiment, Linnett & Tipper (1976) demonstrated that enzymes necessary for the synthesis of peptidoglycan precursors were synthesised at two periods during sporulation. In the first period (Stages II–III), enzymes responsible for making vegetative peptidoglycan precursors were synthesised, but not meso-diaminopimelyl ligase. In a second period of synthesis shortly before cortex appeared (Stage IV), the cortex specific meso-diaminopimelyl ligase was made. During this second period, the lysyl ligase required for vegetative peptidoglycan synthesis was not made. Tipper & Linnett (1976) were then able to show by selective disruption of the sporulating cells that, as predicted, the diaminopimelyl ligase was confined to the mother-cell compartment. L-lysine ligase activity which is not required for cortex synthesis, was found in forespore and mother cell, but its specific activity was four times greater in the former, than in the latter. Other enzymes common to the synthesis of both vegetative and spore peptidoglycan also occurred in both compartments but with similar specific activity. These results together with those of their earlier experiments, allow Linnett & Tipper (1976) to propose that cortex is assembled during the second period of enzyme synthesis while the first period is concerned with the deposition of primordial cell wall.

As was found for Ca^{2+} uptake (Hogarth et al., 1977a), membrane bound electron transport components (Hogarth et al., 1977b) and the distribution of small molecules and enzymes between sporangial compartments (Singh et al., 1977; Andreoli et al., 1973; Freese, 1977), these studies of cortex biosynthesis reveal that the architecture of the sporulating cell can accommodate the differential expression of genes in the two cellular compartments. The work of Andreoli et al. (1975), Chasin & Szulmajster (1969) and Zytkowicz & Halvorson (1972) has also suggested that the enzymes involved in dipicolinic acid synthesis may be confined to the mother-cell compartment. It remains to be seen where the various modifications to transcriptional and translational patterns which occur during sporulation (Szulmajster, 1973; Linn & Losick, 1976; Doi & Sanchez-Anzaldo, 1976) are located within the sporangium.

THE ROLE OF CORTEX
IN SPORE DORMANCY AND RESISTANCE

Data from a number of sources suggests that spores are characterised by a low water content compared to vegetative cells (Gould & Measures, 1977). If this is the case, it might help to explain the unique high heat resistance of bacterial spores. Several mechanisms have been proposed by which the forespore cytoplasm could be dehydrated during morphogenesis and some discussion of these is appropriate here since they are based on the successful architecture of spore specific structures. One such mechanism (Lewis, Snell & Burr, 1960) proposes that dehydration is accomplished by compressive contraction of the cortex during sporulation. If the cortex is considered as a predominantly electronegative polymer, then this contraction could be brought about by cross-linking it with cations, such as Ca^{2+}. In examining the implications of the unique chemistry of spore cortex, Rogers (1977) has drawn attention to two general properties which might be expected to influence its physical properties. Firstly, the very low degree of cross-linking would allow the glycan chains more freedom to move relative to each other and might therefore permit the structure to undergo much greater volume changes in response to pH and ionic changes than would be possible for the highly cross-linked vegetative polymer. Secondly, on the basis of known cortex composition it could possess a much higher negative charge than the peptidoglycan of vegetative cells. As Rogers (1977) points out however, it would be unwise to make too much of this last comparison. While spore cortex contains no accessory polymers, teichoic and teichuronic acids occur in large amounts in vegetative walls and in some organisms these polymers constitute the major source of acidic groups for binding inorganic ions. In support of the notion of a contractile cortex, salt-induced contraction of the vegetative polymer has been observed (Marquis, 1968). Although as Rogers (1977) has pointed out, the extent of this contraction is quite small, the contractility of the poorly cross-linked, more negatively charged spore polymer may prove more impressive.

Evidence that the cortex is linked in some way to spore dormancy and resistance comes from many sources including mutant studies (Pearce & Fitz-James, 1971a, b; Freese, Cole, Klofat & Freese, 1970; Imae & Strominger, 1976) and antibiotic inhibition (Murrell & Warth, 1964). The results of Pearce & Fitz-James (1971b) were especially interesting since they suggested that the role of cortex might not be to produce the dehydrated spore state, but rather to maintain it.

Also on the basis of the known cortex properties, Gould & Dring (1975) have recently proposed a second mechanism by which it may function to dehydrate the spore core. They argue that a polymer of this type is more likely to be expanded than contracted and go on from this to suggest that the role of the cortex is that of an osmoregulatory organelle. Their novel proposal is illustrated in Fig. 5. With its poorly cross-linked, electronegative structure, they suggest that in the absence of high concentrations of cations, spore cortex would be expanded by electrostatic repulsion of adjacent acidic groups. Contraction of this expanded polymer could then be brought about by exposure to high cation concentrations, especially multivalent cations. In the dormant spore they envisage the cortex as electrically neutral and expanded, by virtue of the association between the peptidoglycan carboxyl groups and free positively charged counterions. They propose that this complex of polymer plus counterions will be osmotically active and capable of exerting osmotic pressures in excess of 30 atm. Provided that the osmotic pressure of the spore core is not of the same order, this osmoregu-

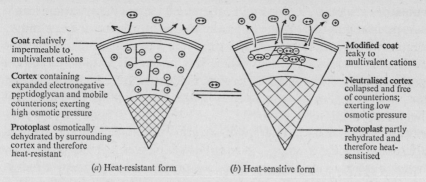

Coat relatively impermeable to multivalent cations

Cortex containing expanded electronegative peptidoglycan and mobile counterions; exerting high osmotic pressure

Protoplast osmotically dehydrated by surrounding cortex and therefore heat-resistant

Modified coat leaky to multivalent cations

Neutralised cortex collapsed and free of counterions; exerting low osmotic pressure

Protoplast partly rehydrated and therefore heat-sensitised

(*a*) Heat-resistant form   (*b*) Heat-sensitive form

Fig. 5. Diagrammatic representation of the osmoregulatory expanded cortex in normal heat-resistant spores (*a*) and in coat-modified spores made heat sensitive by multivalent cations (*b*). −, Negatively charged groups on cortex peptidoglycan. +, monovalent positively charged mobile counterions, + +, multivalent cations.

latory cortex would dehydrate the core and preserve it in this state by osmosis. In fact, the osmotic pressure in the forespore compartment is quite likely to be much less than that of a vegetative cell, since spores contain much reduced levels of soluble low molecular weight solutes (Nelson & Kornberg, 1970). In addition, the high concentration of Ca^{2+} and DPA in spores strongly suggests that many of the remaining pool solutes will be precipitated out of solution (Ellar, 1978*a*; Gould & Dring, 1975). Dielectric studies (Carstensen, Marquis & Gerhardt, 1971) have indicated that spore ions are immobilised in some way and free ions are present in only small amounts. The model of Gould & Dring (1975) is therefore one in which a low osmotic pressure, dehydrated spore cytoplasm is in osmotic equilibrium with a surrounding 'wet' cortex. Among the attractive features of this model, is the fact that it is compatible with the observation that the bulk of spore water is exchangeable with external water (Marshall & Murrell, 1970). The experimental work carried out by Gould & Dring (1975) amounts to strong support for their model.

If we accept that the forespore compartment becomes progressively dehydrated, we might also consider yet another dehydration mechanism (Ellar, 1978*a*) which is based on our accumulated knowledge of the biochemistry and architecture of the sporulating cell and the membranes which enclose the two compartments. This is a mechanism in which cortex is needed to maintain the dehydrated state of spore cytoplasm, but not to create it. In this respect it is compatible with the results of Pearce & Fitz-James (1971*b*) discussed above. In this proposal osmosis remains the driving force for dehydration, but it is the forespore membrane(s) rather than the cortex which is considered to be the osmoregulatory structure. According to this model spore dehydration would be a natural outcome of the macromolecular syntheses and ion movements which occur during sporulation. We have already seen that the levels of free amino acids and other low molecular weight solutes are progressively reduced in the forespore compartment as a result of polymer synthesis, general reduction in metabolic potential, or because of chelation/precipitation with Ca^{2+} or Ca^{2+}–DPA. Added to these observations is the fact that the spore is a cell within a cell, i.e. its external environment is the mother-cell cytoplasm from which it is separated by the two forespore membranes. Conceivably then, the progressive reduction in osmotic pressure in the forespore compartment could lead to a compensatory outflow of water from the forespore cytoplasm. It follows from this model that the role of spore integuments, e.g. cortex, would be to form a rigid casing around the spore protoplast which could resist protoplast expansion and thus oppose the tendency for water to flow back into the spore when it is released from the mother-cell cytoplasm into what would inevitably be a more dilute environment. Spore water content undoubtedly increases rapidly upon germination and therefore the corollary to this model is that initiation of germination would require reduction or elimination of the rigid component of spore integuments. Collapse of an expanded cortex upon exposure to Ca^{2+}, as proposed by Gould & Dring (1975) is one mechanism for bringing this about. Alternatively, the same collapse could be achieved if the spore responds to germinants by initiating the activity of latent enzymes capable of reducing cortex rigidity by selective hydrolysis.

CAPABILITIES OF ISOLATED FORESPORES

Using isolated forespores it has been possible to probe the function of sporulating cell structures in yet another way. Forespores isolated at Stages IV, V and VI from *B. megaterium* have been compared with mature spores in terms of such characteristic spore properties as germinability, response to heat activation and resistance to heat and octanol (Ellar *et al.*, 1975; Stewart & Ellar, 1977). By examining the timing of the appearance of these properties in the forespore during morphogenesis, it was hoped to discover any interdependence of one spore property with another, or any requirement for a particular spore specific structure. For example, in attempting to understand the events during germination, it would be valuable to know if the forespore can exhibit the typical germination response before the synthesis or completion of any of the spore specific structures.

When Stage III and IV forespores were isolated and examined in this way they were found to be osmotically fragile. In the absence of a stabiliser such as sucrose, they swelled rapidly and lysed (Ellar & Posgate, 1974). They were also non-viable in any of the media tested (Ellar *et al.*, 1975; Stewart & Ellar, 1977). Upon addition of germinants to a sucrose-stabilised suspension of Stage IV forespores no response typical of germination could be detected. At this stage these forespores had accumulated one third of their final content of Ca^{2+} and DPA. In contrast, forespores isolated from Stage V onwards were osmotically stable in aqueous buffers and completely viable in normal culture media. Further examination of these forespores revealed that other spore properties appeared at distinct stages. Thus Stage V forespores were resistant to octanol and showed a typical response to germinants with respect to loss of optical density and Ca^{2+}, but were heat sensitive. Stage VI forespores responded similarly to octanol and germinants, but were heat resistant and also showed the normal response to heat activation. These experiments are continuing, but a number of interesting conclusions can be drawn at this point. First we see that although at Stage V, approximately 80% of the total spore Ca^{2+} content has been accumulated by the forespore (C. Hogarth and D. J. Ellar, unpublished observation), it is not yet heat resistant. This could mean that the subsequent 20% is of special importance in establishing heat resistance, or alternatively that the maximum spore content of Ca^{2+} is required. It is also noteworthy in terms of mechanisms for spore dehydration, that cortex synthesis is virtually complete by Stage V in this organism. Secondly, these experiments reveal that from Stage V onwards the forespore is equipped to reinitiate vegetative growth in response to germinants and that this potential develops in the forespore compartment between Stage IV and V. When however intact Stage V sporangia are suspended in a growth medium, sporulation continues and heat resistant free spores are released (D. J. Ellar and G. Denton, unpublished observations). 'Commitment' to sporulation (Freese, 1977) is thus a characteristic of the intact sporangium and not a feature of the forespore. Finally, the properties of the Stage V forespore suggest that two typical germination responses (optical density decrease and Ca^{2+} release) may be dissociable from a third, viz. loss of heat resistance.

SPORE DORMANCY AND RESISTANCE
Spore membranes

The studies of Wilkinson *et al.* (1975) on forespore membrane polarity demonstrated the presence of several enzyme activities in the *outer* membrane of isolated forespores of *B. megaterium*. Tipper & Linnett (1976) have also suggested that a D-glutamic acid-meso-diaminopimelic acid endopeptidase may be located on the outer forespore membrane of *B. sphaericus*. In early studies on the protein and lipid composition of forespore membranes (Ellar & Posgate, 1974) only the inner forespore membrane was examined and compared with the mother cell plasma membrane. The results showed a dramatic shift in membrane phospholipid composition in both sporangial compartments at the time of forespore engulfment (Stage III). From about Stage IV of sporulation, the forespore outer membrane becomes less and less visible in the electron microscope as the spore coats are deposited around it. Nevertheless NADH oxidase remains active in the outer membrane of isolated Stage V forespores of *B. megaterium* (Hogarth *et al.*, 1977*a*). In assessing the role of the various spore specific structures, it is clearly important to know more about the comparative biochemistry of the two forespore membranes and also to discover whether the outer membrane persists as a functional entity in the dormant spore. Significant progress towards these objectives has been made through the development of a method by which the inner spore membrane can be separated from the outer membrane plus spore integuments (Koncewicz, Ellar & Posgate, 1977). The fact that the two membranes differ markedly in lipid composition in *B. megaterium* suggests that this separation is essential in any study of lipid metabolism during sporulation and germination. In dormant spores of *B. megaterium*, 66% of total spore lipid and about 80% of spore phospholipid occur in the spore inner membrane (Koncewicz *et al.*, 1977). By contrast, the outer membrane contains the majority (60%) of the total neutral lipid of the spore and only 20% of spore phospholipid. The spore inner membrane was also found to contain a characteristic red carotenoid which was absent from both the spore outer membrane and the mother-cell membrane. This analysis has been taken further by a comparison of the enzyme complement and polypeptide profile of inner and outer forespore membranes (Crafts-Lighty & Ellar, 1977). Of the total spore particulate activities, the outer membrane fraction contained 38% of the NADH oxidase, 43% of the NADH dehydrogenase, 19% of the NADH–cytochrome *c* reductase and 8% of the NADPH dehydrogenase. A range of other oxidase activities which are absent from the spore inner membrane (Wilkinson & Ellar, 1975) was also not present in the outer membrane. Approximately one third of the total spore cytochrome content was located in the out membrane. No adenosine deaminase or ribosidase activity could be detected in either membrane.

The existence of enzymes in the outer membrane of dormant spores raises questions about their possible role in germination. None of the outer membrane enzymes so far investigated show any greater heat resistance than enzymes of the inner membrane (A. Crafts-Lighty, unpublished observation). One other interesting comparison between the inner and outer spore membranes concerns their lipid content. Calculations based on the absolute amount of lipid by weight in spore and vegetative cell membranes of *B. megaterium* show that while the vegetative membrane contains sufficient phospholipid to cover the membrane surface one and a half times in the bilayer form, the spore inner membrane has sufficient lipid to cover its surface only once as a bilayer (M. A. Koncewicz and D. J. Ellar, in preparation). Direct analyses reveal that phospholipid comprises only 12% by weight of the spore inner membrane compared to a figure of 25% for the vegetative cell membrane. The inner spore membrane therefore appears to be characterised by a very high protein to lipid ratio. In contrast, similar calculations show that the amount of phospholipid in the spore outer membrane is only sufficient to form a monolayer over the membrane surface. Whether these figures reflect any membrane properties which are important to sporulation, dormancy or subsequent germination, remains to be seen. One such property which may be relevant, is the existence of a block in the electron transport chain of the dormant spore inner membrane which is removed in the first five minutes of spore germination (Wilkinson, Ellar, Scott & Koncewicz, 1977). It was seen in an earlier section that sporulation is accompanied by a reduction in cytochrome content and a parallel decline in membrane oxidative capacity. The demonstration by Wilkinson *et al.* (1977) of a block in the chain between NADH dehydrogenase and cytochrome *b* provides one reason for this decline.

Spore germination

Since spore specific structures are involved in the termination of dormancy, it is important to examine their chemistry and architecture with this function in mind. In discussing the spore cortex, we have already dealt with some suggestions by which this structure might play a key role in germination. The work of Setlow (1975*a*, *b*, 1976) and Setlow & Waites (1976) has recently indicated that a number of low molecular weight spore specific basic proteins located in the spore core, may play an important part in germination. Although these proteins are not spore specific structures on the scale of cortex, membranes or spore coats, it is possible that they are involved with other macromolecules in creating molecular aggregates whose properties are important in spore dormancy and resistance. The finding that these proteins bind to purified DNA and to a nuclear body from dormant spores to produce a marked elevation in the DNA melting temperature (Setlow, 1975b), led Setlow to suggest that they may be involved *in vivo* in altering the conformation of spore DNA. The simultaneous appearance in the developing spore of these basic proteins and the acquisition of resistance to ultraviolet radiation, raises the possibility that resistance may result from a DNA conformational change induced by the proteins (Setlow, 1975a, b).

Setlow (1975a; 1976) has also shown that these proteins which comprise 10 to 15% of total spore protein, are rapidly degraded during germination by a spore specific endoprotease. Of the 20% of dormant spore protein which is degraded during germination, three quarters of this is accounted for by the proteolysis of two of these small basic proteins. During sporulation, although these proteins and the specific endoprotease are found together confined to the forespore compartment, no turnover of the proteins was observed (Singh *et al.*, 1977). This suggests first that some mechanism exists within the sporangium to prevent the protease hydrolysing the basic proteins during morphogenesis, and second that initiation of the activity of the protease may be an important part of the spore response to germinants (Singh *et al.*, 1977).

In addition to degradation of these low molecular weight basic proteins, other spore specific structures such as cortex are degraded rapidly during germination (Gould, 1969). The spore inner membrane and primordial wall, however, are destined to serve the emerging new vegetative organism and an examination of these structures during the first minutes after addition of germinants might yield information on the site of the germination response mechanism. One dramatic change which occurs very early in germination is the elimination of the block in the spore inner membrane electron transport chain referred to above (Wilkinson *et al.*, 1977). This results in a 5–10 fold increase in NADH oxidase activity in the first five minutes of germination. Further experiments showed that this increase occurs even when protein synthesis is blocked, and therefore some mechanism of activation must be occurring. Interestingly, no corresponding activation of the outer membrane NADH oxidase activity occurs during germination (A. Crafts-Lighty unpublished,). One possible mechanism for this activation may be found in the membrane conformational changes during the expansion of the spore inner membrane which is believed to occur early in germination. The observation that lipid turnover starts approximately 2 minutes after addition of germinants to *B. megaterium* spores, whereas de-novo lipid synthesis is not detectable for a further 8 minutes (Koncewicz *et al.*, 1977) may also be relevant.

Biophysical studies

Almost by definition, the biologically and commercially important characteristics of the dormant spore are particularly unsuited to investigation by conventional biochemical techniques. Attempts to study the structural and biochemical basis of dormancy and resistance by traditional methods requiring prior cell breakage, are likely to disrupt the cellular architecture which is responsible for creating and maintaining these characteristics. In this area, several non-destructive physicochemical techniques such as nuclear magnetic and electron spin resonance, differential thermal analysis and dielectric studies are beginning to provide important information on the conformation of spore structures, the state of spore water and the degree of mobility of spore ions. Carstensen & Marquis (1975) concluded from dielectric measurements that the ions within dormant spores were almost totally immobilised when compared with vegetative organisms. During heat activation, germination and outgrowth of spores, the internal ion mobility increased. Such restrictions on the mobility of spore core constituents might be anticipated if the presence of Ca^{2+} and DPA causes precipitation of spore constituents as was discussed earlier. Nuclear magnetic resonance spectroscopy of bacterial spores has also begun to yield promising results. When Maeda, Fujita, Sugiura & Koga (1968) examined the spectra of spores and vegetative bacilli with the same water content, the broad water absorption peak seen in vegetative cells was absent from the spores. Recent experiments (Eaton, 1975; M. W. Eaton and D. J. Ellar, in preparation) using similar spectroscopic techniques, have shown that approximately 5% of spore dry weight is composed of non-exchangeable water compared to a figure of 1.4% for bacteria. The measured relaxation time of this non exchangeable water fraction (10^{-2} s at 30 °C) is consistent with its adsorption by macromolecules in a relatively anhydrous environment. Nuclear magnetic resonance was also used to study the changes occurring in spore components during heat activation and germination. At the normal heat activation temperatures (60–70 °C) there was a dramatic appearance of spectral peaks derived from spore lipids. This effect is similar to the changes seen in nuclear magnetic resonance spectra when phospholipids change from the ordered (gel) to the disordered (liquid crystal) state. The absence of these characteristic peaks in the spectrum of spores below heat activation temperatures, suggests that the molecular mobility of the phospholipids is constrained in some way in the spore. Among the mechansims which could restrict lipid mobility in this way is the interaction between multivalent cations such as Ca^{2+} and the polar groups of phospholipids (Ellar, 1978b). The abundance of this cation in spores might favour this mechanism. Studies on the effect of added Ca^{2+} on the nuclear magnetic resonance spectra of sonicated aqueous dispersions of B. megaterium spore lipid (M. W. Eaton and D. J. Ellar, in preparation), showed considerable peak broadening, indicative of a restriction in molecular motion. In related experiments with lipid monolayers, it was found that the contraction of a spore lipid monolayer induced by Ca^{2+} was three times greater than that observed for vegetative lipids. This probably reflects the greatly increased cardiolipin content of B. megaterium spore membranes (Ellar & Posgate, 1974). One particularly interesting result from these monolayer experiments was the finding that when Ca^{2+} and DPA were present in the subphase in equimolar concentrations, Ca^{2+} appeared to interact preferentially with the anionic phospholipids in the monolayer. It now remains to be seen whether a restricted mobility of phospholipids in spore membranes is important in creating or maintaining any of the typical spore properties. In addition we can enquire to what extent a restoration of normal membrane fluidity is involved in the initial germination events. The correlation between the potency of the homologous series of long chain alkyl amines as germinants and their relative effectiveness at displacing calcium from intact spores and spore lipid monolayers, has led to the suggestion that calcium displacement from spore membranes may be an important early event in germination (Ellar, Eaton & Posgate, 1974). The rapid increase in phospholipid mobility occurring during heat activation and germination provides one possible explanation for the electron transport chain activation described above. These early results from the application of physio-chemical methods to probe the architecture of spores are encouraging. Because such methods permit the interior of the spore and spore structures to be studied without loss of spore dormancy, they are likely to become increasingly useful.

The author wishes to acknowledge the contribution of many colleagues who have worked in Cambridge during the past nine years and whose efforts and suggestions are to be found in this survey. He is grateful to the Editors of Biochemical Journal and Nature for permission to reproduce Figs. 2 and 5 respectively Research carried out at Cambridge was supported by the Science Research Council, Medical Research Council, Welcome Foundation and the Broodbank Foundation.

REFERENCES

ANDREOLI, A. J., SUEHIRO, S., SAKIYAMA, D., TAKEMOTO, J., VIVANCO, E., LARA J. C. & KLUTE, M. C. (1973). Release and recovery of forespores from *Bacillus cereus. Journal of Bacteriology*, **115**, 1159–66.

ANDREOLI, A. J., SARANTO, J., BAECKER, P. A., SUEHIRO, S., ESCAMILLA, E. & STEINER, A. (1975). Biochemical properties of forespores isolated from *Bacillus cereus.* In *Spores VI*, pp. 418–24. American Society for Microbiology.

ARONSON, A. I. & FITZ-JAMES, P. C. (1976). Structure and morphogenesis of the bacterial spore coat. *Bacteriological Reviews*, **40**, 360–402.

BRONNER, F., NASH, W. C. & GOLUB, E. E. (1975). Calcium transport in *Bacillus megaterium.* In *Spores V*, pp. 356–61. American Society for Microbiology.

CARSTENSEN, E. L., MARQUIS, R. E. & GERHARDT, P. (1971). Dielectric study of the physical state of electrolytes and water within *Bacillus cereus* spores. *Journal of Bacteriology*, **107**, 106–13.

CARSTENSEN, E. L. & MARQUIS, R. E. (1975). Dielectric and electrochemical properties of bacterial cells. In *Spores VI*, pp. 563–71. American Society for Microbiology.

CHASIN, L. A. & SZULMAJSTER, J. (1969). Biosynthesis of dipicolinic acid in *Bacillus subtilis. Biochemical and Biophysical Research Communications*, **29**, 648–54.

CLARKSON, B. D. (1974). The survival value of the dormant state in neoplastic and normal cell populations. In *Control of Proliferation in Animal Cells*; *Cold Spring Harbor Conferences on Cell Proliferation*, **1**, ed. B. Clarkson & R. Baserga, pp. 945–72. Cold Spring Harbor Laboratory, USA.

CRAFTS-LIGHTY, A. & ELLAR, D. J. (1977). Evidence for the presence of the outer forespore membrane in dormant-spore integuments. *Biochemical Society Transactions*, **5**, 113–16.

CROSS, T., DAVIES, F. L. & WALKER, P. D. (1971). *Thermoactinomyces vulgaris* I. Fine structure of the developing endospores. In *Spore Research 1971*, ed. A. N. Barker, G. W. Gould & J. Wolf, pp. 175–80. London: Academic Press.

DOI, R. H. & HALVORSON, H. (1961). Comparison of electron transport systems in vegetative cells and spores of *Bacillus cereus. Journal of Bacteriology*, **81**, 51–8.

DOI, R. H. & SANCHEZ-ANZALDO (1976). Complexity of protein and nucleic acid acid synthesis during sporulation of Bacilli. In *Microbiology 1976*, ed. D. Schlessinger, pp. 145–63. American Society for Microbiology.

EATON, M. W. (1975). The biochemistry of bacterial sporulation and germination. Ph.D. Thesis, University of Cambridge, England.

EATON, M. W. & ELLAR, D. J. (1974). Protein synthesis and breakdown in the mother-cell and forespore compartments during spore morphogenesis in *Bacillus megaterium. Biochemical Journal*, **144**, 327–37.

EISENSTADT, E. & SILVER, S. (1972). Calcium transport during sporulation in *Bacillus subtilis.* In *Spores V*, pp. 425–33. American Society for Microbiology.

ELLAR, D. J. (1978a). Some strategies of osmoregulation and ion transport in microorganisms. In *Comparative Physiology: Water, Ions and Fluid Mechanics*, ed. K. Schmidt-Neilsen, L. Bolis & S. H. P. Maddrell, pp. 125–49. Cambridge University Press.

ELLAR, D. J. (1978b). Membrane fluidity in microorganisms. In *Companion to Microbiology*, ed. A. T. Bull and P. M. Meadow, in press. Longman. London.

ELLAR, D. J., EATON, M. W. & POSGATE, J. A. (1974). Calcium release and germination of bacterial spores. *Biochemical Society Transactions*, **2**, 947–8.

ELLAR, D. J., EATON, M. W., HOGARTH, C., WILKINSON, B. J., DEANS, J. & LA

NAUZE, J. (1975). Comparative biochemistry and function of forespore and mother cell compartments during sporulation of *Bacillus megaterium* cells. In *Spores VI*. pp. 425–33. American Society for Microbiology.

ELLAR, D. J. & LUNDGREN, D. G. (1966). Fine structure of sporulation in *Bacillus cereus* grown in a chemically defined medium. *Journal of Bacteriology*, **92**, 1748–64.

ELLAR, D. J. & POSGATE, J. A. (1974). Characterisation of forespores isolated from *Bacillus megaterium* at different stages of development into mature spores. In *Spore Research 1973*, ed. A. N. Barker, G. W. Gould and J. Wolf, pp. 21–40. London: Academic Press.

FITZ-JAMES, P. C. (1965). Spore formation in wild and mutant strains of *B. cereus* and some effects of inhibitors. In *Mécanismes de regulation chez les microorganisms; Colloques Internationaux du Centre National de la Recherche Scientifique, Sciences Humaines*, **124**, 529–54.

FITZ-JAMES, P. C. & YOUNG, E. (1969). Morphology of sporulation. In *The Bacterial Spore*, ed. G. W. Gould & A. Hurst, pp. 39–72. New York: Academic Press.

FREESE, E. (1972). Sporulation of bacilli, a model of cellular differentiation. *Current Topics in Developmental Biology*, **7**, 85–124.

FREESE, E. (1977). Metabolic control of sporulation. In *Spore Research 1976*, ed. A. N. Baker, J. Wolf, D. J. Ellar, G. J. Dring & G. W. Gould, in press. London: Academic Press.

FREESE, E. B., COLE, R. M., KLOFAT, W. & FREESE, E. (1970). Growth, sporulation and enzyme defects of glucosamine mutants of *Bacillus subtilis*. *Journal of Bacteriology*, **101**, 1046–62.

FUKUDA, A., GILVARG, C. & LEWIS, J. C. (1969). 4H-pyran-2,6-dicarboxylate as a substitute for dipicolinate in the sporulation of *Bacillus megaterium*. *Journal of Biological Chemistry*, **244**, 5636–43.

GOULD, G. W. (1969). Germination. In *The Bacterial Spore*, ed. G. W. Gould & A. Hurst, pp. 397–444. New York: Academic Press.

GOULD, G. W. (1970). Germination and the problem of dormancy. *Journal of Applied Bacteriology*, **33**, 34–49.

GOULD, G. W. & DRING, G. J. (1975). Heat resistance of bacterial endospores and concept of an expanded osmoregulatory cortex. *Nature, London*, **258**, 402–5.

GOULD, G. W. & MEASURES, J. C. (1977). Water relations in single cells. *Philosophical Transactions of the Royal Society of London B*, **278**, 151–66.

GREENE, R. A., HOLT, S. C., LEADBETTER, E. R. & SLEPECKY, R. A. (1971). Correlation of light and electron microscopic observations of sporulation in *Bacillus megaterium*. In *Spore Research 1971*, ed. A. N. Barker, G. W. Gould & J. Wolf, pp. 161–74. London: Academic Press.

HALVORSON, H. D. & SWANSON, A. (1969). Role of dipicolinic acid in the physiology of bacterial spores. In *Spores IV*, pp. 121–32. American Society for Microbiology.

HANSON, R. S. (1975). Role of small molecules in regulation of gene expression and sporogenesis in bacilli. In *Spores VI*, pp. 318–26. American Society for Microbiology.

HIGGINS, M. L., TSIEN, H. C. & DANEO-MOORE, L. (1976). Organisation of mesosomes in fixed and unfixed cells. *Journal of Bacteriology*, **127**, 1519–23.

HITCHINS, A. D. & SLEPECKY, R. A. (1969). Bacterial sporulation as a modified procaryotic cell division. *Nature, London*, **223**, 804–7.

HOGARTH, C., DEANS, J. A. & ELLAR, D. J. (1977a). Calcium accumulation and membrane morphogenesis in forespore and mother cell compartments during sporulation of *Bacillus megaterium*. In *Spore Research 1976*, ed. A. N. Barker,

J. Wolf, D. J. Ellar, G. J. Dring and G. W. Gould, in press. London: Academic Press.

HOGARTH, C., WILKINSON, B. J. & ELLAR, D. J. (1977b). Cyanide resistant electron transport in sporulating *Bacillus megaterium* KM. *Biochimica et Biophysica Acta*, **461**, 109–23.

HOLT, S. C., GAUTHIER, J. J. & TIPPER, D. J. (1975). Ultrastructural studies of sporulation in *Bacillus sphaericus*. *Journal of Bacteriology*, **122**, 1322–38.

HOLT, S. C. & LEADBETTER, E. R. (1969). Comparative ultrastructure of selected aerobic spore-forming bacteria: a freeze-etching study. *Bacteriological Reviews*, **33**, 346–378.

IMAE, Y. & STROMINGER, J. L. (1976). Relationship between cortex content and properties of *Bacillus sphaericus* spores. *Journal of Bacteriology*, **126**, 907–13.

JONES, C. W., BRICE, J. M., DOWNS, A. J. & DROZD, J. W. (1975). Bacterial respiration-linked proton translocation and its relationship to respiratory chain composition. *European Journal of Biochemistry*, **52**, 265–71.

KEILIN, D. & HARTREE, E. F. (1949). Effect of low temperature on the absorption spectra of haemoproteins; with observations on the absorption spectrum of oxygen. *Nature, London*, **164**, 254–9.

KEYNAN, A. (1968). The initiation of metabolism in metabolically dormant systems and its comparison with egg fertilisation. *Accad. Naz. Lincei*, **104**, 135–45.

KONCEWICZ, M. A., ELLAR, D. J. & POSGATE, J. A. (1977). Metablism of membrane lipids during bacterial spore germination and outgrowth. *Biochemical Society Transactions*, **5**, 118–19.

LA NAUZE, J. M., ELLAR, D. J., DENTON, G. & POSGATE, J. A. (1974). Some properties of forespores isolated from *Bacillus megaterium*. In *Spore Research* 1973 ed. A. N. Barker, G. W. Gould & J. Wolf, pp. 41–6. London: Academic Press.

LEMAN, A. (1973). Interference microscopical determination of bacterial dry weight during germination and sporulation. *Jena Review*, **5**, 263–70.

LEWIS, J. C. (1969). Dormancy. In *The Bacterial Spore*, ed. G. W. Gould & A. Hurst, pp. 301–58. New York: Academic Press.

LEWIS, J. C., SNELL, N. S. & BURR, H. K. (1960). Water permeability of bacterial spores and the concept of a contractile cortex. *Science*, **132**, 544–5.

LINN, T. & LOSICK, R. (1976). The programme of protein synthesis during sporulation in *Bacillus subtilis*. *Cell*, **8**, 103–14.

LINNETT, P. E. & TIPPER, D. J. (1976). Transcriptional control of peptidoglycan precursor synthesis during sporulation in *Bacillus sphaericus*. *Journal of Bacteriology*, **125**, 565–74.

MAEDA, Y., FUJITA, T., SUGIURA, Y. & KOGA, S. (1968). Physical properties of water in spores of *Bacillus megaterium*. *Journal of General and Applied Microbiology*, **14**, 217–26.

MANDELSTAM, J. (1976). Bacterial sporulation: a problem in the biochemistry and genetics of a primitive developmental system. *Proceedings of the Royal Society of London, B*, **193**, 89–106.

MANDELSTAM, J., KAY, D. & HRANUELI, D. (1975). Biochemistry and morphology of Stage I in sporulation of *Bacillus subtilis* cells. In *Spores VI*, pp. 181–8, American Society for Microbiology.

MARQUIS, R. E. (1968). Salt-induced contraction of bacterial cell walls. *Journal of Bacteriology*, **95**, 775–81.

MARSHALL, B. J. & MURRELL, W. G. (1970). Biophysical analysis of the spore. *Journal of Applied Bacteriology*, **33**, 103–29.

MEYER, D. J. & JONES, C. W. (1973). Oxidative phosphorylation in bacteria which contain different cytochrome oxidases. *European Journal of Biochemistry*, **36**, 144–51.

MURRELL, W. G. (1967). The biochemistry of the bacterial spore. *Advances in Microbial Physiology*, **1**, 133–251.

MURRELL, W. G. (1969). Chemical composition of spores and spore structures. In *The Bacterial Spore* ed. G. W. Gould & A. Hurst, pp. 215–73. New York: Academic Press.

MURRELL, W. G. & WARTH, A. D. (1964). Composition and heat resistance of bacterial spores. In *Spores III*, pp. 1–24. American Society for Microbiology.

NELSON, D. L. & KORNBERG, A. (1970). Biochemical studies of bacterial sporulation and germination. XVIII. Free amino acids in spores. *Journal of Biological Chemistry*, **245**, 1128–37.

PEARCE, S. M. & FITZ-JAMES, P. C. (1971*a*). Sporulation of a cortexless mutant of a variant of *Bacillus subtilis*. *Journal of Bacteriology*, **105**, 339–48.

PEARCE, S. M. & FITZ-JAMES, P. C. (1971*b*). Spore refractility in variants of *Bacillus cereus* treated with actinomycin D. *Journal of Bacteriology*, **107**, 337–44.

PIGGOT, P. J. & COOTE, J. G. (1976). Genetic aspects of bacterial endospore formation. *Bacteriological Reviews*, **40**, 908–62.

ROGERS, H. J. (1977). Peptidoglycans (mucopeptides), structure, form and function. In *Spore Research 1976*, ed. A. N. Barker, J. Wolf, D. J. Ellar, G. J. Dring & G. W. Gould, in press. London: Academic Press.

RYTER, A. (1965). Étude morphologique de la sporulation de *Bacillus subtilis*. *Annales de l'Institut Pasteur*, **108**, 40–60.

RYTER, A., BLOOM, B. & AUBERT, J. P. (1966). Localisation intracellulaire des acides ribonucléiques synthetisés pendent la sporulation chez *Bacillus subtilis*. *Compte rendue Hebdomadaire des Séances de l'Académies des Sciences D*, **262**, 1305–7.

RYTER, A., IONESCO, H. & SCHAEFFER, P. (1961). Étude au microscope éléctronique de mutants asporogènes de *Bacillus subtilis*. *Compte rendue Hebdomadaire des Séances de l'Académies des Sciences D*, **252**, 3675–7.

SCHAEFFER, P., IONESCO, H., RYTER, A. & BALASSA, G. (1963). La sporulation de *Bacillus subtilis*: *étude génétique et physiologique*. *Colloques Internationaux du Centre National de la Recherche Scièntifique, Sciences Humaines*, 529–44.

SETLOW, P. (1975*a*). Identification and localisation of the major proteins degraded during germination of *Bacillus megaterium* spores. *Journal of Biological Chemistry*, **250**, 8159–67.

SETLOW, P. (1975*b*). Purification and properties of some unique low molecular weight basic proteins degraded during germination of *Bacillus megaterium* spores. *Journal of Biological Chemistry*, **250**, 8168–73.

SETLOW, P. (1976). Purification and properties of a specific proteolytic enzyme present in spores of *Bacillus megaterium*. *Journal of Biological Chemistry*, **251**, 7853–62.

SETLOW, P. & PRIMUS, G. (1975). Protein metabolism during germination of *Bacillus megaterium* spores. I. Protein synthesis and amino acid metabolism. *Journal of Biological Chemistry*, **250**, 623–30.

SETLOW, P. & WAITES, W. M. (1976). Identification of several unique low-molecular-weight basic proteins in dormant spores of *Clostridium bifermentans* and their degradation during spore germination. *Journal of Bacteriology*, **127**, 1015–17.

SILVER, S., TOTH, K. & SCRIBNER, H. (1975). Facilitated transport of calcium by cells and subcellular membranes of *Bacillus subtilis* and *Escherichia coli*. *Journal of Bacteriology*, **122**, 880–5.

SINGH, R. P., SETLOW, B. & SETLOW, P. (1977). Levels of small molecules and enzymes in the mother cell and forespore of sporulating *Bacillus megaterium*. *Journal of Bacteriology*, **130**, 1130–8.

SLEPECKY, R. A. (1972). Ecology of bacterial sporeformers. In *Spores V*, pp. 297–313. American Society for Microbiology.

STEWART, G. S. A. B. & ELLAR, D. J. (1977). Germination and subsequent outgrowth capacity of isolated forespores compared with mature spores of *Bacillus megaterium* KM. *Biochemical Society Transactions*, **5**, 116–18.

SZULMAJSTER, J. (1973). Inititation of bacterial sporogenesis. In *Microbial Differentiation*; *Symposia of the Society for General Microbiology*, **23**, ed. J. M. Ashworth & J. E. Smith, pp. 45–83. Cambridge University Press.

TIPPER, D. J. & GAUTHIER, J. J. (1972). Structure of the bacterial endospore. In *Spores V*, pp. 3–12. American Society for Microbiology.

TIPPER, D. J. & LINNETT, P. E. (1976). Distribution of peptidoglycan synthetase activities between sporangia and forespores in sporulating cells of *Bacillus sphaericus*. *Journal of Bacteriology*, **126**, 213–21.

TOCHIKUBO, K. (1971). Changes in terminal respiratory pathways of *Bacillus subtilis* during germination, outgrowth and vegetative growth. *Journal of Bacteriology*, **108**, 652–61.

WALKER, P. D. (1970). Cytology of spore formation and germination. *Journal of Applied Bacteriology*, **33**, 1–12.

WARTH, A. D. & STROMINGER, J. L. (1969). Structure of the peptidoglycan of bacterial spores: occurrence of the lactam of muramic acid. *Proceedings of the National Academy of Sciences, USA*, **64**, 528–35.

WARTH, A. D. & STROMINGER, J. L. (1972). Structure of the peptidoglycan from spores of *Bacillus subtilis*. *Biochemistry*, **11**, 1389–95.

WEBER, M. M. & BROADBENT, D. A. (1975). Electron transport in membranes from spores and from vegetative and mother cells of *Bacillus subtilis*. In *Spores VI*, pp. 411–17. American Society for Microbiology.

WILKINSON, B. J., DEANS, J. A. & ELLAR, D. J. (1975). Biochemical evidence for the reversed polarity of the outer membrane of the bacterial forespore. *Biochemical Journal*, **152**, 561–9.

WILKINSON, B. J. & ELLAR, D. J. (1975). Morphogenesis of the membrane bound electron transport system in sporulating *Bacillus megaterium* KM. *European Journal of Biochemistry*, **55**, 131–9.

WILKINSON, B. J., ELLAR, D. J., SCOTT, I. R. & KONCEWICZ, M. A. (1977). Rapid, chloramphenicol-resistant activation of membrane electron transport on germination of *Bacillus* spores. *Nature, London*, **266**, 174–6.

WOODS, D. D. (1965). The architecture of the microbial cell. In *Function and Structure in Micro-organisms*; *Symposia of the Society for General Microbiology*, **15**, ed. M. R. Pollock & M. H. Richmond, pp. 1–7. Cambridge University Press.

YOUNG, I. E. & FITZ-JAMES, P. C. (1959a). Chemical and morphological studies of bacterial spore formation. I. The formation of spores in *Bacillus cereus*. *Journal of Biophysical and Biochemical Cytology*, **6**, 467–81.

YOUNG, I. E. & FITZ-JAMES, P. C. (1959b). Chemical and morphological studies of bacterial spore formation. II. Spore and parasporal protein formation in *Bacillus cerus* var. Alesti. *Journal of Biophysical and Biochemical Cytology*, **6**, 483–98.

ZYTKOVICZ, T. H. & HALVORSON, H. O. (1972). Some characteristics of dipicolinic acid less mutant spores of *Bacillus cereus*, *Bacillus megaterium* and *Bacillus subtilis*. In *Spores V*, pp. 49–52. American Society for Microbiology.

THE GAS VESICLES OF AQUATIC PROKARYOTES

A. E. WALSBY

Marine Science Laboratories,
Menai Bridge, Anglesey LL59 5EH, Wales, UK

INTRODUCTION

The gas vesicle is the submicroscopic component of gas vacuoles, which provide aquatic prokaryotes with buoyancy. It is probably the simplest of all organelles, being made up of a single protein which forms a monolayer enclosing a hollow space into which gases diffuse. In a sense the creation, maintenance and regulation of the space can be regarded as the *direct* function of the gas vesicle. The *biological* functions suggested for gas vesicles all depend on the space itself, rather than on the membrane which encloses it. The functions of the gas vesicle should, therefore, be considered at these two levels.

GAS VESICLE STRUCTURE

In this section I will summarise the basic information on gas vesicle structure which I have reviewed previously (Walsby, 1972a, 1975) and will consider in more detail the recent developments and aspects in need of further investigation.

Gas vacuoles and gas vesicles

Gas vacuoles were discovered at the end of the last century. Observed under the light microscope as objects of irregular outline and high refractility they could be readily distinguished from all other refractile granules by their disappearance on being subjected to a hydrostatic pressure of several bars. The term 'gas vacuole' was coined by Klebahn (1895) who showed that gas could be obtained from them. However, the first description of these structures was by Winogradsky (1888) who with great foresight referred to them as 'hohlungen' (hollow cavities). The early work on gas vacuoles of blue-green algae is reviewed by Fogg (1941).

The gas vesicles which make up gas vacuoles were discovered by Bowen & Jensen (1965) using conventional electron microscopy of thin-sectioned blue-green algae. They demonstrated that the disappearance of the vacuoles under pressure was explained by the constituent vesicles collapsing flat and thereby losing their gas-filled, hollow spaces. Similar gas vesicles have now been demonstrated in many blue-green algae, photosynthetic bacteria, halobacteria and a wide variety of heterotrophic bacteria from various taxonomic orders (see Smith & Peat, 1967a; Cohen-Bazire, Kunisawa & Pfennig, 1969; Walsby, 1972a, 1974a; Caldwell & Tiedje, 1975; Dubinina & Kuznetsov, 1976). They are clearly homologous structures in all of these groups. Their shape is typically cylindrical with conical ends. Gas vesicles of blue-green algae are of about 70 nm diameter and of mean length 400 nm (maximum length about 1 μm), though rather longer ones occur in *Nostoc muscorum* (Waaland & Branton, 1969). Those in the green photosynthetic bacteria are very similar, while the vesicles in the purple photosynthetic sulphur bacteria and in many heterotrophic bacteria are rather wider (100 to 120 nm) and shorter (300 nm maximum). The halobacterial gas vesicles are exceptional in being much wider (up to 300 nm) and usually lemon-shaped, though cylindrical ones also occur (Larsen, Omang & Steensland, 1967; Stoeckenius & Kunau, 1968). Dubinina & Kuznetsov (1976) have recently described a bacterium with gas vesicles only 30 nm wide.

A constant feature of all gas vesicles is that they are made up of ribs, 4.5 nm wide, which lie normal to the long axis of the structure (Plate 1). It seems likely that these ribs represent turns of a shallow spiral rather than stacks of hoops (Jost & Jones, 1970). The ribbed construction is visible from both the inside (in freeze-fractured specimens) and the outside (by negative staining, metal shadowing and freeze-etching of isolated vesicles). They are also clear in thin-sectioned material when the section goes parallel to the long axis. A rather prominent rib is seen at the middle of the structure (Waaland & Branton, 1969). The gas vesicle may be made of two identical halves which are joined back-to-back at this rib (Walsby, 1972a). Occasionally the two halves separate when the vesicle is collapsed in a drying droplet (Walsby & Buckland, 1969).

The gas vesicle membrane is 2 nm thick according to X-ray diffraction studies (see below); electron microscopy tends to give an overestimate of this dimension (see Walsby, 1972a).

Chemical composition

Gas vesicles are easily isolated and purified in an intact state but unusual precautions must be taken to avoid exposing them to pressures which might cause their collapse during the process (Walsby, 1974b). The cells are lysed by chemical (Larsen *et al.*, 1967), osmotic-shrinkage

(Walsby & Buckland, 1969) or enzymatic (Jones & Jost, 1970) procedures. The gas vesicles are then separated by centrifugation which causes them to float to the supernatant surface, but under conditions calculated to avoid supercritical collapse pressures (Walsby, 1974b).

Chemical analysis of gas vesicles isolated from two blue-green algae (Walsby & Buckland, 1969; Jones & Jost, 1970), two halobacteria (Krantz & Ballou, 1973; Falkenberg, 1974) and a heterotrophic bacterium (Konopka, Lara & Staley, 1977) shows that protein is the sole organic constituent. Small quantities of phosphate are also recorded in two of the preparations (Krantz & Ballou, 1973; Konopka et al., 1977) but are not present in the others. In each case, lipids were said to be undetectable (<1%) as were carbohydrates, with one possible exception (Krantz & Ballou, 1973). Several workers have pointed out that gas vesicle 'membranes' are very different from typical cell unit membranes containing lipid bilayers. Smith, Peat & Bailey (1969) drew a parallel between the proteinaceous coats of the cylindrical viruses and gas vesicles, which remains apt despite the fact that it was originally based on mistaken evidence (see Falkenberg, Buckland & Walsby, 1972). It does not necessarily follow that because gas vesicles resemble viral structures they are derived from them, and it certainly cannot be concluded on this basis that they are pathogenic inclusions (Fjerdingstad, 1972).

Evidence for a single protein, and its molecular weight

Jones & Jost (1971) found a single mobile protein, MW 14300, by polyacrylamide–urea gel electrophoresis of the gas vesicles from *Microcystis aeruginosa*. Similar results were obtained for the gas vesicles of *Anabaena flos-aquae* in which the empirical amino acid composition suggested a MW of 15100 (Falkenberg et al., 1972). As discussed previously (Walsby, 1972a, 1975), the evidence presented did not rigorously preclude the presence of other non-mobile proteins and other molecular weight values. However, further confirmatory evidence has accumulated. Falkenberg (1974) has produced evidence vindicating the molecular weight estimates obtained by the polyacrylamide–urea gels used. They gave an identical value (15100) for the anabaena vesicle protein. Weathers, Jost & Lamport (1977) carried out quantitative amino acid sequence analysis of the microcystis protein which provides additional support. Assuming only one residue of the least abundant amino acid (tryptophan) per protein molecule they calculated an empirical MW of 7300. However, 89 non-overlapping residues were separated and partially sequenced, accounting for a MW of 9800, and the yield of the N-terminal amino acid, alanine, indicated a MW for the protein of 11440 (calculated from their data assuming 100% recovery of the terminal alanine derivative). They therefore concluded that there must be two tryptophan residues, giving a MW of 14600, close to the value given by electrophoresis. There is, however, a confusing conflict with a view expressed elsewhere in their paper that the protein might exist as a 7000 MW unit which aggregates during electrophoresis, and a reference to unpublished data suggesting the existence of a species of 21 500 MW.

The only other conflicting evidence on molecular weight comes from the recent work of Konopka *et al.* (1977) on gas vesicles isolated from the heterotrophic bacterium *Microcyclus aquaticus*. They obtained an estimate of 50000 MW by polyacrylamide gel electrophoresis of gas vesicles suspended in 0.1% sodium dodecyl sulphate, conditions usually considered as giving a dependable correlation between mobility and size. They mention, however, that proteins with such a high degree of hydrophobicity may give anomalous migration rates. Moreover, it can be seen from their figure that a large proportion of the protein remained at the origin of their electrophoresis gel so that, again, the mobile band may represent a minor component or a contaminant. Konopka *et al.* (1977) have demonstrated, by an immunological technique, that the microcyclus protein is similar to that of the blue-green algae, and this conclusion is supported by similarities in their amino acid composition.

Examples with more than one protein

In some other organisms it has been suggested that there may be more than one gas vesicle protein. The only case where this is proven is in *Halobacterium* sp. strain 5 (of Larsen *et al.*, 1967) where Falkenberg (1974) found two components, A of MW 12900 representing about 20% of the total protein, and B of MW 15100 representing the remaining 80%. Falkenberg speculated on the smaller molecule being a specific component of the conical or cylindrical regions of the vesicle but pointed out that the ratio between these regions (1.2 to 1) was very different from the ratio of the two proteins (4.0 to 1). Although the gas vesicles of halobacteria are atypical in respect of their shape, size, strength and amino acid composition, it has not been proved that they also differ in needing two proteins to produce them. It might be that they could be made with one, but that the organism has two proteins which function equally to produce gas vesicles. Moreover, it has not been demonstrated that the two proteins are present in the same individual vesicles

Table 1. *Amino acid composition of gas vesicle proteins from five aquatic prokaryotes*

Mole %

Amino acid	(a) Anabaena flos-aquae		(b) Microcystis aeruginosa	(c) Microcyclus aquaticus	(d) Halo-bacterium halobium	(e) Halobacterium sp. strain 5 Protein A		Protein B	
Alanine	15.9	(23)	15.2	18.9	13.4	14.3	(18)	17.9	(26)
Valine	12.6	(18)	15.0	21.6	16.4	16.3	(20)	15.9	(23)
Glutamic acid	12.4	(18)	9.2	8.1	14.1	13.4	(17)	14.4	(21)
Isoleucine	10.1	(14)	10.3	4.1	5.9	5.6	(7)	5.4	(8)
Leucine	10.0	(14)	9.3	9.9	10.6	9.5	(12)	9.0	(13)
Serine	9.9	(14)	9.3	7.1	6.2	7.4	(9)	5.5	(8)
Aspartic acid	5.6	(8)	6.3	7.1	8.1	7.3	(9)	6.7	(10)
Threonine	4.9	(7)	4.5	3.6	4.4	4.8	(6)	4.2	(6)
Lysine	4.7	(7)	4.2	4.8	3.2	3.4	(4)	2.3	(4)
Glycine	4.3	(6)	4.5	6.2	6.3	6.4	(8)	6.1	(9)
Arginine	4.2	(6)	4.0	4.9	4.3	3.5	(4)	4.3	(6)
Tyrosine	2.8	(4)	2.4	2.5	1.5	1.7	(2)	1.5	(2)
Proline	1.4	(2)	5.6	1.2	1.7	1.8	(2)	2.6	(4)
Tryptophan	0.7	(1)	n.d.	0	0.4	n.g.	(1)	n.g.	(1)
Phenylalanine	0.6	(1)	0.4	0	1.6	1.7	(2)	1.5	(2)
Histidine	0	—	tr.	tr.	1.6	1.6	(2)	1.4	(2)
Methionine	0	—	0	0	0.4	0	—	0	—
Cystine	0	—	0	0	0	0	—	0	—
Amide N	5.0	(7)	n.d.	n.d.	n.d.	n.d.		n.d.	

n.d. = not determined; n.g. = not given; tr = trace.

Data from (a) Falkenberg *et al.* (1972); (b) Weathers *et al.* (1977); (c) Konopka *et al.* (1977); (d) Krantz & Ballou (1973); (e) Falkenberg (1974). All figures adjusted to one decimal place.

Figures given in brackets are numbers of amino acid residues per protein molecule, assuming a molecular weight of 13000–15000.

If gas vesicle isoproteins are produced, the redundancy can perhaps be explained by the abundance of gas vesicle proteins in cells. Jones & Jost (1970) estimated that gas vesicles account for about 3.5% of the total protein in *M. aeruginosa*, and Walsby (1977a) commented that they are one of the 10 most abundant proteins in *A. flos-aquae*. They are therefore produced at least 100 times more rapidly than the average cell protein and transcription of the relevant messenger RNA may be a rate limiting step in their formation (Walsby, 1976). Mutants which are partially defective in gas vesicle production have been isolated from both *A. flos-aquae* and the heterotroph *Prosthecomicrobium pneumaticum*, and it has been suggested that they may have lost one copy of the gas vesicle gene (Walsby, 1976, 1977a). These mutants produce gas vesicles too slowly to render the growing cells buoyant and would be selected out in their natural habitats (see below p. 349).

Konopka, Staley & Lara (1975) suggested that *M. aquaticus* might

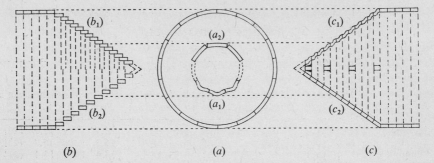

Fig. 1. A representation of the possible, different geometrical relationships between the protein molecules in the gas vesicle membrane. It is assumed that the molecules are in their relaxed form and make perfect contact with their neighbours in the cylindrical section; they must either be twisted or make imperfect contact in the conical end sections.

(a) Viewed in cross-section. The width of the particles has been greatly exaggerated (more than 5-fold) to emphasise the necessary twisting (a_1) or imperfect contact (a_2) in the cone section. The distortion increases towards the centre.

(b) and (c) Viewed in longitudinal section. In each case the 2 nm thickness and 4.6 nm width of the molecules, and the 70 nm diameter of the gas vesicle cylinder are reproduced to scale, and the correct 70° end angle (Jost & Jones, 1970) of the cone is used. In (b) two alternatives are shown; the molecules are not distorted but their mutual contacts are. In (b_1) they overlap, generating a reduced rib periodicity of 2.86 nm (=2 nm/tan 35°). In (b_2) the rib periodicity is maintained at 4.6 nm and this results in loss of contact between adjacent ribs, apparently ruling out this alternative. If the molecules are to touch only at the corners (as shown in the first molecule in the cone in (b_2)) this will result in an end cone angle of 47° (2 × arc tan 2 nm/4.6 nm), smaller than observed.

In (c_1) the molecules are twisted into s-bends which permits the area of contact between adjacent ribs to be maintained planar to the rib. In (c_2) the molecules are not twisted and the area of contact now lies at an angle of 145° to the rib. This will result in a progressive decrease in the angle between the end faces of adjacent molecules in the same rib, that will impair the contact with molecules in the next rib (not shown in this simplified diagram). In both (c_1) and (c_2) the rib periodicity is reduced to 3.77 nm (= 4.6 nm × cos 35°) in the cone.

Because the circumference of each successive rib in the cone decreases by 16.6 nm (= 3.77 nm × 2 π tan 35°) the molecules occupy spaces which become increasingly wedge-shaped towards the end of the cone (examples shown on the principal axis in (c)) resulting in loss of contact between neighbours in the same rib. This problem does not occur in (b).

have more than one gas vesicle protein on other grounds; they isolated a mutant which produced only 'juvenile' gas vesicles with opposing conical end caps. The simplest explanation for this would be that the cylindrical part is made from a different protein which has mutated to a non-functional form. There are, however, other possibilities. For example the mutant may lack a co-factor needed to assemble the protein in the cylindrical mode. An explanation is possible with a single protein even if no co-factors are required. The wild-type protein would possess the flexibility to accommodate the different relationships with its neighbours in the cone and cylinder, as summarised in Fig. 1. Possibly the mutant form of the protein lacks the flexibility to form both sorts of relationship (the straight and sloping in Fig. 1(a). As shown in Fig.

1(*b*), the angle subtended by the particles will also differ markedly in different parts of the cone, the difference being particularly marked at its tip. I would like to suggest that the diameter of the gas vesicle is determined by these mutual relationships. Thus the protein is bent out of its most relaxed form in the tighter circles of the cones, but achieves its minimum free energy in circles having the diameter of the central cylinder.

Konopka *et al.* (1977) have subsequently reported that they have been able to detect only a single protein in gas vesicles they isolated from *M. aquaticus*.

GAS VESICLE PROTEIN STRUCTURE AND ARRANGEMENT

Accepting that the gas vesicle is built from a single protein, the complete solution of its structure will require determination of the protein's amino acid sequence, knowledge of its secondary and tertiary structure, how it is arranged in the membrane, and the quasi-equivalent arrangements which accommodate the different relationships in the cone and cylinder.

Amino acid composition of the protein

The amino acid composition of the gas vesicles isolated from five species of prokaryotes is given in Table 1. The gas vesicles of the blue-green algae were more than 97% pure according to a ^{14}C cross-labelling technique. The purity of the other preparations was said to be good on the basis of both electron microscopy, which indicated the absence of contaminating structures, and chemical analysis, which showed the absence of lipids and nucleic acids.

General features common to all of the gas vesicle proteins are: (1) the high proportion, $> 50\%$, of hydrophobic amino acids; (2) the low proportion of aromatic amino acids; and (3), with one possible exception, the absence of sulphur amino acids. Cystine is definitely absent in each case, so that disulphide bridges are precluded from the gas vesicle protein structure.

As is to be expected, the gas vesicle proteins of the two blue-green algae are closely similar to one another, as are those of the two halo-bacteria. The differences between components A and B from *Halo-bacterium* sp. strain 5 could be explained by the loss of 23 residues and a mutation at one additional point. The vesicle protein of the hetero-trophic bacterium *M. aquaticus* is distinctive in the relatively higher proportions of alanine and valine, the most abundant amino acids in each protein; the remaining amino acids show roughly equal numbers of affinities with the other two groups.

As Konopka *et al.* (1977) have pointed out, the composition of the gas vesicle protein from these diverse prokaryotic groups appears to be quite highly conserved, as expected from the number of structural constraints on the molecule.

Amino acid sequence

The sequencing of proteins that are as hydrophobic as the gas vesicle protein is a formidable task owing to the insolubility of many of the peptides generated by its digestion. Nevertheless, Weathers *et al.* (1977) have isolated more than half of the expected peptides, seven by treatment with trypsin and two with N-bromosuccinimide, and have obtained sequence information on them. Some of the sequences overlapped but the relative positions of others in the protein has not yet been determined. Two features are of particular interest. First, there is a long stretch of 15 aliphatic residues representing one of the most hydrophobic stretches of a protein yet described. The significance of this is discussed below (p. 342). Secondly, there is an octapeptide (–ala–glu–ala–val–gly–leu–thr–glu–) which is said to repeat three times in the molecule. They suggest that this serves as a 'structural building block for the membrane'. The possible locations of the different sequenced peptides demonstrates that there are separate polar and nonpolar regions in the primary structure. This correlates with the protein having different hydrophilic and hydrophobic surfaces (see below, p. 338).

Secondary structure

Infrared spectroscopy of films of microcystis gas vesicles indicated the presence of α-helix, random-coil and β-sheet conformations in the protein (Jones & Jost, 1971). Similar conclusions were drawn from infrared spectroscopy and circular dicroism studies on the gas vesicles of *A. flos-aquae* (Buckland, 1971). Information on the orientation of the β-sheet from X-ray diffraction studies is given below.

Arrangement of the protein in the membrane

It is reasonable to assume that the protein molecule is arranged in a repeating pattern along the ribs of the gas vesicle, but it is not clear whether it corresponds to any of the particles claimed to be seen under the electron microscope (see the discussion in Walsby, 1972*a*). A particularly unsatisfactory feature of these putative particles is the irregularity of their periodicity; optical diffraction analysis of electron microscope images of single vesicles reveals no clear reflection in the plane of the ribs (D. Branton & A. E. Walsby, unpublished).

Jost and Jones (1970) suggested on the basis of electron microscopy of shadowed specimens that there were globular particles repeating at intervals of 2.8 to 3.5 nm along the ribs. Assuming a mean value of 3 nm they pointed out that the volume of these particles would correspond with that of the microcystis gas vesicle protein (MW 14 300). Further corroboration of this model was sought by Jost, Jones and Weathers (1971) by showing that the number of such particles required to cover the area of an average gas vesicle corresponded with the number of molecules per vesicle calculated independently from determinations of the number of vesicles in a sample of known dry weight. As pointed out previously (Walsby, 1972a), the latter measurement, which relates a volume of protein to an area of membrane, establishes an independent estimate of membrane mean thickness of about 1.7 nm but does not indicate how the protein is arranged. Another estimate of the mean thickness has been obtained from measurements of the weight of protein in a gas vesicle suspension in which the volume of gas-filled space has been determined by manometric methods; the value is 1.9 nm for the anabaena gas vesicle (Walsby, 1972a).

X-ray and neutron diffraction studies

Recent X-ray diffraction studies by Blaurock and his collaborators have helped to bridge the information from electron microscopy and protein chemistry of the membrane. The diffraction patterns generated by a preparation of collapsed halobacterium gas vesicles, dried down to form a sheet which was placed edge on to the X-ray beam, showed a strong reflection with a Bragg spacing of 4 nm normal to the sheet. It was argued that this represented the stacking periodicity of the flattened, double membrane envelopes (Blaurock & Wober, 1976). The mean thickness of the membrane is, therefore, half this figure, 2 nm (as confirmed by the density measurements given above). The thickness of the membrane is unlikely to be uniform, however, as electron microscopy clearly shows both surfaces to be corrugated. Moreover, differences between the X-ray diffraction patterns of intact and collapsed membranes show that the ribs which form these corrugations intermesh at the gas-facing surfaces when the structure collapses. Within the thickness of the membrane are two layers of β-sheet with a centre-to-centre distance of 1 nm. The presence of the sheet is indicated by diag-

nostic Bragg spacings of 0.47 and 0.49 nm. The electron densities of
the two surfaces of the membrane are unequal. This is compatible with
the idea of hydrophobic amino acids (with their side chains richer in
the less dense C and H) being at the inner surface and the hydrophilic
amino acids (with side chains richer in the more dense O and N) being
more abundant at the outer surface (Blaurock & Wober, 1976). Analysis
of the collapsed gas vesicle membranes from the blue-green alga *A. flos-
aquae* showed that the membrane structure is identical in profile
(Blaurock & Walsby, 1976).

It was possible to orientate the longer gas vesicles of *A. flos-aquae*
by drying a thick paste of the intact structures in aqueous suspension,
down to a fibre suspended between two points. The vesicles tended to
align themselves with their long axes parallel to that of the fibre. Diffraction of this fibre held normal to the X-ray beam gave a pattern made
up of arcs subtending an angle of about 44° at the centre, indicating the
degree of orientation achieved. From the positions of the centres of
these arcs a reciprocal lattice was calculated which gave further information on the structure (Blaurock & Walsby, 1976). The β-sheet was again
indicated by the 0.47 nm Bragg spacing between adjacent parallel or
antiparallel chains, which were now seen to run at an angle of 35° to
the long axis of the vesicle. A spacing of 0.6 to 0.7 nm in the direction
of these chains was observed which corresponds to the repeating distance of the dipeptide unit in the chain. The distance between pairs of
chains, about 0.95 nm normal to the chains, gave a projected interval
of about 1.14 nm along the rib. It is tentatively suggested that the unit
cells in adjacent ribs are offset by 0.3 nm, i.e. the line between their
centres forms an angle of 86° to the rib. The volume of the unit cell
indicates a molecular weight of 7800, about half of that indicated by
electrophoresis and amino acid composition. It is thus possible that the
molecule comprises two similar parts, each made of two sets of two
chains in two layers, as indicated in Fig. 2. It is impossible to distinguish
the direction (left or right) of the inclined β-sheet from our data, as the
direction at the back of the vesicle will appear opposite to that at the
front. (Hence, the two sides of the diffraction pattern are mirror images).
It is also impossible to distinguish whether the two layers of β-sheet
within the membrane run in the same direction, but if they do, the ridges
formed by the protruding amino acid side chains in one layer would fit
between the grooves left between the corresponding ridges in the other
layer. This interlocking would stiffen the structure and it might explain
the rigidity of the gas vesicle and its resistance to collapse under pressure
(Blaurock & Walsby, 1976).

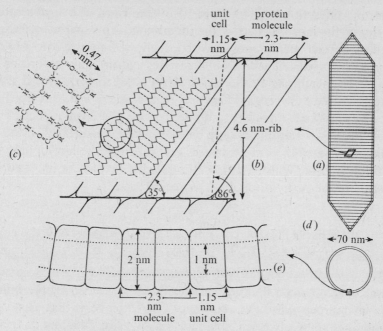

Fig. 2. The proposed structure of the blue-green algal gas vesicle, based on the results of chemical analysis and X-ray diffraction studies. (*a*) Longitudinal section of a vesicle showing the 2 nm thick wall and 4.6 nm wide ribs. (*b*) A section of a rib showing the 1.15 nm wide unit cells sloping at an angle of 35°. The protein molecule (2.3 nm wide) contains two unit cells. Its polypeptide chain is wound into eight parallel lengths in two layers (only one layer of β-sheet shown). There are hydrogen bonds between the adjacent lengths within each cell, between adjacent cells and between adjacent molecules, as shown. Note that the 35° angle of the peptide chains results in the dipeptide residues in adjacent cells being in horizontal register. Note also that the 86° angle made by the shortest (dashed) line between unit cells in adjacent ribs results in the unit cells being in alternate register. (*c*) A section of the β-sheet, here shown to be antiparallel (not yet proven). Note that the side groups (**R**) form ranks normal to the direction of the peptide chains, as mentioned in the text. (*d*) Cross-section of the vesicle. (*e*) A section of the vesicle wall showing the 1.15 nm wide unit cell and the planes of the two layers of β-sheet (dotted lines 1 nm apart). (Drawn after figure in Walsby, 1977*c*.)

Further information on the structure has been given by neutron diffraction studies on anabaena gas vesicles suspended in H_2O and ^2H_2O (D. L. Worcester, A. E. Blaurock & A. E. Walsby, in preparation; see Worcester, 1975, for a preliminary account). Because hydrogen and deuterium have neutron scattering amplitudes of different signs, a mixture (8% ^2H_2O) can be prepared with zero scattering density. When intact gas vesicles are suspended in such a mixture only the protein membrane scatters (as it has gas on one side and the zero scattering water on the other) and its thickness can be determined directly. The value given, 2±0.4 nm, again confirms that the thickness has been correctly interpreted from the electron diffraction patterns.

Neutron diffraction studies have also confirmed the low scattering density of the inner surface of the membrane. An experiment in which dried, collapsed membranes have been allowed to swell in ^2H_2O has also been carried out which confirms the hydrophobic nature of only one of the membrane surfaces. As ^2H_2O is taken up, there is a large increase in the first order diffraction of a 4 nm repeat, but not in the second order (i.e. equivalent to the first order of a 2 nm repeat). Hence the deuterated water enters only between alternate surfaces of the 2nm membrane. Since the outer surfaces are known to be hydrophilic from partition studies (Walsby, 1971), it is concluded that the inner surfaces must therefore be hydrophobic (Worcester, 1975).

THE FUNCTION OF CREATING, MAINTAINING AND REGULATING THE HOLLOW SPACE

Creating the space

It was once assumed that gas vesicles must be created by the inflation of a membrane with gas. However, the discovery that gas vesicle membranes are freely permeable to gases (Walsby, 1969) made this idea untenable. It was proposed that the hollow space in the gas vesicle must be a product of the way the structure was assembled from its constituent particles (of protein). The proteins would aggregate to form a cluster and would be mutually oriented in such a way that as more particles were added, a space would form within the cluster. The erection of the structure would require energy but gas would enter the hollow space by passive diffusion from the surrounding aqueous solution (Walsby, 1969). The contrasting hydrophilic and hydrophobic surfaces of the gas vesicle protein (see below) would be important in producing the required orientation of the protein during its assembly (Walsby, 1971, 1972a).

Waaland & Branton (1969) provided evidence of the de-novo production of gas vesicles by following the stages in their formation in *Nostoc muscorum*. They found conditions for inducing gas vesicle formation in this alga and demonstrated that the first new vesicles which appeared were small biconical structures. The successive stages indicated that these vesicles grew to a width of 70 nm and then expanded by the formation and elongation of the central cylindrical portion. Analysis of their micrographs gives an initial growth rate of 100 nm h^{-1}, equivalent to the addition of 22 ribs per hour, or the incorporation of 35 protein molecules (each measuring 2.3 nm along the rib), per minute. The growth rate subsequently declined.

After gas vesicles have been collapsed by pressure, new ones appear in the cells (Bowen & Jensen, 1965) but again the permeability of the membranes to gas precludes their being formed by re-inflation of the collapsed vesicles (Walsby, 1969). Lehmann & Jost (1971) confirmed this by following the stages of gas vesicle development in the manner described above; their figures indicate a similar gas vesicle assembly rate in *Microcystis aeruginosa*. A similar but more detailed study was carried out by Konopka *et al.* (1975) on *Microcyclus aquaticus*. From their plot of increase in gas vesicle area with time I calculate an incorporation rate of 34 molecules per minute. The similarity in growth rates in these three organisms may indicate that there is an intrinsic assembly time required for the insertion of a protein into a vesicle, about 1.7 seconds. This suggests, then, that proteins are incorporated at a localised point. Waaland & Branton (1969) thought that the growing point might be the distinctive central rib. This seems the most likely possibility; addition of particles to the ends of the structure would result in continuous rearrangement of the particles in the cones. Lehmann & Jost (1971) have suggested that the assembly kinetics of the gas vesicle cone and cylinder may be different, but the evidence for this is as yet inconclusive (see Walsby, 1972*a*; Konopka *et al.*, 1975).

Konopka *et al.* (1975) demonstrated by the use of inhibitors that gas vesicle formation was immediately dependent on protein synthesis, probably on the synthesis of the vesicle protein itself. Perhaps the assembly of the gas vesicle protein goes hand-in-glove with its synthesis; cells which have been prevented from forming gas vesicles, by being grown under pressure, do not appear to accumulate a store of gas vesicle precursors in a form which can be subsequently assembled when the pressure is released (A. E. Walsby & D. Branton, unpublished).

Enlargement of a gas vesicle takes place against pressures acting on it and therefore requires energy (Walsby, 1969). The source of this energy may be simply the difference in free energy of the vesicle protein in its unassembled state (with high energy values at the hydrophobic surfaces in contact with water) and the assembled state (with these surfaces hydrophobically bonded together and facing the gas-filled space). Several observations demonstrate an energy requirement: gas vesicle formation is inhibited by lack of oxygen and by dinitrophenol in halobacteria (Larsen *et al.*, 1967), by lack of light in a blue-green alga (Lehmann & Jost, 1971), and by KCN in *M. aquaticus* (Konopka *et al.*, 1975). In each case the energy requirement might be that needed for protein synthesis, which is dependent on ATP.

To determine the full requirements for gas vesicle assembly it would be useful to have a defined cell-free assembly system. Lehmann & Jost (1972) have published a preliminary report which suggests that concentrated slurries of broken cells remain effective in producing gas vesicles for a short period of time.

Maintenance of the hollow space

The problems of maintaining the hollow space are basically those of resisting collapse and keeping out water.

The gas vesicle membrane is rigid, permeable to gases and impermeable to water (Walsby, 1969, 1971). The structure has thus been likened to a porous pot from which liquid water is prevented from entering by surface tension acting at the wall of the pores through which the gas passes. In such a system hydrostatic pressures acting on the outer surface are borne principally by the wall itself and are not transmitted to the gas inside. The pressure of the gas is, at the rapidly achieved equilibrium, the same as the pressure of gas dissolved in the surrounding medium. These are the essential properties of the gas vesicle which have to be reconciled with its molecular structure.

Resisting collapse

The regular, round cross-sectional shape of gas vesicles, reminiscent of the form of submersible vessels and commercial gas containers, is obviously an adaptation to pressure. The gas vesicle will withstand moderate pressures with very little elastic (i.e. reversible) compression. A rise in hydrostatic pressure of 1 bar on the structure results in a barely detectible ($<0.1\%$) volume decrease according to light-scattering measurements. At certain pressures, however, the vesicle collapses irreversibly. When this happens the cylindrical portion flattens to a rectangular envelope; for geometrical reasons, splits occur between this and the end cones as they flatten separately. The collapsed double membranes are presumably cracked along their edges. It has been concluded that the gas escapes by diffusion through the permeable membrane, as it collapses, and dissolves without forming a bubble, though larger bubbles may subsequently form as a result of the ensuing supersaturation of the suspending water (Walsby, 1971).

The critical pressure (p_c) of a gas vesicle is the minimum pressure difference across the membrane which results in its collapse. It is determined from the pressure required in excess of the ambient pressure after the suspension has been equilibrated with air by shaking (Walsby,

1971). The collapse of the gas vesicles can be followed by nephelometry (Walsby, 1973). Critical pressures vary between individual vesicles from the same cell and the mean values show differences between species. The p_c varies from 0.6 to 1.6 bars in halobacteria, from 2 to 5 bars in certain photosynthetic sulphur bacteria (Walsby, 1971), 4 to 7 bars in a heterotrophic bacterium (Walsby, 1976), 4 to 8 bars in green bacteria (A. E. Clark & A. E. Walsby, unpublished) and over a similar range, or in some cases up to 10 bars, in many freshwater, planktonic blue-green algae (Walsby, 1971; Reynolds, 1973; Walsby & Klemer, 1974; E. H. Randall & A. E. Walsby, unpublished). In each example the range of values of p_c encountered is just sufficient to withstand the pressures normally experienced by the gas vesicles. These arise principally from the head of overlying water in the natural habitat and from the cell turgor pressure. Thus the halobacteria with weak vesicles apparently inhabit the surfaces of brine pools and have no turgor pressure, while planktonic blue-green algae may inhabit deeper lakes and have a cell turgor pressure of from 2 to 5 bars (Walsby, 1971). This economy of strength, in which gas vesicles are no stronger than they have to be, is in some cases a feature of a gas vesicle regulation mechanism (see below). It was also thought to be explained in mechanical terms: as the diameter of a hollow structure increases it becomes more efficient at enclosing space but its critical pressure decreases, assuming the thickness and strength of its wall remains constant. Until recently a general inverse correlation was found between the size and strength of gas vesicles in different organisms (from the wide, weak vesicles of halobacteria to the strong, narrow ones of the blue-green algae) indicating that the strength of the membrane protein might be about the same in each case. However, I have recently found that the marine blue-green alga *Trichodesmium erythraeum* has gas vesicles of exceptional strength, p_c 10 to 18 bars (Walsby, 1977b) despite their normal width (Van Baalen & Brown, 1969). Those of other species, *T. contortum* and *T. thiebautii* are even stronger, p_c 20 to 45 bars (Walsby, 1977b and in preparation) but the vesicle width has not yet been determined. The great strength of vesicle in these species is obviously an adaptation to the considerable depth at which these algae occur (at least 200 m; see Carpenter & McCarthy, 1975).

Obviously the strength of the gas vesicle membrane is dependent on the forces holding the constituent protein molecules together (Buckland & Walsby, 1971). Hydrophobic bonding seems likely in view of the high proportion of non-polar amino acids. Hydrogen bonds may link the β-sheets of adjacent molecules as well as those within molecules. Other features which might explain the rigidity of the protein membrane are discussed above (p. 336, and see Blaurock & Walsby, 1976).

In addition to hydrostatic pressures, interfacial tension at the highly curved surface of the gas vesicle in contact with water would tend to collapse the structure. Experiments in which gas vesicles are partitioned between oil and water demonstrate that their outer surfaces are hydrophilic. This evidently decreases the interfacial tension to a negligible value as addition of surface active agents to gas vesicles in aqueous suspension does not bring about any improvement in their critical pressure (Walsby, 1971).

Keeping out water

The gas vesicle membrane evidently serves to keep liquid water out of the hollow space, while at the same time admitting gas. What is the molecular basis of this semi-permeability?

First, it is unlikely that gases pass through the membrane by dissolving in it and diffusing out the other side as this would infer mobility of part of the membrane protein molecule. It is more reasonable to assume that gases pass through pores. We do not know if the pores are within or between the protein molecules but the latter seems more likely (see Walsby, 1972a). By manometric techniques and investigation of gas vesicle collapse it has been demonstrated that gas vesicle membranes are permeable to nitrogen, oxygen, argon (Walsby, 1969) hydrogen, carbon dioxide, carbon monoxide and methane (Walsby, 1971). It has not been possible to determine whether they are permeable to water molecules in the vapour phase but this seems possible because the pores must accommodate methane which is a larger molecule than water and also carbon monoxide which, like water, is polar (though it has a dipole moment 16 times smaller than water). On these grounds the polarity of the pore is unimportant (cf. Weathers *et al.*, 1977).

Liquid water can be kept out of the structure by surface tension if the pore has a hydrophobic lining or opens to a surface which is hydrophobic (Stoeckenius & Kunau, 1968; Walsby, 1969). Since the inner surface of the gas vesicle is now known to be hydrophobic (see p. 338) the hydrophobicity of the pore itself remains an open question. Perhaps the sequence of 15 hydrophobic amino acid residues, which Weathers *et al.* (1977) discovered in the gas vesicle protein, is located at the inner surface of the membrane rather than at the surface of the pore as they suggest. If this sequence forms part of the β-sheet (see Fig. 2b) its length, about 5 nm, could just be accommodated in a single run across the 4.5 nm rib if inclined at the observed angle of 35° (4.5 nm/cos 35° = 5.5 nm).

The permeability of the gas vesicle membrane to gas is probably incidental, and of no biological importance (see p. 352). There is no particular advantage in having a membrane which is impermeable to gases and it may not be feasible to construct such a membrane from a protein monolayer. The only structural advantage in having a gas permeable membrane rather than an impermeable one is that it will normally result in there being gas in the structure providing a supporting pressure of 1 bar (Walsby, 1971) although the energetic advantage of this is probably unimportant.

Regulation of the hollow space

Regulation of the amount of hollow space provided by gas vesicles is important in their main biological function, to provide buoyancy. The relative gas vesicle content of a cell can be determined by the respective rates at which intact gas vesicles and other cell material are produced and destroyed. Thus the trend for gas-vacuolation of various bacteria and blue-green algae to increase in the late exponential or stationary phase of growth (e.g. Larsen *et al.*, 1967; Smith & Peat, 1967*b*; Lehmann & Jost, 1971; Walsby, 1976) may be explained by gas vesicle production continuing after cell growth has been limited by factors not affecting protein synthesis.

Regulation of gas vesicle formation rates

The majority of planktonic blue-green algae seem to produce gas vesicles constitutively and although their gas vesicle content is known, in some instances, to vary according to defined conditions (Walsby, 1969; Meffert, 1971), it is not known whether they vary the rate of forming them; the same is true for the planktonic bacteria. There are a few blue-green algae which form gas vesicles only in hormogonia (Canabaeus, 1929; Singh & Tiwari, 1970) or under certain conditions (Waaland & Branton, 1969; Waaland, Waaland & Branton, 1971). These are evidently examples of on–off differentiation rather than regulation involving feedback.

The only organism in which regulation of gas vesicle formation has been investigated is the heterotrophic bacterium *M. aquaticus* strain M1. Van Ert & Staley (1971) showed that it formed gas vesicles only under certain culture conditions. Konopka (1977) has recently found that the synthesis of gas vesicle protein by this strain is inhibited by L-lysine. L-threonine and L-cystine also produced partial inhibition, probably by causing the accumulation of lysine, as all three amino acids are part of the same branched metabolic pathway. Konopka (1977) suggested that a decrease in the intracellular level of lysine might occur under unfavourable growth conditions. In natural waters the resulting increase in gas vesicle production might result in the organism floating up into more favourable conditions.

Regulation by collapse of intact vesicles

Gas vesicles in blue-green algae and bacteria are normally subjected to a cell turgor pressure of several bars. When this is removed by placing the cells in a hypertonic sucrose solution the true critical pressure of the gas vesicles may be determined. The difference between the critical pressure observed after and before suspending the cells in sucrose can therefore be used to give a measure of the turgor pressure (Walsby, 1971). In blue-green algae the turgor pressure may rise above the critical pressure of the weakest gas vesicles present, causing them to collapse (Walsby, 1971). Dinsdale & Walsby (1972) showed that there is a consistent rise in turgor pressure when the blue-green algae *A. flos-aquae* is transferred from a low to a high light intensity, and that this rise brings about collapse of sufficient of the alga's gas vesicles to make it negatively buoyant. They showed that the rise in turgor pressure was dependent on photosynthesis; up to half of the rise may be accounted for by the increased accumulation of low molecular weight products of photosynthesis dissolved in the cell water (Grant & Walsby, 1977), the balance presumably resulting from increased accumulation of inorganic ions from the culture medium. We have now demonstrated a similar turgor pressure rise in 6 other species of blue-green algae (E. H. Randall & A. E. Walsby, unpublished). Such a response appears to be important in regulating gas-vacuolation in these organisms. In order to work, it depends on the critical pressure of individual gas vesicles varying within each cell (Walsby, 1971). The cause of this variation is not known.

Regulation of gas vesicles by collapse appears a wasteful process, though no more so than the inactivation of enzymes and mRNA which features in the control of metabolic pathways. Although the collapsed vesicles cannot be reinflated it is possible that their proteins are recycled or are broken down to their constituent amino acids.

Gas vesicles are not regulated by collapse under turgor pressure in all organisms. It is out of the question in *Trichodesmium* where the gas vesicles are so strong (Walsby, 1977b), in *Halobacterium* where turgor pressure is absent (Walsby, 1971), and it seems not to occur in the heterotrophic bacterium *Prosthecomicrobium pneumaticum* (Walsby, 1976). The capacity for it exists in photosynthetic green bacteria as the ranges of turgor pressure and gas vesicle critical pressure encountered overlap, but it is not clear whether it actually operates (A. E. Clark & A. E. Walsby, unpublished).

THE BIOLOGICAL FUNCTIONS OF GAS VESICLES

All of the biological functions so far suggested for gas vesicles depend on the presence of the hollow space rather than of the membrane *per se*. They are providing (1) buoyancy, (2) light shielding, (3) channels for gas diffusion, (4) inert space and (5) gas-storage space. In my view the first is universal and usually the only function of gas vesicles; the second is possible, of restricted occurrence and then dependent on the first, the third is unlikely, the fourth very unlikely, and the fifth is impossible.

Buoyancy

Buoyancy has no relevance in stirred laboratory cultures and this seems to prompt biologists whose experience is limited to such environments to look for functions of gas vesicles which would more directly affect the metabolism of isolated cells. In the section below I summarise the evidence for gas vesicles providing buoyancy, and discuss its importance in the natural habitats of gas-vacuolate organisms.

Evidence for buoyancy

Klebahn (1895) proved that gas vacuoles provide buoyancy in blue-green algae which form waterblooms, by showing that algae which initially floated in water, immediately sank after their gas vacuoles were destroyed by pressure. This simple experiment has been repeated with many other gas-vacuolate organisms with the same result (see Walsby, 1972*a*, for references). Klebahn (1895, 1922) also showed by direct measurement that the specific gravity of the cells immediately increased as their gas vacuoles were destroyed; the gas vacuole space occupied 0.8% of the cell volume, 0.7% was required to give the cells the same density as water. This experiment has been repeated by Walsby (1970) and Reynolds (1972, 1973) but they obtained larger estimates (2.1%, 1.0%, and 2.4% respectively) to provide neutral buoyancy. I suspect that the lower values here result from an overestimate of cell volume by including interstitial water.

Efficiency at providing buoyancy

Gas vesicles provide buoyancy more efficiently than any liquid or solid substance. The density of a gas vesicle from *A. flos-aquae* is about 0.126 g cm^{-3} (Walsby, 1972*a*). Thus 1 g of gas vesicles occupies about 7.94 cm^3 and provides upthrust of 6.94 g in fresh water. Protein has an energy content of about 23.6 kJ g^{-1} (calculated from figures of S. Brody, quoted by Crisp, 1971) so that the energy required to provide 1 g of upthrust is 3.4 kJ. By contrast, lipids, the lightest solids or liquids likely to be accumulated by living cells, have a density of 0.86 g cm^{-3} or more (Sargent, 1976) and an energy content of 39.5 kJ g^{-1} (see Crisp, 1971). The equivalent upthrust therefore requires lipid which occupies over 6 times as much volume and takes more than 70 times as much energy to produce.

In considering the space required to provide buoyancy it may be more appropriate, for certain purposes, to consider the density of the gas vacuole as a whole. The interstitial cytoplasm-filled space between the constituent vesicles occupies about 9% of the total space in the gas vacuole of blue-green algae and results in the vacuole's overall density being 0.22 g cm^{-3} (Walsby, 1975). The wider gas vesicles of halobacteria are less dense than those of blue-green algae, only 0.082 g cm^{-3}, but they pack together less tightly (Walsby, 1972*a*). Gas bubbles are, of course, even less dense than gas vesicles and take less energy to produce. However, they are ruled out for reasons of instability. First, the excess pressure inside a bubble, resulting from surface tension, varies inversely with its diameter (it is 1.5 bars for a bubble of 1 μm). Secondly, the gas inside would be further compressed by cell turgor pressure (up to 5 bars inside blue-green algae). To maintain such a bubble, the surrounding cytoplasm would have to sustain gases in solution at several times the saturation concentration, which is quite out of the question in tiny cells suspended in water and dependent on diffusion for the exchange of metabolic O_2 and CO_2. Similar considerations rule out other types of gas-filled structures such as those with limited permeability and those having flexible walls. The rigid but permeable gas vesicle seems to provide the ideal solution (see Walsby, 1972*b*, for discussion of this topic).

Evaluation of the buoyancy role from ecological and physiological studies

Our thesis is that the buoyancy provided by gas vacuoles enables an organism to stay at the most favourable depth for its growth and perhaps enables it to move to that depth from other positions. Implicit in this idea is some aspect of buoyancy regulation by gas vacuoles, particularly when the organism is to be suspended at an intermediate depth. The speed with which an object sinks increases as the square of its radius (Stokes' Law); therefore large organisms must regulate their buoyancy more accurately than small ones, to stay within a given depth range. An organism which moves to the preferred depth must regulate its buoyancy with respect to factors which give information on whether it is above or below that depth. These will almost certainly be physico-chemical factors, such as light intensity, temperature, pH, and concentrations of oxygen, sulphide and other dissolved substances, which affect growth and which form stable gradients with depth.

To test our thesis it will be necessary (a) to show by culture studies that the conditions in the water occupied in nature by the organism are suitable for its growth; (b) to demonstrate that the accumulation or regulation of gas vacuoles is necessary to explain the organism's observed stratification or movement; (c) to demonstrate, where appropriate, the existence of a buoyancy-regulating mechanism which accounts for the observed movements; and (d) to bring together these observations by experiments in the natural habitat. The details will be different for different organisms. Although studies with no organism are complete in the sense of satisfying all these criteria, plausible accounts can be constructed in several cases.

Certain planktonic blue-green algae form populations which peak at the metalimnion of temperate lakes which stratify in summertime (e.g. Zimmermann, 1969). Extrapolating from culture studies on blue-green algae, though not planktonic species, it is proposed that the moderate light intensities, reduced oxygen tensions and higher concentrations of mineral nutrients, which prevail in the metalimnion, are particularly favourable to growth (Fogg, 1969). It has been proposed that the response in which gas vesicles are collapsed by the rising turgor pressure that is developed under increasing light intensity, permits blue-green algae to regulate their position on vertical light gradients in natural waters (Walsby, 1971).

Walsby & Klemer (1974) found that *Oscillatoria agardhii*, which formed a metalimnetic population in a stratified lake, had just sufficient gas vacuoles to render it neutrally buoyant. Without its gas vacuoles the alga sank at 0.8 m per day. Determinations *in situ* showed that the algal population grew at less than 5% per day. Without the buoyancy provided by gas vacuoles, losses from the metalimnetic population by sinking would have exceeded gains from growth. Walsby & Klemer

(1974) argued that in order to stratify the alga must actively regulate its buoyancy, and they provided evidence that it does from experiments in which the alga was suspended in bottles held at different depths in the lake. The buoyancy appeared to be determined by the alga's response to the daily mean light intensity, possibly by the turgor rise mechanism. The response to light was determined by an interaction with the concentration of a nutrient (combined N) as predicted for this mechanism (Dinsdale & Walsby, 1972). These results suggest that the alga might be growing at the most favourable depth, though this remains to be proved. In short term exposures the alga photosynthesises more efficiently nearer the surface (Baker, Brook & Klemer, 1969) but these higher rates would probably not be sustained for long in the nutrient-poor waters of this lake.

In less stable lakes, gas-vacuolate blue-green algae stratify when calm conditions permit. In such cases aggregation of cells or filaments into colonies enables the alga to move more rapidly to its preferred depth. In some instances there is evidence that large-colony forms respond quickly enough to migrate up and down on a daily pattern, in response to the varying light intensity. Under certain combinations of conditions the algae may form surface waterblooms that may result from failure of the buoyancy regulation mechanism and be detrimental to the alga. These aspects are discussed in detail by Reynolds & Walsby (1975).

As yet, no direct measurements on growth rate or buoyancy have been performed on gas-vacuolate bacteria in natural waters. However, a study of the photosynthetic and heterotrophic gas-vacuolate bacteria in Crose Mere, Shropshire, suggests that these anaerobic bacteria overwinter in the bottom mud and then invade the successive layers of overlying water as they become anaerobic. The gas vacuoles may be important in floating the bacteria upwards (A. E. Clark & A. E. Walsby, in preparation). It is possible that the bacteria are able to regulate their buoyancy as they don't accumulate in the overlying oxygenated layers. Alternatively, the gas vacuoles may simply provide the cells with approximately neutral buoyancy which permits the bacteria to form stable populations at certain depths where conditions allow growth. The smaller bacteria sink rather slowly (3 cm day^{-1}) even with their gas vacuoles collapsed, so accurate regulation of their gas-vacuolation is unnecessary. The ability to occupy intermediate depths is of obvious advantage to photosynthetic bacteria which require light in combination with anaerobic conditions. It also opens up an alternative niche for anaerobic heterotrophs which might otherwise sink to the overcrowded mud surface.

Petter (1932) suggested that gas vacuoles are important to halobacteria in floating them to the surface of brine pools where the availability of oxygen would be higher for these obligately aerobic organisms. Other gas-vacuolate aerobes should benefit in the same way, though no case has yet been directly investigated.

Evidence from competition experiments

It may be difficult to make observations in natural situations which test the role of gas vacuoles in providing buoyancy. An alternative approach is to study the responses of gas-vacuolate organisms in model environments which reproduce the relevant features of the natural habitat. The selective advantages of gas vacuoles are best assessed in competition experiments between the gas-vacuolate organism and its non-vacuolate mutant. If the advantage accrues from buoyancy, it should not show up in stirred cultures.

I isolated a mutant, nearly devoid of gas vacuoles, of the aerobic heterotroph *P. pneumaticum* described by Staley (1968). In shaken culture it showed the same exponential growth rate as the vacuolate wild-type and when mixed together in shaken culture, the ratio of the wild-type to the mutant did not significantly change. However, when the two were grown together in standing culture the wild-type, which floated at the surface, outgrew the mutant, which sank to the bottom. The results of other experiments supported the idea that the wild-type gained its advantage from the greater availability of oxygen at the culture surface (Walsby, 1976). The gas-vacuolate wild-type would probably enjoy similar advantages in stagnant pools, as Petter (1932) suggested.

M. J. Booker and I (manuscript in preparation) have isolated a gas-vacuoleless mutant from *A. flos-aquae*, a blue-green alga from Lake Windermere. The growth rates of the wild-type and mutant are about the same when the two are grown together in a stirred culture. However, in a large, thermally stratified water column, which we constructed to model a freshwater lake (Booker, Dinsdale & Walsby, 1976), we found that the wild-type, which remained suspended in the illuminated 'epilimnion' of the column, outgrew the mutant, which sank into the darker hypolimnion. Our column is shallower (2.2 m) than Lake Windermere, and the light gradient compressed by the more concentrated algal population. However, it seems likely that the non-vacuolate form would be at a similar disadvantage in lakes where its high sinking rate (0.4 m day^{-1}) would also lead to its more rapid disappearance from the euphotic zone.

Besides showing the advantages conferred by the gas vacuole buoyancy, these experiments also demonstrate that gas vacuoles have no direct effect on the intrinsic growth rate of these organisms (at least under the conditions investigated). Das & Singh (1976) have recorded that a gas-vacuoleless mutant of another alga grows slightly more slowly than its vacuolate wild-type but it is not clear if they took precautions to prevent the mutant from settling out.

Evidence for the buoyancy role from the occurrence of gas vacuoles

Gas vacuoles have now been recorded in nearly a hundred different prokaryotes representing 15 genera of blue-green algae, 9 genera of photosynthetic sulphur bacteria, 16 named genera of heterotrophic bacteria, and a number of unnamed forms distinguished on morphological characters. (For lists and references see Walsby, 1972a, 1974a; Caldwell & Tiedje, 1975; Dubinina & Kuznetsov, 1976). The buoyancy provided by gas vacuoles is relevant to microorganisms only when they are suspended in water. It is therefore highly significant that with only one apparent exception, all of these organisms are aquatic. (The exception is a bacterium with gas-vacuolate caps to its spores, found in soil from a rice field (Krasil'nikov, Duda & Pivovarov, 1971) which would be periodically flooded)). Further corroborative evidence for the importance of gas vacuole buoyancy is that gas-vacuolate forms are abundant amongst the prokaryotes occupying non-turbulent water bodies. Thus, not only are all the constitutively gas-vacuolate blue-green algae planktonic but all of the dominant planktonic forms have gas vacuoles (Reynolds & Walsby, 1975). According to the results of an extensive survey of British water bodies which we have just completed, gas-vacuolate bacteria are found in all lakes which form an anaerobic hypolimnion in summer (A. E. Clark & A. E. Walsby, in preparation). In some such lakes gas-vacuolate forms appear to dominate the bacterial plankton (Lauterborn, 1915; Skuja, 1956, 1964; Walsby, 1974a; Caldwell & Tiedje, 1975; Dubinina & Kuznetsov, 1976). About half the planktonic photosynthetic bacteria are gas-vacuolate; the others are flagellates. Hardly any gas-vacuolate prokaryotes possess flagella (Cohen-Bazire et al., 1969). This suggests that the gas vesicle and the flagellum provide alternative solutions to the problem of positioning a prokaryote cell in the vertical water column. Flagella are, of course, also effective in the horizontal plane.

Other functions

A number of other functions for gas vesicles have been suggested and these are discussed below. All of these functions, if discharged, would benefit aquatic and non-aquatic microorganisms alike. I view the absence of gas vesicles in non-aquatic prokaryotes as an indication that they do not perform any of these other functions efficiently enough to ensure their natural selection; i.e. the cost of producing gas vesicles outweighs the benefit they bring. However, in aquatic organisms where they are already present, providing buoyancy, gas vesicles might be put to additional uses with little additional cost.

Light shielding

Isolated gas vesicles suspended in water show very strong light scattering on account of their small size and the large refractive index difference between their gas spaces and the surrounding liquid. At wavelengths over 450 nm, the amount of light scattered by the gas space varies approximately as λ^{-4}, as expected for Rayleigh scattering (Walsby, 1970). The apparent absorption spectrum for the gas spaces in gas vacuoles in intact cells of *A. flos-aquae* is fairly flat over the visible spectrum, which suggests that the vesicles act together like mirrors, reflecting light (independently of wavelength) rather than scattering it (Shear & Walsby, 1975). In *N. muscorum* the apparent absorption spectrum is basically similar but has distinct troughs at the absorption maxima of the main photosynthetic pigments which demonstrates that the gas vacuoles prevent part of the incident illumination from reaching the photosynthetic lamellae (Waaland et al., 1971). It is not possible to determine the extent of light shielding quantitatively by this method, but an estimate can be obtained by comparing the rates of photosynthesis at limiting light intensities in cells with intact and collapsed gas vesicles; the rate should be higher in cells with the collapsed vesicles. In one organism investigated in this way, *A. flos-aquae*, the increase after collapsing the vesicles is only 4%, which is probably not significant (Shear & Walsby, 1975). It should be pointed out that light shielding by gas vesicles is only of selective advantage if it protects the cells containing the structures. Porter & Jost (1973, 1976) suggested a light-shielding effect by gas vacuoles in *Microcystis aeruginosa*. However, this seems to occur only in dense suspensions, and has consequently been interpreted as 'sacrificial shading' (Walsby, 1972a, Shear & Walsby, 1975) in which cells at the surface provide increased shading for those deep in the culture. In a competitive situation non-vacuolate mutants would benefit equally from this protection and no selection would result.

To obtain any significant light shielding *within* a cell may require the gas vesicle to be situated at the cell periphery (Van Baalen & Brown, 1969; Waaland *et al.*, 1971). Where gas vesicles are already present, providing buoyancy, this would only require storing the appropriate information and would not entail further expenditure of energy. The absence of gas vesicles from prokaryotes occupying terrestrial habitats such as rock surfaces which are subject to high insulation argues against their being efficient light shields.

Gas diffusion channels

Taylor, Lee & Bunt (1973) have made the suggestion that the peripheral, cylindrical gas vacuoles of *Trichodesmium* may provide a channel for the rapid diffusion of oxygen away from the central cytoplasm, perhaps providing sufficiently microaerophilic conditions for nitrogen fixation to occur in the centre of the cell.

The rate of diffusion of a solute in a liquid varies as the square of distance. Small molecules like gases diffuse very rapidly over distances equal to the radii of prokaryotic cells (usually $< 2\ \mu\text{m}$) and it is therefore unlikely that gas vacuoles have any importance in ventilating the cytoplasm. It is just possible, however, that they could perform such a function in those organisms like *Trichodesmium* which form large colonies.

E. J. Carpenter & I (manuscript in preparation) have found that collapsing the gas vesicles of *Trichodesmium* does bring about a decrease in nitrogenase activity. However, this reduction was observed with separated filaments as well as intact colonies, and also under anaerobic incubation conditions. It is unlikely, therefore that gas diffusion pathways are implicated. The fact that a similar reduction was observed in photosynthetic carbon fixation suggests that it may have resulted simply from a disturbance of the cellular architecture on collapsing the gas vesicles.

Providing inert space

It has been suggested that by occupying space gas vesicles produce an increase in the ratio of cell surface area to cytoplasm volume (Houwink, 1956) and that they bring about an increase in the concentration of cell solutes (M. Jost quoted by Wolk, 1973). These changes can be construed to affect cell growth or metabolism. A similar function could be imputed to any of the other cell components. Greater increases could be achieved by decreasing the cell size (Walsby, 1972a).

Gas storage

Gas vesicles cannot perform the function of gas storage as they are far too permeable to gases (Walsby, 1969).

I wish to thank Dr D. L. Worcester for supplying material prior to its publication. I also wish to thank Dr A. E. Blaurock for many useful discussions on gas vesicle structure, and also for suggesting improvements in Fig. 2.

REFERENCES

BAKER, A. L., BROOK, A. J. & KLEMER, A. R. (1969). Some photosynthetic characteristics of a naturally occurring population of *Oscillatoria agardhii* var. *isothrix*. *Limnology and Oceanography*, **14**, 327–33.

BLAUROCK, A. E. & WALSBY, A. E. (1976). Crystalline structure of the gas vesicle wall from *Anabaena flos-aquae*. *Journal of Molecular Biology*, **105**, 183–99.

BLAUROCK, A. E. & WOBER, W. (1976). Structure of the wall of *Halobacterium halobium* gas vesicles. *Journal of Molecular Biology*, **106**, 871–88.

BOOKER, M. J., DINSDALE, M. T. & WALSBY, A. E. (1976). A continuously monitored column for the study of stratification by planktonic organisms. *Limnology and Oceanography*, **21**, 915–9.

BOWEN, C. C. & JENSEN, T. E. (1965). Blue-green algae: fine structure of the gas vacuoles. *Science*, **147**, 1460–2.

BUCKLAND, B. A. (1971). Studies on gas vacuoles in blue-green algae and bacteria. Ph.D. Thesis, University of London, England.

BUCKLAND, B. & WALSBY, A. E. (1971). A study of the strength and stability of gas vesicles isolated from a blue-green alga. *Archiv für Mikrobiologie*, **79**, 327–37.

CALDWELL, D. E. & TIEDJE, J. M. (1975). A morphological study of anaerobic bacteria from the hypolimnia of two Michigan lakes. *Canadian Journal of Microbiology*, **21**, 362–76.

CANABAEUS, L. (1929). Über die Heterocysten und Gasvakuolen der Blaualgen und ihre Beziehung Zueinander. *Pflanzenforschung, Jena*, **13**, 1–48.

CARPENTER, E. J. & MCCARTHY, J. J. (1975). Nitrogen fixation and uptake of combined nitrogenous nutrients by *Oscillatoria* (*Trichodesmium*) *thiebautii* in the western Sargasso Sea. *Limnology and Oceanography*, **20**, 389–401.

COHEN-BAZIRE, G., KUNISAWA, R. & Pfennig, N. (1969). Comparative study of the structure of gas vacuoles. *Journal of Bacteriology*, **100**, 1049–61.

CRISP, D. J. (1971). Energy flow measurements. In *Methods for the Study of Marine Benthos*, IBP Handbook No. 16, ed. N. A. Holme & A. D. McIntyre, pp. 197–279. Oxford: Blackwell.

DAS, B. & SINGH, P. K. (1976). Isolation of gas vacuole-less mutants of the blue-green alga *Anabaenopsis raciborskii*. *Archives of Microbiology*, **111**, 195–6.

DINSDALE, M. T., & WALSBY, A. E. (1972). The interrelations of cell turgor pressure, gas vacuolation, and buoyancy in a blue-green alga. *Journal of Experimental Botany*, **23**, 561–70.

DUBININA, G. A. & KUZNETSOV, S. I. (1976). The ecological and morphological characteristics of microorganisms in Lesnaya Lamba (Karelia). *Internationale Revue der gesamten Hydrobiologie*, **61**, 1–19.

FALKENBERG, P. (1974). Kjemisk karakterisering av gassvakuole membranen hos halobakterier. Thesis. Institutt for Teknisk Biokjemi Norges. Tekniske Høgskole, Universitetet i Trondheim.

FALKENBERG, P., BUCKLAND, B. & WALSBY, A. E. (1972). Chemical composition of gas vesicles isolated from *Anabaena flos-aquae*. *Archiv für Mikrobiologie*, **85**, 304–9.

FJERDINGSTAD, E. (1972). Gas-vacuoles and other viruslike structures in blue green algae. *Schweizerische Zeitschrift für Hydrologie*, **34**, 135–54.

FOGG, G. E. (1941). The gas vacuoles of the *Myxophyceae* (*Cyanophyceae*). *Biological Reviews*, **16**, 205–17.

FOGG, G. E. (1969). The physiology of an alga nuisance. *Proceedings of the Royal Society of London B*, **173**, 175–89.

GRANT, N. G. & WALSBY, A. E. (1977). The contribution of photosynthate to turgor pressure rise in the planktonic blue-green alga *Anabaena flos-aquae*. *Journal of Experimental Botany*, **28**, 409–15.

HOUWINK, A. L. (1956). Flagella, gas vacuoles and cell wall structure in *Halobacterium halobium*; an electron microscope study. *Journal of General Microbiology*, **15**, 146–50.

JONES, D. D. & JOST, M. (1970). Isolation and chemical characterization of gas vacuole membranes from *Microcystis aeruginosa* Kuetz. emend. Elenkin. *Archiv. für Mikrobiologie*, **70**, 43–64.

JONES, D. D. & JOST, M. (1971). Characterization of the protein from gas vacuole membranes of the blue-green alga *Microcystis aeruginosa*. *Planta*, **100**, 277–87.

JOST, M. & JONES, D. D. (1970). Morphological parameters and macromolecular organization of gas vacuole membranes of *Microcystis aeruginosa* Kuetz. emend. Elenkin. *Canadian Journal of Microbiology*, **16**, 159–64.

JOST, M., JONES, D. D. & WEATHERS, P. J. (1971). Counting of gas vacuoles by electron microscopy in lysates and purified fractions of *Microcystis aeruginosa*. *Protoplasma*, **73**, 329–35.

KLEBAHN, H. (1895). Gasvakuolen, ein Bestandteil der Zellen der Wasserblute-bildenden Phytochromaceen. *Flora* (*Jena*), **80**, 241–82.

KLEBAHN, H. (1922). Neue Untersuchungen uber die Gasvakuolen. *Jahrbuch für wissenschaftliche Botanik*, **61**, 535–89.

KOLKWITZ, R. (1928). Über Gasvakuolen bei Bakterien. *Bericht der Deutschen Botanischen Gesellschaft*, **46**, 29–34.

KONOPKA, A. E. (1977). Inhibition of gas vesicle production in *Microcyclus aquaticus* by L-lysine. *Canadian Journal of Microbiology*, **23**, 363–8.

KONOPKA, A. E., LARA, J. C. & STALEY (1977). Isolation and characterization of gas vesicles from *Microcyclus aquaticus*. *Archives of Microbiology*, **112**, 133–40.

KONOPKA, A. E., STALEY, J. C. & LARA (1975). Gas vesicle assembly in *Microcyclus aquaticus*. *Journal of Bacteriology*, **122**, 1301–9.

KRANTZ, M. J. & BALLOU, C. E. (1973). Analysis of *Halobacterium halobium* gas vesicles. *Journal of Bacteriology*, **114**, 1058–67.

KRASIL'NIKOV, N. A., DUDA, V. I. & PIVOVAROV, G. E. (1971). Characteristics of the cell structure of soil anaerobic bacteria forming vesicular caps on their spores. *Microbiology*, **40**, 592–97.

LARSEN, H., OMANG, S. & STEENSLAND, H. (1967). On the gas vacuoles of the halobacteria. *Archiv für Mikrobiologie*, **59**, 197–203.

LAUTERBORN, R. (1915). Die sapropelische Lebewelt. *Verhandlungen des Natur-historische-Medizinischen Vereins zu Heidelberg*, N.F., **3**, 396–480.

LEHMANN, H. & JOST, M. (1971). Kinetics of the assembly of gas vacuoles in the blue-green alga *Microcystis aeruginosa* Kuetz. emend. Elenkin. *Archiv für Mikrobiologie*, **79**, 59–68.

LEHMANN, H. & JOST, M. (1972). Assembly of gas vacuoles in a cell free system of the blue-green alga *Microcystis aeruginosa*, Kettz. emend. Elenkin. *Archiv für Mikrobiologie*, **81**, 100–2.

MEFFERT, M. E. (1971). Cultivation and growth of two planktonic *Oscillatoria* species. *Mitteilungen der internationale Vereinigung für theoretische und angewandte Limnologie*, **19**, 189–205.

PETTER, H. F. M. (1932). Over roode en andere bakterien van getzouten visch. Doctoral Thesis, University of Utrecht, Holland.

PORTER, J. & JOST, M. (1973). Light-shielding by gas-vacuoles in *Microcystis aeruginosa*. *Journal of General Microbiology*, **75**, xxii (Abstract).

PORTER, J. & JOST, M. (1976). Physiological effects of the presence and absence of gas vacuoles in the blue-green alga *Microcystis aeruginosa* Kuetz. emend. Elenkin. *Archives of Microbiology*, **110**, 225–31.

REYNOLDS, H. C. S. (1972). Growth, gas vacuolation and buoyancy in a natural population of a blue-green alga. *Freshwater Biology*, **2**, 87–106.

REYNOLDS, C. S. (1973). Growth and buoyancy of *Microcystis aeruginosa* Kutz. emend. Elenkin in a shallow eutrophic lake. *Proceedings of the Royal Society of London B*, **184**, 29–50.

REYNOLDS, C. S. & WALSBY, A. E. (1975). Water blooms. *Biological Reviews (Cambridge)*, **50**, 437–81.

SARGENT, J. R. (1976). The structure, metabolism and function of lipids in marine organisms. In *Biochemical and Biophysical Perspectives in Marine Biology*, **3**, ed. D. C. Malins & J. R. Sargent, pp. 150–212. London: Academic Press.

SHEAR, H. & WALSBY, A. E. (1975). An investigation into the possible light-shielding role of gas vacuoles in a planktonic blue-green alga. *British Phycological Journal*, **10**, 241–51.

SINGH, R. N. & TIWARI, D. N. (1970). Frequent heterocyst germination in the blue-green alga *Gloeotrichia ghosei* Singh. *Journal of Phycology*, **6**, 172–6.

SKUJA, H. (1956). Taxonomische und Biologische Studien über das Phytoplankton Schwedischer Binnengewasser. *Nova Acta Regiae Societatis Scientiarum Upsaliensis*, Ser. IV, Vol. 16, No. 3.

SKUJA, H. (1964). Grundzuge der Algen flora und Algen vegetation der Fjeldgegenden um Abisco in Schwedisch–Lappland. *Nova Acta Regiae Societatis Scientiarum Upsaliensis*, Ser. IV, Vol. 18, No. 3.

SMITH, R. V. & PEAT, A. (1967a). Comparative structure of the gas-vacuoles of blue green algae. *Archiv für Mikrobiologie*, **57**, 111–22.

SMITH, R. V. & PEAT, A. (1967b). Growth and gas vacuole development in vegetative cells of *Anabaena flos-aquae*. *Archiv für Mikrobiologie*, **58**, 117–26.

SMITH, R. V., PEAT, A. & BAILEY, C. J. (1969). The isolation and characterization of gas-cylinder membranes and α-granules from *Anabaena flos-aquae* D 124. *Archiv für Mikrobiologie*, **65**, 87–97.

STALEY, J. T. (1968). *Prosthecomicrobium* and *Ancalomicrobium*: new prosthecate freshwater bacteria. *Journal of Bacteriology*, **95**, 1921–42.

STOECKENIUS, W. & KUNAU, W. H. (1968). Further characterization of particulate fractions from lysed cell envelopes of *Halobacterium halobium* and isolation of gas vacuole membranes. *Journal of Cell Biology*, **38**, 336–57.

TAYLOR, B. F., LEE, C. C. & BUNT, J. S. (1973). Nitrogen fixation associated with the marine blue-green alga, *Trichodesmium*, as measured by the acetylene-reduction technique. *Archiv für Mikrobiologie*, **88**, 205–12.

VAN BAALEN, C. & BROWN, R. M., JR. (1969). The ultrastructure of the marine blue-green alga, *Trichodesmium erythraeum*, with special reference to the cell walls, gas vacuoles and cylindrical bodies. *Archiv für Mikrobiologie*, **69**, 79–91.

VAN ERT, M. & STALEY, J. T. (1971). Gas-vacuolated strains of *Microcyclus aquaticus*. *Journal of Bacteriology*, **108**, 236–40.

WAALAND, J. R. & BRANTON, D. (1969). Gas vacuole development in a blue-green alga. *Science*, **163**, 1339–41.

WAALAND, J. R., WAALAND, S. D. & BRANTON, D. (1971). Gas vacuoles. Light shielding in blue-green algae. *Journal of Cell Biology*, **48**, 212–15.

WALSBY, A. E. (1969). The permeability of blue-green algal gas-vacuole membranes to gas. *Proceedings of the Royal Society of London*, B **173**, 235–55.

WALSBY, A. E. (1970). The nuisance algae: curiosities in the biology of planktonic blue-green algae. *Water Treatment and Examination*, **19**, 359–73.

WALSBY, A. E. (1971). The pressure relationships of gas vacuoles. *Proceedings of the Royal Society of London B*, **178**, 301–26.

WALSBY, A. E. (1972a). Structure and function of gas vacuoles. *Bacteriological Reviews*, **36**, 1–32.

WALSBY, A. E. (1972b). Gas-filled structures providing buoyancy in photosynthetic organisms. In *The Effects of Pressure on Organisms*; *Symposia of the Society for Experimental Biology*, **26**, ed. M. A. Sleigh & A. G. MacDonald, pp. 233–50. Cambridge University Press.

WALSBY, A. E. (1973). A portable apparatus for measuring relative gas vacuolation, the strength of gas vacuoles, and turgor pressure in planktonic blue-green algae and bacteria. *Limnology and Oceanography*, **18**, 653–58.

WALSBY, A. E. (1974a). The identification of gas vacuoles and their abundance in the hypolimnetic bacteria of Arco Lake, Minnesota. *Microbial Ecology*, **1**, 51–61.

WALSBY, A. E. (1974b). The isolation of gas vesicles from blue-green algae. In *Methods in Enzymology XXXI, Biomembranes Part A*, ed. S. Fleischer & L. Packer, pp. 678–86. New York: Academic Press.

WALSBY, A. E. (1975). Gas vesicles. *Annual Review of Plant Physiology*, **36**, 427–39.

WALSBY, A. E. (1976). The buoyancy-providing role of gas vacuoles in an aerobic bacterium. *Archives of Microbiology*, **109**, 135–42.

WALSBY, A. E. (1977a). Absence of gas vesicle protein in a mutant of *Anabaena flos-aquae*. *Archives of Microbiology*, **114**, 167–70.

WALSBY, A. E. (1977b). Gas vacuoles in sea sawdust. *British Phycological Journal*, **12**, 123.

WALSBY, A. E. (1977c). The gas vacuoles of blue-green algae. *Scientific American*, **237** (2), 90–7.

WALSBY, A. E. & BUCKLAND, B. (1969). Isolation and purification of intact gas vesicles from a blue-green alga. *Nature, London*, **224**, 716–7.

WALSBY, A. E. & KLEMER, A. R. (1974). The role of gas vacuoles in the micro-stratification of a population of *Oscillatoria agardhii* var. *isothrix* in Deming Lake, Minnesota. *Archiv für Hydrobiologie*, **74**, 375–92.

WEATHERS, P. J., JOST, M. & LAMPORT, D. T. A. (1977). The gas vacuole membrane of *Microcystis aeruginosa*. A partial amino acid sequence. *Archives of Biochemistry and Biophysics*, **178**, in press.

WINOGRADSKY, S. (1888). *Zur morphologie und Physiologie der Schwefelbacterien*. Liepzig: Arthur Felix.

WOLK, C. P. (1973). Physiology and cytological chemistry of blue-green algae. *Bacteriological Reviews*, **37**, 32–101.

WORCESTER, D. L. (1975). Neutron diffraction studies of biological membranes and membrane components. In *Neutron Scattering for the Analysis of Biological Structures*; *Brookhaven Symposia in Biology* no. 27, pp. III, 37–57.

ZIMMERMANN, U. (1969). Ökologische und physiologische Untersuchungen und der planktonische Blaualge *Oscillatoria rubescens* D.C. unter besonderer Beruck-sichtingung von Licht und Tempteratur. *Schweizerische Zeitschrift für Hydro-logie*, **31**, 1–58.

PLATE 1

EXPLANATION OF PLATE

Electron micrograph of the prosthecate bacterium *Prosthecomicrobium pneumaticum* Staley showing the gas vesicles (whose inner, gas-facing surfaces are exposed) mainly fractured in longitudinal section. (magnification 60000 ×). The inset shows one of the gas vesicles at higher magnification (210000 ×). Note the distinctive central rib. The rib periodicity appears smaller in the cone, as predicted in the legend of Fig. 1, but this may be an artifact generated by the viewing angle. Micrographs by D. Branton and A. E. Walsby, from Walsby (1975).

PLATE 1

INDEX

A protein of bacteriophages, 33, 42, 46, 47, 57

N-acetylglucosaminidase, in autolysis of peptidoglycans, 256, 258, 259, 263

N-acetylmuramic acid-L-alanine amidase, in autolysis of peptidoglycans, 256, 259, 263; mutant of *B. licheniformis* deficient in, 264

Acholeplasma laidlawii, cell membrane of, 204, 213

adenylate cyclase, membrane-associated multifunctional system, 126

affinity labelling, of 30S ribosome subunit, 61, 62, 63, 82

L-alanine, as germinant for spores, 297

alkaline phosphatase, periplasmic enzyme in *E. coli*, 240

amide groups, in peptidoglycan of myco-bacteria, 179; essential in structure of immunostimulant, 187

amines, long-chain alkyl: as germinants for spores, 320

amino acids: of gas vesicle proteins, 331; of peptidoglycans, 142–3; sequences of, in *E. coli* lactose repressor protein, 48–9, in *E. coli* ribosome proteins, 2, 8, and in gas vesicle proteins, 334

Anabaena flos-aquae, gas vesicles of: buoyancy provided by, 346; light inten-sity and, 344; mutant defective in, 349; protein of, 329, 331, 336–7

anaerobic bacteria, with gas vesicles, 348

antibiotics: affecting translation, 73–4, 84–5; cell wall of Gram-negative bacteria often impermeable to, 122, 131; β-lactam, inhibit transpeptidation in peptidoglycan synthesis, 151; β-lactam, mycobacteria resistant to, 180

antibodies, used to locate specific proteins: on interface of 30S and 50S ribosome subunits, 87; on 30S subunit, 62

antigens, in bacterial cell membranes, 209–15

arabinoglycan: mycolate of, linked to pepti-doglycan, in cell walls of mycobacteria, 178, 180–2

arginine, not synthesised in forespores, 303

Arthrobacter crystallopoietes: markers on surface of, 159; peptidoglycan of, 143

Arthrobacter spp., dimycolates of trehalose in, 184

aspartate-β-semialdehyde dehydrogenase, defective in stable L-forms of bacilli, 250

ATP synthesis, forespores lose capacity for, 306

ATPase: as chemoreceptor for divalent metal ions in bacterial chemotaxis, 126; complex of 5S RNA and three ribo-somal proteins as, 83–4; on cytoplasmic side of cell membrane of *M. lysodeikticus*, 211, 212, 214–15; in *E. coli* cell membrane, 214; particles of, 205, 215–16; of forespore membrane, 302; proton-translocating, of a thermophilic bacterium, 216

aurintricarboxylic acid, inhibits binding of mRNA to ribosomes, 44

autolysins, acting on peptidoglycans, 152; inhibit reversion of protoplasts of *B. subtilis*, 255–6; septa of dividing cells fail to split in mutants deficient in, 161, 167

axial filaments of spirochaetes, *see Lepto-spira*

bacilli: L-forms of, 250; turnover of pepti-doglycans in, 152; *see also individual species*

Bacillus amyloliquefaciens, extrusion of exoenzymes by, 239

Bacillus cereus: NADH-flavoprotein oxi-dase in spores of, 307; peptidoglycan of, 142; surface markers on, 159

Bacillus licheniformis: autolysin-deficient mutant of, 161; autolysins in protoplasts of, 263; excretion of penicillinase by, *see* penicillinase; helical growth of mutants of, 166; L-forms of, 250; peptidoglycan of, 143, 147, 150; polyglycerol phosphate produced by, 243; proton translocation in vegetative forms of, 308; reversion of L-forms and protoplasts of, 251, 255, 256–7, 258–9, 260–1, 265; teichuronic acid of, 147

Bacillus megaterium: N-acetylmuramyl-tri-peptide from, as immunostimulant, 187; cytochrome *c* in mother-cell and spore membranes of, not in vegetative cells, 308; forespores of, 302, 314–15, 316; lipids in membranes of spores of, 316; mutant of, unable to synthesise dipico-linic acid, 304; phospholipid in cell membrane of, 208; protoplasts of, 249; reversion of protoplasts of, 251, 255, 259

Bacillus sphaericus, peptidoglycan of, 300, 309

Bacillus stearothermophilus: binding of bacteriophage RNA to rRNA of, 42, 43, 46, 47; initiation factors in, 38; peptidyl transferase of, 82–3; reconstitution of 50S ribosome subunit of, 12

Bacillus subtilis: autolysins in protoplasts of, 260; binding of bacteriophage RNA to rRNA of, 46, 47; cell-wall growth in, 159–65; cytochrome *c* lacking in vegetative cells of, 308; helical growth of, in magnesium deficiency, 166; levansucrase of, as possible phospholipoprotein, 243; penicillinase-releasing protease of, 244; peptidoglycan of, 143, 144, 147, (in spore cortex) 309, (turnover of) 152, 153–4; polyribitol phosphate formed by mutant of, 243, protein synthesis in, can be initiated without formylation, 30; protoplasts and L-forms of, 249–50; reversion of L-forms of, to bacilli, in contact with fibrous solids, 251–3; reversion of protoplasts of, 253–5, 256, 259–60, 265; synthesis of peptidoglycan during reversion in, 256, 258–9; teichoic acid of cell wall of, in binding of bacteriophage, 153

bacteriophage BF23: binds to receptor for vitamin B12, 121, 125, 126; uptake of, independent of *tonB* function, 126

bacteriophage λ: carrying flagellar genes from *E. coli*, 274–5, 275–6; carrying gene for IF3, 58; receptor protein for, also involved in maltose transport, 120, 122, 123, 128

bacteriophage φ80: *tonB* function required for irreversible binding of, and entry of DNA into cell, 124

bacteriophage φX174, adheres to cell wall above contacts with cell membrane, 127

bacteriophage Qβ: initiation sequence for replicase of, 35; mutant of, in coat-protein cistron, 59

bacteriophage Σ15: adheres to cell wall above contacts with cell membrane, 127–8

bacteriophage SP50: wall teichoic acid of *B. subtilis*, and binding of, 153

bacteriophage TuI: *E. coli* mutants resistant to, lack binding protein for, 113

bacteriophages: binding of, triggers release of nucleic acid and its transport through cell wall and cell membrane, 123; cells became temporarily leaky on infection with, 127

bacteriophages, RNA: binding of RNA of, to bacterial rRNA, 42, 43, 46, 47; binding of, requires IF3, 55; binding sites for, on *E. coli* cell wall, 113, 115; proteins of *E. coli* cell wall serve for both binding of, and transport of substrates, 120–2; RNA of, codes for A protein, coat protein, and replicase, 33, 42; transfer of genes for binding of, from *E. coli* to *S. typhimurium* and *P. mirabilis*, 119

bacteriophages T1 to T7: adhere to cell wall

above contacts with cell membrane, 127; T1 requires *ton B* function for irreversible binding, and entry of DNA into cell, 124; T7 mutant in initiator region, 60

bacteriorhodopsin, in purple membrane of halobacteria, 225, 226, 227–8

bacterioruberin, in cell membrane of halobacteria, 225

bicyclomycin, causes decrease of lipoprotein in *E. coli*, 118

biotin-N-hydroxy-succinimide ester, lysine reagent: labelling of purple membrane with, 228

blue-green algae, gas vesicles of, 327, 382, 329; formed by some species in certain conditions only, 343; present in all planktonic forms, 350; in stratification, 348; turgor pressure and, 344

calcium: accumulation of, in forespores, 300, 303–6; effect of, on nuclear magnetic resonance spectra of lipids from spores and vegetative cells, 320; energy-dependent efflux mechanism for, in vegetative bacilli, 303; release of, from germinating spores, 315

carboxypeptidases: and peptidoglycans, 142, 144, 157; used in study of interpeptide linkages in peptidoglycan, 179–80

cardiolipin (diphosphatidyl glycerol): in cell membranes of micrococci, 207, 208; in membranes of *B. megaterium* spores, 320

cardiolipin synthetase of bacterial cell membrane, inhibited by mesosome fraction, 216–17

carotenoid, in inner membrane of *B. megaterium* spores, 316

Caulobacter crescentus: binding of bacteriophage RNA to rRNA of, 46, 47; initiation factors in, 38

cell cycle, incorporation of cell-wall components of Gram-negative bacteria during short period of, 128

cell division, creating forespore, 296

cell (cytoplasmic, inner, plasma) membrane: antigens and enzymes of, 209–15; dissociation and reconstitution of, 215–16; chemical components of, 206–9; flagella and, 271, 278; of forespores (two, with opposite polarity), 299, 300, 302; of Gram-negative bacteria, 11, 206, 207; of Gram-positive bacteria, 141, 206; in prokaryotes, performs functions of eukaryote organelles, 209; structure of, 203–6

cell walls (outer membranes): components of, synthesised in cell membrane, 209; freeze-fracture plane in, 205; *see also under* Gram-negative bacteria, Gram-positive bacertia, mycobacteria cerulenin, inhibits synthesis of phospholipoproteins, 243

chemoreceptors, in periplasmic space or on outer surface of cell membrane, 278, 279

chemotaxis in bacteria: ATPase in, 126; flagellar rotation and, 271, 272; gene products involved in, in cell membrane and cytoplasm, 278, 280; mutants deficient in, 279–80; scheme for flow of information through system for, 281

chloramphenicol: binding of, to ribosome proteins, 84

chromosome of bacteria: change in conformation of, at onset of sporulation, 298, 299; division of, at formation of forespore, 300

citrate, in transport of iron, 121, 122, 128

cloacin DF13: removes 3′-terminal sequence of 16S rRNA, and inhibits initiation, 64

coat protein of bacteriophages, 33; mutant in cistron for, 59–60; ribosome recognition of initiator site for, 35, 46, 47, 57

codons: for initiation, 35, 37, 41; for termination, 35–6, 88; two-letter, 73

colicins: B, D, and Ia, tonB-dependent killing of E. coli by, 125; E3, removes 3′-terminal sequence of 16S rRNA, and inhibits initiation, 43, 44–5, 64; translocated by protein of E. coli cell wall simultaneously with substrates and bacteriophages, 120–2, 123, 125, 126

concanavalin A: binds to glucosyl substituents of teichoic acid, 154; in study of antigens of M. lysodeikticus membrane, 211

conjugation of E, coli, inhibited by isolated protein and lipopolysaccharide from cell wall, 114

cord factors, of mycobacteria (dimycolates of trehalose), 183–4; anti-tumour activity of, 190, 191; induce non-specific resistance to infections, 189

cortex of spores: degraded on germination, 318; in spore dormancy and resistance, 311–14; structure and synthesis of, 309–11, 315

Corynebacterium spp.: covalent skeleton of, 181–2; dimycolates of trehalose in, 184

covalent skeleton, of cell walls of mycobacteria, 178–81; of allied genera, 181–2; as immunostimulant, 186

Cristispira, mode of swimming of, 291

cross-linking: for locating sites of initiation components on ribosome 30S subunit, 61, 62–3; of peptidoglycan in mycobacteria, 179–50, in reverting protoplasts, 256, 260, and in spores and vegetative cells, 309, 312; of proteins of ribosomes, 6–8, 65–6, 86; of rRNA with ribosome proteins, 16, 19–21

cyanide: changes in sensitivity to, of NADH oxidase of B. megaterium during sporulation, 307, 308, 309

cytochrome b: block between NADH dehydrogenase and, in inner membrane of spores, 317

cytochrome b1: cell cycle, and synthesis of, 128

cytochrome c: lacking in membranes of vegetative bacilli, 308; in membranes of forespore and mother cell of B. megaterium, 307, 308; synthesis of, in developing spore, 301

cytochromes: in cell membrane of M. lysodeikticus, 216; during sporulation, 307

cytoplasm, products of flagellar genes associated with, 278, 280

dehydration of spores: occurs in forespores? 313–14; possible role of cortex in, 311–13

diaminopimelic acid: present in spore cortex of B. sphaericus, not in vegetative cell, 310; protoplasts and L-forms of B. subtilis unable to synthesise, 253; in reverting L-forms, 251

diaminopimelyl ligase, cortex-specific enzyme: appears in mother cell during sporulation, 310–11

dielectric measurements, on bacterial spores, 319

dihydrofolate reductase, 30

dihydro-orotate dehydrogenase, in cell membrane of E. coli, 214

dihydrobenzoate, precursor of enterochelin: E. coli mutant requiring, 124

dinitrophenol, inhibits iron-enterochelin transport, 125

diphosphatidyl glycerol, see cardiolipin

dipicolinic acid, accumulates in forespore as chelate with calcium, 300, 303–6

DNA of spores: conformation of, altered by low-molecular-weight basic proteins of spore core? 318

DNA synthesis, not involved in conditioning of protoplasts for reversion, 255

dormancy, in spores and other cells, 296–7; cortex of spores in, 311–14; lipids and enzymes of membranes in, 316–17

electron microscopy: of attachment of antibodies to ribosome proteins, 3–5, 65, 66, 86; of cell membrane, 203–5; detection by, of surface markers on bacteria, 155, 159; of freeze-dried E. coli, 175, 316; negative staining for, 205–6; of protein

electron microscopy: (*contd.*)
   enclosing gas vesicles, 332–3; of purple
   membrane of halobacteria, 227–8
electron transport chain, in cell membranes
   of Gram-negative bacteria, 124
elongation, in protein synthesis, 68, 69;
   elongation factors in, 68–70, 79–81;
   hypotheses on cycle of, 74–6; peptidyl
   transferase in, 78–9; selection of tRNA
   in, 72–4; ternary complex of tRNA,
   elongation factor, and GTP in, 70–2,
   76–8; translocation in, 81–2
elongation factor EF-G, 68–9, 79–81; catalyses translocation, 79; complex of ribosome and, has GTPase activity, 79, 83;
   ribosomal proteins and, 83
elongation factor EF-Ts, 68, 69; catalyses
   reutilisation of EF-Tu in formation of
   ternary complex, 68, 70
elongation factor EF-Tu, 68–70, 78; binding of EF-G and, mutually exclusive, 80;
   binds to both 2′- and 3′-deoxyadenosine,
   71; catalyses binding of aatRNA to A
   site of ribosome, 68; complex of, with
   aatRNA and GTP, 70–2; GTPase
   activity associated with, 71–2, 76, 77, 83;
   ribosome proteins and, 83, 85; stimulates
   translation at low concentrations of
   magnesium, 77–8
endoplasmic reticulum, rough: in formation
   of proteins to be excreted, 231–2
enterochelin, in transport of iron, 121, 124–
   5, 128
Escherichia coli: antigens in cell membrane
   and cell wall of, 214; cell wall of, 117, 119,
   122; electron microscopy of cell membrane and ATPase particles of, 205–6;
   freeze–fracture plane of cell membrane of,
   204; genes for cell-wall proteins of, transferred to *Proteus* and *Salmonella*, 119,
   130; map of location of flagellar-associ-
   ated genes in, 271–2, and identification of
   products of, 274–7; mutants of, defective
   in motility, 272, with decrease in lipoprotein, 118, and lacking proteins of cell
   wall, 116, 117, 118–19; periplasmic
   alkaline phosphatase of, 240; preparation
   of protoplasts of, 212; proteins of flagella
   of, 271; reversion of spheroplasts of, 262,
   263; subunit of peptidoglycan of cell
   wall of, active as immunostimulant, 186;
   width of cell of, does not increase with
   growth in length, 163
N-ethyl-maleimide, for probing conformational changes in ribosomes, 9, 15
exosporium, 301

fatty acids, in cell wall of Gram-negative
   bacteria, 112, 113

ferritin-labelled antibody, for identifying
   components of cell membranes, 215, 240;
   on reverting protoplasts, 259, 260, 261
flagella, bacterial, 271–2, 285; genes related
   to, in *E. coli* chromosome map, 273;
   products of genes related to, 274–7, (functions) 278–82, (location in cell) 277–8
flagellin, of *E. coli*, 271, 278
forespores, 296; capabilities of isolated,
   314–15; membranes of, 296, 299, 300,
   301–2; methods for isolating, 295, 302;
   transport between mother cell and, 302–3
formaldehyde, cross-linking of proteins to
   RNA by means of, 20
N-formyl-tetrahydrofolate, donor of formyl
   residue, 29
formylation of Met-tRNA^{fMet} for initiation
   of protein synthesis, 29; not required for
   formation of pre-initiation complex on
   30S ribosome subunit, 39; as regulatory
   device in translation of polycistronic
   messengers, 41; some bacteria able to
   initiate protein synthesis without, 30
freeze-etching and freeze-fracture studies
   of cell membranes, 203–4
fusidic acid, stabilises complex of EF-G and
   GDP with ribosome, 80, 81

Gaffkya homari, peptidoglycan synthesis in,
   151
galactose-binding protein: cell cycle, and
   synthesis of, 128
gas vacuoles, provide aquatic prokaryotes
   with buoyancy, 327, 346
gas vesicles, composed of protein enclosing
   hollow space into which gases diffuse,
   327–8; buoyancy as function of, 345–50,
   (possible other functions) 351–2; creation,
   maintenance and regulation of hollow
   space in, 338–45; protein of, 327, 328–33;
   structure and arrangement of protein of,
   333–8
GDP, binds more tightly than GTP to EF-
   Tu, 69
germinability of spore, 296–7; acquisition
   of, by forespore, 296, 300, 301, 315
germination of spores, 317–19
glucosamine: *N*-acetyl, in teichoic acids,
   147; often *N*-acetylated, in peptidoglycans, 142
glutamate dehydrogenases (two), in cell
   membrane of *E. coli*, 214
glycanases, 143
glycans, in peptidoglycans of bacterial cell
   wall: length of chains of, 143–4, (in
   reverting protoplasts) 257–8, 260; in
   mycobacteria, 178, 179
glycerol-3-phosphate dehydrogenase, in cell
   membrane of *E. coli*, 214

glycerophosphate transport system: cell cycle, and synthesis of, 128

glycolipids, in bacterial cell membranes, 207

N-glycolylmuramic acid, in peptidoglycans of mycobacteria, 142, 179

glycoproteins: of cell walls of halobacteria, 225; less abundant in bacterial than in animal cell membranes, 207

Gram-negative bacteria: anti-tumour activity of endotoxin from, with cord factor from mycobacteria, 190, 191; cell membranes of, 11, 206, 207; cell walls of, *see next entry*; protoplasts and spheroplasts of, 212, 249; *see also individual species*

Gram-negative bacteria, cell walls of, 111–12, 130–2; chemical composition of, 112–16; multiple functions of proteins of, 120–6; mutants with additional protein in, 119–20, and lacking protein in, 116–19; as permeability barrier to hydrophilic substances, 120; phospholipase A as only enzyme located exclusively in, 202; structure of peptidoglycan of, 142; synthesis of, 127–30

Gram-positive bacteria: cell membranes of, 141, 206, 208; cell walls of, *see next entry*; initiation without formylation in, 30; *see also individual species*

Gram-positive bacteria, cell walls of, 167–8; chemical composition and morphology of, 140–2; growth of, 154–61; helical growth of, 165–7; models for growth of, 161–5; peptidoglycans of, 142–8, 186; proteins of, *see* proteins of cell walls of Gram-negative bacteria; synthesis of polymers of, 148–51; turnover of constituents of, 151–4

GTP: binding sites on IF2 for, 53, 83; in formation of initiation complex, 28, 39, 40, 52, 53; in elongation cycle, 69–72, 80; hydrolysis of, on release of EF-G from ribosome, 81–2, on release of EF-Tu from ribosome, 77, on release of IF2 from initiation complex, 54, on stable binding of aatRNA, 76, and in translocation, 90

GTPase: associated with EF-Tu, 71–2, 76, 77, 83; associated with complex of ribosome and EF-G, 79, 83; associated with complex of ribosome and IF2, 54, 83; complex of 5S rRNA and three ribosome proteins as, 83–4

halobacteria, gas vesicles of, 328, 239, 349

Halobacterium halobium, purple membrane of, 225–8

Halobacterium sp.: protein from gas vesicles of, 330, 331, 333; turgor pressure absent in, 344

halophilic bacterium: lacking transformylase, able to initiate protein synthesis without formylation, 30

heat resistance, of spores, 300, 301, 315; of enzymes of inner and outer forespore membranes, 317; lost on germination, 315

helical growth, of bacteria in some conditions, 165–7

hook structure: of axial filaments of *Leptospira*, 286; of flagella of *E. coli*, 271, (genes for) 273, 277, 281

hydrogen bonding, in peptidoglycans, 145

immunoaffinoelectrophoresis, 211

immunodepression, by derivatives of immunostimulant, 188

immunoelectrophoresis, crossed: for analysis of membrane antigens of bacteria, 210, 212, 213

immunostimulant, in killed mycobacteria, 177–8; identification of substance involved (adjuvant), 185–6; minimum structure for (*N*-acetylmuramyl-L-alanyl-D-isoglutamine: MDP), 186–7; relation between structure and activity of, 187–8; target cells for, 188

initiation codon, 35, 37, 41

initiation complex, in protein synthesis, 27–8, 89; components of, (initiation factors) 36–8, 50–2, (ribosomes) 30–2, 41–9, (mRNA) 32–6, 48–50, (tRNA) 28–30, 39–41; formation of, 52–9, 89; location of components of, on 30S ribosome subunit, 61–5; location and structural studies of, correlated, 65–8; mutations affecting, 59–60

initiation factor IF1, 36–7; in formation of initiation complex, 28, 52; and IF2, 53, 54, 56; increases rate of dissociation of 70S ribosomes, 51, 52; requirements for, with different mRNAs, 55

initiation factor IF2 (two forms), 37–8; in binding of formylated Met-tRNA, 28, 52–3, 57; has stronger affinity for 70S ribosome than for 30S subunit, 52; recycling of, 54, 56; release of, 53; ribosome proteins and, 83; ribosome-associated GTPase activity of, 54, 83; sites for binding GTP on, 53, 83

initiation factor IF3, 38; and binding of formylated Met-tRNA, 54, 57; binds to 30S ribosome subunit and prevents reassociation, 51, 52; binds to 3′-terminus of both 16S and 23S rRNA, 58, 86; different mRNAs and requirements for, 55–7, 58–9; λ bacteriophage carrying structural gene for, 58; mutant with thermolabile, 60; promotes recycling of IF2, 56; test for activity of, 55 initiation factors, 50–2; binding sites for, on 30S ribosome subunit, 63–4, 67; promote binding of formylated Met-tRNA, inhibit binding of unformylated Met-tRNA, 39; recycling of, 53–4; *see also individual factors*

iodination, catalysed by lactoperoxidase: to identify components of cell membrane, 214–15; to probe conformational changes in ribosomes, 9, 15

ions, immobilised in spores, 319

iron, systems for transport of: in complex with citrate, 121, 122, 128; in complex with enterochelin, 121, 124–5, 128

kasugamycin: acts on 3′-terminus of 16S rRNA, inhibiting initiation, 43, 86; structure of rRNA in mutants resistant to, 85

kethoxal, reagent binding to non-paired guanine bases in RNA, 13, 14; sites sensitive to, as candidates for functional importance, 14–15

kirromycin: blocks elongation by inhibiting release of EF-Tu, 72; induces ribosome-independent GTPase on EF-Tu, 72, 83

L-forms of bacteria: conversion of protoplasts to, 249–50; reversion of, to bacteria, *see under* protoplasts; stable (unable to revert, owing to mutation), 250–1

β-lactamase, in cell membrane, 209

lactate dehydrogenase, in cell membrane of *E. coli*, 214

Lactobacillus acidophilus, turnover of peptidoglycan in, 152

lactose operon of *E. coli*: formylation, and polarity of expression of genes of, 41

lactose repressor protein of *E. coli*, gene for: initiator codon in, 35, preceded by sequence complementary to 16S rRNA, 48; restarting sites in, after chain termination at nonsense mutation, 48–9

lectins, in study of antigens of membrane of *M. lysodeikticus*, 211

Leptospira interrogans, axial filaments of (under external sheaths), 285–6; in basic movements, 287–8; in movement in semisolid medium, 291–2; in non-translational movement, 288–9; in translational movement, 289–91

levansucrase, of *B. subtilis*, as possible phospholipoprotein, 243

lipids: in cell membranes, 203, 207, 208; in cell walls of Gram-negative bacteria, 207, and of mycobacteria, 178, 182–3; in inner and outer membranes of *B. megaterium* spores, 316, 319; in purple membrane of halobacteria, 225, 227

lipomannan: in mesosomes, 216; succinylated, in cell membrane of micrococci, 208, 211

lipopolysaccharide, in cell walls of Gram-negative bacteria, 112, 214; appears on cell walls at sites of adhesion to cell membrane, and spreads over whole surface, 127; not affected in mutants lacking proteins of cell wall, 116, 117; ring of basal structure of flagella associated with, 271; synthesised in cell membrane, 209, 213

lipoprotein: of cell membrane of *E. coli*, 214; of cell wall of *E. coli*, 114, 128, 214; linked to murein, 114, 117–18, 129–30; mutants lacking, 117, 118–19, and with reduced amounts of, 118; in-vitro translation of mRNA for, 128–9

lipoteichoic acid, in cell membrane of Gram-positive bacteria, 208

lysine: in bacteriorhodopsin, 228; inhibits synthesis of protein of gas vesicles, 343–4

lysyl ligase, during sporulation, 310, 311

magnesium: in association and dissociation of ribosome subunits, 31, 32; in binding of formylated and non-formylated Met-tRNA, 39; in binding of GTP or GDP to EF-Tu, 69; teichoic and teichuronic acids and, in cell walls of Gram-positive bacteria, 148, 167–8

malate dehydrogenase, in cell membranes, 214; on cytoplasmic side of membrane in *M. lysodeikticus*, 211, 212; in outer membrane of forespore, 302

maltose: enzymes transporting and catabolising, in *E. coli*, associated with λ-binding protein, 120, 122–3; λ-binding protein induced in presence of, 128; specific channel through cell wall of Gram-negative bacteria for maltodextrins and, 121, 122

maturation protein, of RNA bacteriophages, 33; initiator codon for, 35

MDP, minimal adjuvant structure, for immunostimulant activity by cell wall preparations from mycobacteria, 187–8; induction of non-specific resistance to infection by, 189

mecillinam, induction of *E. coli* spheroplasts by, 262, 263

membranes: cell, cytoplasmic, plasma, *see* cell membrane; outer, *see* cell walls

mercaptobutyrimidate, for cross-linking of proteins, 7

mesosomes: as infoldings of cell membrane, in Gram-positive bacteria, 201–2; lack enzymes, 216–17; mark position of future septum, 141, 168; membrane-bound penicillinase and penicillinase-

mesosomes: (*contd.*)
  releasing protease located near, 236–7; negative staining to show, 205; sporulation septum associated with, 300
metabolism, not detectable in spores, 296
methionine, not synthesised in forespores, 303
methyl-accepting chemotaxis protein, 276, 277
methylation and demethylation, in chemotactic system, 276, 279, 280
micrococci: cardiolipin as chief phospholipid of, 207; succinylated lipomannan in cell membranes of those species of, devoid of lipoteichoic acid, 208; *see also individual species*
Micrococcus luteus (*M. lysodeikticus*): antigens and enzymes of cell membrane of, 207, 208, 210–12, 214–15; negative staining of cell membrane, ATPase particles, and mesosomes of, 205; peptidoglycan synthesis by isolated cell walls and membranes from, 150, 151; percentage of peptidoglycan in cell walls of, 141; protein synthesis in, can be initiated without formylation, 30
Micrococcus lysodeikticus, see *M. luteus*
Micrococcus roseus: subunit of peptidoglycan of, active as immunostimulant, 186
Microcystis aeruginosa, gas vesicles of: production of, 339, (mutant defective in) 331; protein of, 329, 331
Microcyclus aquaticus, gas vesicles of: production of, 339; protein of, 330, 331, 333–4; synthesis of protein of, inhibited by lysine, 343–4
mitogenic activity, for lymphocytes, of soluble peptidoglycan fragments, 188
motility, bacterial, 271, 285; mutants of *E. coli* defective in, 272
muramic acid in peptidoglycans, 142, 179; carboxyl groups of, substituted by peptide, 144; lactam of, in *B. sphaericus* spores, 309; *N*-glycolyl-, in mycobacteria, 142, 179
murein, *see* peptidoglycan
mutants, for identifying whether a component of a system is necessary, 89
mycobacteria, cell walls of, 178–9; arabinogalactan of, 180–2; cord factors of, 183–4; lipids of, 182–3; peptide chains of (nonpeptidoglycan), 178, 184; peptidoglycan of, 142, 179–80; polyglutamic acid of, 184–5
mycobacteria, immunostimulant properties of cell walls of, 177–9; homologous and heterologous resistance to infections induced by, 189; minimum structure involved in immuno-stimulation (MDP), 185–8; prevention, suppression, and regression of tumours induced by, 189–91
Mycobacterium spp.; structures of different mycolic acids of, 181
mycolic acids: combined with arabinogalactan in cell walls of mycobacteria, 178, 180, 181, in *Nocardia* and *Corynebacterium*, 181–2, unambiguous generic identification by means of, 182

NADH-cytochrome *c* reductase, in membranes of *B. megaterium* forespores, 316
NADH dehydrogenase: block between cytochrome *c* and, in spores, 317, removed on germination, 318; in cell membrane of *E. coli*, 214; in membranes of forespores, 302, 316–17; two forms of, on cytoplasmic side of cell membrane of *M. lysodeikticus*, 211, 212
NADH-flavoprotein oxidase, in spores of *B. cereus*, 307
NADH oxidase, of *B. megaterium*: change in cyanide-sensitivity of, during sporulation, 308, 309; in dormant spores, 307; in forespore membranes, 316; of inner and outer membranes of spores, on germination, 318
neamine, ribosome proteins altered in *E. coli* mutant resistant to, 85
Neisseria gonorrhoeae, crossed immunoelectrophoresis of membranes of, 213
neomycin, affects translation, 74
neutron-scattering, neutron diffraction: to measure distances between centres of mass of pairs of deuterated ribosome proteins, 5–6; to measure thickness of protein layer around gas vesicle, 337–8
Nocardia spp.: covalent skeleton of, 181–2; dimycolates of trehalose in, 184
Nostoc muscorum, gas vesicles of, 328, 338
nuclear magnetic resonance spectroscopy, of bacterial spores, 319, 320
nucleosides: specific channels for, through cell walls of Gram-negative bacteria, 121, 122
nucleotides: sequences of, in MS2 bacteriophage, 83–4, in mRNA of lactose repressor gene, 48–9, in rRNA, 14, and in tRNAs, 29; untranslated, in RNA bacteriophages, 36

octanol: acquisition of resistance to, by forespore, 300, 315
oil, in anti-tumour activity of cell-wall preparations from mycobacteria, 191
optical density of spores, during germination, 315
Oscillatoria agardhii, gas vacuoles and buoyancy of, 347

penicillin: inhibits transpeptidation in peptidoglycan synthesis, 151; prevents reversion of L-forms, 250

penicillinase, of *B. licheniformis*, 232, 243; membrane-bound and exoenzyme forms of, 130, 232–3; membrane-bound, located near mesosomes, 236–7; phospholipopeptide portion of, 233–5, 241, 242; release of membrane-bound, to yield exoenzyme, 235–6; synthesis of, 237–8, 241–2, (site of) 239–40, (source of phosphatidyl serine for), 238–9, 241

peptides (non-peptidoglycan), in cell walls of mycobacteria, 178, 184; involved in immunostimulant properties, 184

peptidoglycan (murein): between cell membrane and cell wall, in *E. coli*, 203; between two membranes of forespores, 300; in cell wall and cell membrane of Gram-negative bacteria, 111, 113, 116; in cell walls of Gram-positive bacteria, 141, (conformation of) 144–5, (polymers attached to) 146–8, (structure of) 142–4; in cell walls of mycobacteria, 178, (immunostimulant fraction of) 186–8, (structure of), 179–80; incorporation of, into cell wall, 128; lipoprotein linked to, 114, 117–18, 129–30; ring of basal structure of flagella associated with? 271; synthesis of (defective in stable L-forms) 250, (during conditioning of protoplasts for reversion) 255, 256, (during reversion) 256–8, 259–63; usually lacking in sporulation septum, 300

peptidyl hydrolase, formylated Met-tRNA resistant to, 29

peptidyl transferase, at A and P sites of ribosomes, in elongation of proteins, 78–9; antibiotics affecting, 84–5; capable of forming both peptide and ester bonds, 79; proteins at sites of, 85, 86; puromycin as acceptor-substrate for, 77, 78

permeases, for cell membranes: genes for, 121

phosphatidylethanolamine, principal phospholipid of many bacterial cell membranes, 207; in inner and outer layers of cell membrane in *B. megaterium*, 208

phosphatidyl glycerol, in inner and outer layers of cell membrane of *M. lysodeikticus*, 208; di-, *see* cardiolipin

phosphatidylserine, at terminus of phospholipopeptide moiety of penicillinase, 233, 234–5; source of, in synthesis of penicillinase, 238–9, 241

phospho - *N* - acetylmuramyl - pentapeptide translocase, in peptidoglycan synthesis: defect of, in stable L-forms, 250, 251

6-phosphogluconate dehydrogenase, in cell membrane of *E. coli*, 214

phospholipase A, in cell wall of Gram-negative bacteria, 202

phospholipids: of bacterial cell membranes, 207–8; in cell walls of Gram-negative bacteria, 112–13; changes in, in heat activation of spores, 319–20; in membranes of *B. megaterium* spores, 316, 317, and vegetative cells, 317; in mutants lacking cell-wall protein, 116, 117, 130

phospholipopeptide portion of penicillinase, 233–5, 241, 242

phospholipoprotein: levansucrase of *B. subtilis* as? 243; membrane-bound penicillinase as, 234, 237

phosphotransferase, membrane-associated multifunctional system, 126

photosynthetic bacteria: gas vesicles of, 328; planktonic forms of, either have gas vesicles, or are flagellate, 350

polyglutamic acid, partly amidated, in cell walls of mycobacteria, 178, 179, 184–5; related to virulence? 185

polyglycerol phosphate, produced by *B. licheniformis*, 243

polyribitol phosphate, produced by *B. subtilis*, 243

polysaccharides, capsular: synthesised in cell membrane, 209

pressure: critical, causing collapse of gas vesicles, 340–1

Prosthecomicrobium pneumaticum, 345; mutant of, defective in gas vesicle production, 331, 349

protease: in cell membrane and cell wall of *E. coli*, 214; of spores, degrading basic proteins on germination, 318

protease, penicillinase-releasing, 235–6, 241, 242; located near mesosomes, 236–7; present in organisms not producing penicillinase, 243–4

protein synthesis, in bacteria, 88–90; in cell membrane, 209; in conditioning of protoplasts for reversion, 255; elongation in, *see* elongation; gas-vesicle formation immediately dependent on, 339; inhibition of, causes thickening of cell wall, 140; initiation of, *see* initiation; translocation in, *see* translocation

proteins: of cell membranes of bacteria, 206, 209–15; of cell walls of Gram-negative bacteria, *see next entry*; crystalline, external to cell wall, of some bacterial species, 202; excretion of, from cells, 231–2; of flagella, 271; of gas vesicles, *see under* gas vesicles; low-molecular-weight basic, in spore cores, 317–18; methyl-accepting chemotaxis, 276, 277; of ribosomes, *see* proteins of ribosomes; of RNA viruses, 33; *see also*

proteins: (*contd.*)
glycoproteins, lipoproteins, phospholipoproteins
proteins of cell walls of Gram-negative bacteria, 112–14, 120, 130; changes in, associated with bacteriophage infection, 119; interactions between, 115; loss of, balanced by increase of phospholipid, 116, 117, 130; multiple functions of (receptor and translocation), 120–6, 131
proteins of ribosomes (S1 to S21 in 30S subunit, L1 to L34 in 50S subunit), 1–3; amino-acid sequences of, 2, 8; binding sites on rRNA for, 17–19; chemical modification to identify, 8–9; cross-linking of, 6–8, 65; cross-linking of rRNA and, 16, 19–21; electron microscopy of specific antibodies attached to, 4–5; energy transfer between fluorescent markers of, to estimate distances between, 9–10; at interface between ribosome subunits, 87; involved with components of elongation process, 85–6; in model for 50S subunit, 86–8; reconstitution of subunits from rRNA and, *see under* ribosome subunits; S1, 16, 42, 48, 57–8; S12, 42, 48; shapes of, shown by neutron scattering, 5–6; at sites for mRNA, 63; streptomycin labelling of, 84–5
Proteus mirabilis: genes for *E. coli* cell-wall proteins transferred to, and expressed, 119, 130; reversion of spheroplasts and L-forms of, 251
proton magnetic resonance spectra, in study of initiator site recognition, 45
proton pump: light-driven, creating electrochemical gradient across purple membrane of halobacteria, 226–7
protoplasts: of *B. licheniformis*, penicillinase synthesis by, 237; of Gram-negative bacteria, difficult to prepare, 212; of *M. luteus*, antigens on surface of, 211; reversion of L-forms and, to bacteria, 249–51, (autolysins in) 263–5, (conditioning for) 253–5, (environment and) 251–3, (synthesis of wall during) 255–9, (ultrastructural studies of) 259–63
Pseudomonas aeruginosa: 3′-terminal sequence of 16S rRNA of, 46, pairs with sequence of R17 bacteriophage RNA, 47
puromycin, analogue of aatRNA, 39; aminoacyl derivatives of, 40; enters peptidyl-transferase centre on ribosome, 77, 78
purple membrane, of halobacteria, 225–8
pyridine nucleotide transhydrogenase, lacking in membranes of vegetative bacilli, 308

release factors in ribosomes: RF1 and RF2, recognising termination codons, and RF3, stimulating RF1 and RF2, 88
replicase gene of bacteriophage RNA, 49–50; ribosome recognition of initiation site for, 42, 46, 47; recognition dependent on initiation factors, 57
retinal, in purple membrane of halobacteria, 225
reversion inhibitory factor, of *B. subtilis*, 256, 264
reversion of protoplasts and L-forms, *see under* protoplasts
ribonuclease: digestion of protein-RNA complexes by, to show RNA regions 'protected' by protein, 17, 33, 35; susceptibility to, of rRNA in 30S subunit and 70S ribosome, 58
ribosome subunits, 30S and 50S, associating to 70S whole, 2; binding sites for mRNA on 30S, 63, 65–6, 66–7; conformational changes of 30S on association, 9, 15; kinetics of association and dissociation of, 31–2; location on 30S of components of initiation complex, 61–5; and on 50S of translation components, 82–6; location and structural studies correlated, for translation components, 65–8; and for initiation components, 86–8; models for, (30S) 4, 67, (50S) 4, 87; normal equilibrium towards 70S affected by initiation factors, 28, 50–2; reactions of 30S with mRNA and with 50S are mutually exclusive, 58–9; reassembly of, from proteins and rRNA, (30S) 10–11, 21, (50S) 12; reassembly with one protein lacking, or substituted, 61–2; recognition by 30S, of initiation sites on mRNA, 13, 41–2, 57
ribosomes (of *E. coli*), 2, 30–2; A site in, 68, 72–6, 78; binding of elongation complex to, 68, 76–8; binding sites on, for RNA bacteriophages, 33, 34, 35, 36, 37; bound to cell membrane during synthesis of membrane proteins, 128–9, and of penicillinase, 239, 241; conformation of isolated, dependent on method of preparation and conditions, 62; mutations of, 62, 74; P site in, 39, 68, 78; proteins of, *see* proteins of ribosomes; subunits of, *see* ribosome subunits; two classes of, A and B, 30–1, 32, 50; two-state equilibrium of, formyl group selecting more active conformation? 40
mRNA: binding sites for, on ribosome 30S subunit, 63, 65–6, 66–7; component of initiation complex, 28, 32–6, 52, 63; from different sources, and requirements for IF3, 55–7, 58–9; estimate of errors made

368   INDEX

mRNA: (contd.)
in decoding of, *in vivo*, 74; for penicillinase, 234, 242; recognition of initiation
sites on, by 30S subunit, 13, 41–2, 57;
rRNA 3′-terminus in recognition of
initiation sites on, 42–5, 46, 65; of RNA
viruses, coding for three proteins, 33, 34,
35–6; secondary structure of, as control
in initiation, 49–50; strength of interaction of rRNA and, and efficiency of
initiation, 48–9
rRNA (16S in 30S ribosome subunit, 5S
and 23S in 50S subunit), 1, 2, 6; binding
sites on, for ribosome proteins, 17–19;
complex of 5S with three ribosome proteins has GTPase and ATPase activity,
83–4; cross-linking of, in 30S subunit, 15;
cross-linking of proteins and, 16, 19–21;
functions of, in protein synthesis, 13;
involved in binding of mRNA, and in
association of ribosome subunits, 89; in
mutants resistant to kasugamycin, 85;
nucleotide sequences of, 14; pairing of
16S, with 23S or mRNA, 58–9; reassembly of ribosome subunits from
proteins and, *see under* ribosome subunits; secondary structure of, 14–15;
strength of interaction of mRNA and,
and efficiency of initiation, 48–9; tertiary
structure of 16S, 15; 3′-terminus of
16S, on interface between ribosome subunits, 44, and in recognition of mRNA
initiator sites, 42–5, 46, 65
tRNA: with attached phosphatidyl serine,
in extracts of cells of *B. licheniformis*, 238;
binding of, to 5S rRNA of 50S ribosome
subunit, 13; formylation in function of,
39–41; ribosome proteins and EF-Tu in
binding of, 82–3, 85; selection of, in
elongation of protein, 72–6
aatRNA: binding of, to ribosome A site,
72–6; in complex with EF-Tu and GTP in
elongation of proteins, 68, 69–72, 76–8;
puromycin as analogue of, 39
tRNA^fMet (initiator tRNA), 28; nucleotide
sequence of, 29
tRNA^Met (elongation tRNA), 28; binding
of 39; nucleotide sequence of, 29
fMet-tRNA^fMet: binding of, to 30S ribosome
subunit, 57, 64–5, 68; in initiation complex,
28, 35, 40; nucleotide sequence of, 29
Met-tRNA^fMet: binding of, in pre-initiation complex, 39–40; does not react with
EF-Tu/GTP, 71; formylation of, 29,
30
Met-tRNA^Met, 40
RNA synthesis, involved in conditioning of
protoplasts for reversion, 255
tRNA synthetase, 41

Salmonella abortesqui: mutants of, with
flagellar defects, 281
Salmonella typhimurium, cell wall of: *E.
coli* cell-wall protein incorporated into
119; impermeable to hydrophilic substances above certain molecular weight,
122; mutants in, lacking lipoprotein,
118, and lacking protein, 116, proteins of,
115; vesicle reconstruction from, 122
Sarcina, extrusion of protease by, 239
septa: in dividing cells of Gram-negative
bacteria, 139; mesosomes mark future
position of? 141; new material incorporated into cell wall in region of, 155, 158;
in reverting protoplasts, 261
signal peptides, at N-terminus of nascent
secretory proteins in eukaryotes, 129,
241–2
singlet–singlet energy transfer, between two
fluorescent dyes labelling two ribosome
proteins, for determination of distance
between them, 9–10
solid fibrous environment, in reversion of
protoplasts, 251–3
spectinomycin, ribosome proteins altered in
E. coli mutants resistant to, 85
spheroplasts, 249, 262
Spirillum volutans: movement of, in 5%
gelatin, 292
spirochaetes, motion of, 285–6
spores, bacterial: biophysical studies of,
319–20; cortex of, 309–11, (in dormancy
and resistance) 311–14; dormancy and
germinability of, 296–7; formation of,
see sporulation; germination of, 317–19;
membranes of, 301–3, (in dormancy and
resistance), 316–17; *see also* forespores
sporulation, 296, 298–301; calcium movement during, 303–6; electron transport
system during, 307–9; initiated by lack
of an essential metabolite, 296, 298;
phospholipoproteins in? 244
staphylococci, peptidoglycan of, 142
Staphylococcus aureus: L-forms of, 250,
251; length of teichoic acids of, 147;
peptidoglycan of, 142, 143, 147, (immunostimulant activity of subunit of) 186
Staphylococcus epidermidis: N-acetylmur-
amyl tripeptide from, as immunostimulant, 187; peptidoglycan of, 143
streptococci: pole formation in, 156–7; no
turnover of peptidoglycans in, 152; sites
of wall growth in, 154–6
Streptococcus faecalis: cell-wall growth in,
151, at sites of division, 155, 158, 262;
freeze-fracture plane of cell membrane of,
204; peptidoglycan of, 142, 143; protein
synthesis in, can be initiated without
formylation, 30; quantitative model for

Streptococcus faecalis: (contd.)
  growth of, using analogue rotation technique, 156–9; secondary expansion of cells of, after cell separation, 162; septum formation in, 162; splitting of septum fails if autolysin content is reduced, 161
Streptococcus faecium: L-forms of, 250, 251; reversion of L-forms of, 251, 261–2, 263; septum formation in, 139
Streptococcus haemolyticus, new wall incorporated at sites of incipient septa in, 154
Streptococcus lactis, length of teichoic acids of, 147
Streptococcus pneumoniae, new wall incorporated at sites of incipient septa in, 154
Streptococcus pyogenes, L-forms of, 250, 251
streptomycin: acts on 3′-terminus of 16S rRNA, and inhibits initiation, 43; affects translation, 73–4; binding of, inhibited by antibodies to some ribosome proteins, 84–5; ribosome proteins labelled by, 84–5
succinate dehydrogenase: in cell membrane of *E. coli*, 214; on cytoplasmic side of cell membrane of *M. lysodeikticus*, 211, 212, 216
sulphydryl groups, in EF-Tu and EF-Ts, 70

tartaric acid: *bis*-amide derivatives of, for cross-linking of proteins, 7
teichoic acids, in cell walls of Gram-positive bacteria, 141, 146, 147; adsorption of bacteriophage SP50 to *B. subtilis* dependent on, 153; chain lengths of, 147–8; not found in spore cortex, 309–10, 312; replaced by teichuronic acids in cell walls of bacilli during phosphate limitation, 152; and supply of magnesium to cell wall, 148, 167–8; synthesised during reversion of *B. licheniformis* protoplasts, 257
teichuronic acids, in cell walls of Gram-positive bacteria, 141, 146; chain lengths of, 147; not found in spore cortex, 312; and supply of magnesium to cell wall, 167–8; synthesised during reversion of *B. licheniformis* protoplasts, 257
temperature jump relaxation methods, in study of initiator site recognition, 45
termination, of peptide chains in protein synthesis: codons for, 35–6, 88
termination factor, releases peptide chain without releasing ribosome from mRNA, 41
tetracycline: blocks labelling of ribosome protein L16, 82; inhibits binding of

aatRNA to A site of ribosome, 39, 76–7
thiostrepton, and A site of ribosome, 82
tonB function, between cell wall and cell membrane of *E. coli*, 121, 124, 126; mediates energy function between the two, 125
transformylase, 29, 30
translocation, in elongation of proteins, 81–2, 90; catalysed by EF-G, 79
transpeptidase, of mycobacteria, 180
trehalose dimycolates (cord factors), in mycobacteria and allied genera, 183–4; anti-tumour activity of, 190, 191; induce non-specific resistance to infections, 189
Trichodesmium spp., gas vesicles of, 341, 344, 352
trimethoprim, inhibits dihydrofolate reductase, 30
tuberculosis: protection against, by factor from mycobacteria, 177, 189
tumours: prevention, suppression, or regression of, caused by factors from mycobacteria, 177, 189–91
turgor pressure of cell, and gas vesicles, 341, 344

ultra-violet light: acquisition of resistance to, by forespores, 318; cross-linking effected by, in RNA, 15, and between RNA and protein, 20; high levels of, cause unfolding of ribosome subunits, 20

vitamin B_{12}: specific channel for, through cell wall of Gram-negative bacteria, 121, 122, 125–6

water: content of, in spores and vegetative cells, 311; content of, in spores, increases rapidly on germination, 314; exchangeable with external water, in spores, 313; nonexchangeable, in spores and vegetative cells, 319; protein round gas vesicles impermable to, 340
wax D, of mycobacteria, autolysis product of cell wall, 182–3; immunostimulant properties of, 186
'wobble pairs', in binding of aatRNA to A site of ribosome, 72, 73

X-ray studies: on oriented fibres of peptidoglycan, 145; on protein enclosing gas vesicles, 335–7; on ribosome proteins, 3

zymogram staining, of immunoprecipitates, for identification of enzymes, 210, 211, 214